T0201953

Contemporary Scientific Realism

Contemporary Scientific Realism

The Challenge from the History of Science

Edited by

TIMOTHY D. LYONS AND PETER VICKERS

OXFORD
UNIVERSITY PRESS

OXFORD
UNIVERSITY PRESS

Oxford University Press is a department of the University of Oxford. It furthers
the University's objective of excellence in research, scholarship, and education
by publishing worldwide. Oxford is a registered trade mark of Oxford University
Press in the UK and certain other countries.

Published in the United States of America by Oxford University Press
198 Madison Avenue, New York, NY 10016, United States of America.

© Oxford University Press 2021

Library of Congress Control Number: 2021934069
ISBN 978-0-19-094681-4

DOI: 10.1093/oso/9780190946814.001.0001

1 3 5 7 9 8 6 4 2

Printed by Integrated Books International, United States of America

Contents

1

History and the Contemporary Scientific Realism Debate

Timothy D. Lyons[1] and Peter Vickers[2]

[1]Department of Philosophy
Indiana University–Purdue University Indianapolis
tdlyons@iu.edu
[2]Department of Philosophy
Durham University
peter.vickers@durham.ac.uk

The scientific realism debate began to take shape in the 1970s, and, with the publication of two key early '80s texts challenging realism, Bas van Fraassen's *The Scientific Image* (1980) and Larry Laudan's "A Confutation of Convergent Realism" (1981), the framework for that debate was in place. It has since been a defining debate of philosophy of science. As originally conceived, the scientific realism debate is one characterized by dichotomous opposition: "realists" think that many/most of our current best scientific theories reveal the truth about reality, including *unobservable* reality (at least to a good approximation); and they tend to justify this view by the "no miracles argument," or by an inference to the best explanation, from the success of scientific theories. Further, many realists claim that scientific realism is an empirically testable position. "Antirealists," by contrast, think that such a view is lacking in epistemic care. In addition to discussions of the underdetermination of theories by data and, less commonly, competing explanations for success, many antirealists—in the spirit of Thomas Kuhn, Mary Hesse, and Larry Laudan—caution that the history of science teaches us that empirically successful theories, even the very best scientific theories, of one age often do not stand up to the test of time.

The debate has come a long way since the 1970s and the solidification of its framework in the '80s. The noted dichotomy of "realism"/"antirealism" is no longer a given, and increasingly "middle ground" positions have been explored. Case studies of relevant episodes in the history of science show us the specific ways in which a realist view may be tempered, but without necessarily collapsing into a full-blown antirealist view. A central part of this is that self-proclaimed "realists," as well as "antirealists" and "instrumentalists," are exploring historical cases in order to learn

Timothy D. Lyons and Peter Vickers, *History and the Contemporary Scientific Realism Debate* In: *Contemporary Scientific Realism*. Edited by: Timothy D. Lyons and Peter Vickers, Oxford University Press. © Oxford University Press 2021.
DOI: 10.1093/oso/9780190946814.003.0001

the relevant lessons concerning precisely what has, and what hasn't, been retained across theory change. In recent decades one of the most important developments has been the "*divide et impera* move," introduced by Philip Kitcher (1993) and Stathis Psillos (1999)—possibly inspired by Worrall (1989)—and increasingly embraced by numerous other realists. According to this "selective" strategy, any realist inclinations are directed toward only those theoretical elements that are really doing the inferential work to generate the relevant successes (where those successes are typically explanations and predictions of phenomena). Such a move is well motivated in light of the realist call to *explain* specific empirical successes: those theoretical constituents that play no role, for instance, in the reasoning that led to the successes, likewise play no role in explaining those successes. Beyond that, however, and crucially, the *divide et impera* move allows for what appears to be a testable realist position consistent with quite dramatic theory-change: the *working parts* of a rejected theory may be somehow retained within a successor theory, even if the two theories differ very significantly in a great many respects. Even the working parts may be retained in a new (possibly approximating) form, such that the retention is not obvious upon an initial look at a theoretical system but instead takes considerable work to identify. This brings us to a new realist position, consistent with the thought that many of our current best theories may one day be replaced. Realism and antirealism are no longer quite so far apart, and this is progress.

A major stimulus for, and result of, this progress has been a better understanding of the history of science. Each new case study brings something new to the debate, new lessons concerning the ways in which false theoretical ideas have sometimes led to success, or the ways in which old theoretical ideas "live on," in a different form, in a successor theory. At one time the debate focused almost exclusively on three famous historical cases: the caloric theory of heat, the phlogiston theory of combustion, and the aether theory of light. An abundance of literature was generated, and with good reason: these are extremely rich historical cases, and there is no simple story to tell of the successes these theories enjoyed, and the reasons they managed to be successful despite being very significantly misguided (in light of current theory). But the history of science is a big place, and it was never plausible that all the important lessons for the debate could be drawn from just three cases. This is especially obvious when one factors in a "particularist" turn in philosophy of science where focus is directed toward particular theoretical systems. These days, many philosophers are reluctant to embrace grand generalizations about "science" once sought by philosophers, taking those generalizations to be a dream too far. Science works in many different ways, in different fields and in different contexts. It follows that the realism debate ought to be informed by a rich diversity of historical cases.

It is with this in mind that the present volume is put forward. In recent years a flood of new case studies has entered the debate, and is just now being worked through. At the same time it is recognized that there are still many more cases out there, waiting to be analyzed in a particular way sensitive to the concerns of the realism debate. The present volume advertises this fact, introducing as it does several new cases as bearing on the debate, or taking forward the discussion of historical cases that have only very recently been introduced. At the same time, the debate is hardly static, and as philosophical positions shift this affects the very *kind* of historical cases that are likely to be relevant. Thus the work of introducing and analyzing historical cases must proceed hand in hand with philosophical analysis of the different positions and arguments in play. It is with this in mind that we divide the present volume into two parts, covering "Historical Cases for the Debate" and "Contemporary Scientific Realism," each comprising of seven chapters.

Many of the new historical cases are first and foremost challenges to the realist position, in that they tell scientific stories that are apparently in tension with even contemporary, nuanced realist claims. The volume kicks off in Chapter 2 with just such a case, courtesy of Dana Tulodziecki. For Tulodziecki, the miasma theory of disease delivered very significant explanatory and predictive successes, while being radically false by present lights. Further, even the parts of the theory doing the work to bring about the successes were not at all retained in any successor theory. Thus, Tulodziecki contends, the realist must accept that in this case false theoretical ideas were instrumental in delivering successful explanations and predictions of phenomena.

This theme continues in Chapter 3, with Jonathon Hricko's study of the discovery of boron. This time the theory in question is Lavoisier's oxygen theory of acidity. Hricko argues that the theory is not even approximately true, and yet nevertheless enjoyed novel predictive success of the kind that has the power to persuade. Just as Tulodziecki, Hricko argues that the realist's *divide et impera* strategy can't help—the constituents of the theory doing the work to generate the prediction cannot be interpreted as approximately true, by present lights.

We meet with a different story in Chapter 4, however. Keith Hutchison provides a new case from the history of thermodynamics, concerning the successful prediction that pressure lowers the freezing point of water. Hutchison argues that although, at first, it seems that false theoretical ideas were instrumental in bringing about a novel predictive success, there is no "miracle" here. It is argued that the older Carnot theory and the newer Clausius theory are related in intricate ways, such that in some respects they differ greatly, while in "certain restricted situations" their differences are largely insignificant. Thus the realist may find this case useful in her bid to show how careful we need to be when we draw antirealist morals from the fact that a significantly false theory enjoyed novel predictive success.

Chapter 5 takes a slightly different approach, moving away from the narrow case study. Stathis Psillos surveys a broad sweep of history ranging from

Descartes through Newton and Einstein. We have here a "double case study," considering the Descartes–Newton relationship and the Newton–Einstein relationship. Psillos argues against those who see here examples of dramatic theory-change, instead favoring a limited retentionism consistent with a modest realist position. Crucially, Psillos argues, there are significant differences between the Descartes–Newton relationship and the Newton–Einstein relationship, in line with the restricted and contextual retention of theory predicted by a nuanced, and epistemically modest, contemporary realist position.

Chapter 6—courtesy of Eric Scerri—introduces another new historical case, that once again challenges the realist. This time the theory is John Nicholson's atomic theory of the early 20th century, which, Scerri argues, "was spectacularly successful in accommodating as well as predicting some spectral lines in the solar corona and in the nebula in Orion's Belt." The theory, however, is very significantly false; as Scerri puts it, "almost everything that Nicholson proposed was overturned." Hence, this case is another useful lesson in the fact that quite radically false scientific theories *can* achieve novel predictive success, and any contemporary realist position needs to be sensitive to that.

Chapter 7 turns to theories of molecular structure at the turn of the 20th century. Amanda Nichols and Myron Penner show how the "old" Blomstrand-Jørgensen chain theory was able to correctly predict the number of ions that will be dissociated when a molecule undergoes a precipitation reaction. While *prima facie* a challenge to scientific realism, it is argued that this is a case where the *divide et impera* strategy succeeds: the success-generating parts of the older theory are retained within the successor, Werner's coordination theory.

The final contribution to the "History" part of the volume—Chapter 8—concerns molecular spectroscopy, focusing on developments in scientific "knowledge" and understanding throughout the 20th century and right up to the present day. Teru Miyake and George E. Smith take a different approach from the kind of historical case study most commonly found in the realism debate. Siding with van Fraassen on the view that Perrin's determination of Avogadro's number, so commonly emphasized by realists, does not prove fruitful for realism, and focusing on diatomic molecules, they emphasize instead the extraordinary amount of evidence that has accumulated after Perrin and over the past ninety years for various theoretical claims concerning such molecules. Taking van Fraassen's constructive empiricism as a foil, they indicate that, in this case, so-called realists and antirealists may really differ very little when it comes to this area of "scientific knowledge."

The second part of the volume turns to more general issues and philosophical questions concerning the contemporary scientific realist positions. This part kicks off in Chapter 9 with Mario Alai's analysis of the *divide et impera* realist strategy: he argues that certain historical cases no longer constitute counterexamples when hypotheses are *essential* in novel predictions. He proposes refined criteria of essentiality, suggesting that while it may be impossible to identify exactly which components are essential in *current* theories, recognizing in hindsight those which were *not* essential in past theories is enough to make the case for deployment realism.

In Chapter 10, Kyle Stanford makes the case that so-called realists and instrumentalists are engaged in a project together. For Stanford, these traditionally diametrically opposed protagonists are now working together to "actively seek to identify, evaluate, and refine candidate indicators of epistemic security for our scientific beliefs." As philosophy of science, and the realism debate in particular, becomes increasingly "local," Stanford also argues for "epistemic guidance intermediate in generality" between the sweeping generalizations of 1970s and '80s realism, and a radical "particularism" where any realist claim should always be specific to one particular theory, or theoretical claim.

Chapter 11 turns to the relationship between the realist's *divide et impera* strategy and the structural realist position. James Ladyman argues that structural realism is not a form of selective realism (or at least doesn't have to be). For Ladyman, structural realism represents a departure from standard scientific realism, not a modification of it. He also argues that scientific realists face ontological questions (not only epistemic ones), and he defends a "real patterns" approach to what he calls the "scale relativity of ontology." This allows for equally "realist" claims to be made at the level of fundamental physics and at the macroscopic level.

In Chapter 12, Jennifer Jhun explores the realism debate in a different territory. In particular, she considers the possibility of taking a structural realist attitude toward macroeconomic theory. Taking the consumption function as a case study, she argues that a better take on macroeconomic theory involves a compromise between structural realism and instrumentalism. For Jhun, when it comes to economics (at least), "theories are instruments used to find out the truth."

In Chapter 13, Ludwig Fahrbach considers a prominent antirealist argument against the claim that realism can be defended against the pessimistic meta-induction (PMI) by invoking the exponential growth of scientific evidence. The antirealist response to this defense depends on the claim that realists could have said the same thing in the past. He introduces this antirealist response as the "PMMI," the pessimistic meta-meta-induction. Fahrbach's challenge focuses on a particular weak spot common to both the traditional PMI and the PMMI. Thus realists unimpressed by the traditional PMI will not be moved by the new PMMI.

Chapter 14 turns to another important aspect of the modern realism debate: the use of "radically false" theoretical assumptions such as infinite limits in many contemporary, highly successful theories. Patricia Palacios and Giovanni Valente note how "infinite idealizations" misrepresent the target system, and sometimes it appears that the introduction of such blatant falsity is necessary to achieve empirical success. Focusing on various examples in physics, such as classical and quantum phase transitions as well as thermodynamically reversible processes, Palacios and Valente propose a realist response to such cases.

Last but not least, in Chapter 15 Anjan Chakravartty tackles the standard model of particle physics, and in particular what the realist might say about representations of fundamental particles. Introducing the realist "tightrope," Chakravartty discusses the trade-off between committing to too little, and committing to too much. But in the end, he argues, this is a tightrope that is not too thin to walk.

These articles have been carefully collected for this volume over many years, and in particular during the Lyons/Vickers 2014–18 Arts and Humanities Research Council (AHRC) project "Contemporary Scientific Realism and the Challenge from the History of Science." The project enjoyed seven major events over its lifetime, out of which the fourteen substantive chapters of this volume ultimately grew. The seven events were:

(i)   "The History of Chemistry and Scientific Realism," a two-day workshop held at Indiana University–Purdue University Indianapolis, United States, December 6–7, 2014.

(ii) "The History of Thermodynamics and Scientific Realism," a one-day workshop held at Durham University, UK, on May 12, 2015.

(iii) "Testing Philosophical Theories against the History of Science," a one-day workshop held at the Oulu Centre for Theoretical and Philosophical Studies of History, Oulu University, Finland, on September 21, 2015.

(iv) "Quo Vadis Selective Scientific Realism?"—a symposium at the biennial conference of the European Philosophy of Science Association, Düsseldorf, Germany, September 23, 2015.

(v)   "Contemporary Scientific Realism and the Challenge from the History of Science," a three-day conference held at Indiana University–Purdue University, United States, February 19–21, 2016.

(vi) "Quo Vadis Selective Scientific Realism?"—a three-day conference held at Durham University, UK, August 5–7, 2017.

(vii) "The Structure of Scientific Revolutions," a two-day workshop held at Durham University, UK, October 30–31, 2017.

The editors of this volume owe a great debt to the participants of all of these events, with special thanks in particular to those who walked this path with us a little further to produce the fourteen excellent chapters here presented. We are also grateful to the unsung heroes, the many anonymous reviewers, who not only helped us to select just which among the numerous papers submitted would be included in the volume but also provided thorough feedback to the authors, helping to make each of the chapters that did make the cut even stronger. The volume has been a labor of love; we hope that comes across to the reader.

References

Kitcher, P. (1993). *The Advancement of Science*, Oxford: Oxford University Press.
Laudan, L. (1981). "A Confutation of Convergent Realism." *Philosophy of Science* 48: 19–49.
Psillos, S. (1999). *Scientific Realism: How Science Tracks Truth*, London: Routledge.
Van Fraassen, B. (1980). *The Scientific Image*, Oxford: Oxford University Press.
Worrall, J. (1989). "Structural Realism: The Best of Both Worlds?" *Dialectica* 43: 99–124.

PART I
HISTORICAL CASES
FOR THE DEBATE

2

Theoretical Continuity, Approximate Truth, and the Pessimistic Meta-Induction: Revisiting the Miasma Theory

Dana Tulodziecki

Department of Philosophy
Purdue University
dtulodzi@purdue.edu

2.1 Introduction

The pessimistic meta-induction (PMI) targets the realist's claim that a theory's (approximate) truth is the best explanation for its success. It attempts to do so by undercutting the alleged connection between truth and success by arguing that highly successful, yet wildly false theories are typical of the history of science. There have been a number of prominent realist responses to the PMI, most notably those of Worrall (1989), Kitcher (1993), and Psillos (1999). All of these responses try to rehabilitate the connection between a theory's (approximate) truth and its success by attempting to show that there is some kind of continuity between earlier and later theories, structural in the case of Worrall and theoretical/referential in the cases of Kitcher and Psillos, with other responses being variations on one of these three basic themes.[1]

In this paper, I argue that the extant realist responses to the PMI are inadequate, since there are cases of theories that were both false and highly successful (even by the realist's own, more stringent criteria for success) but that, nevertheless, do not exhibit any of the continuities that have been

---

[1] It is worth pointing out that this is the case even for the (very) different structural realisms that abound. In particular, even ontic structural realism, which differs substantially from Worrall's epistemic version, shares with Worrall's approach an emphasis on mathematical continuities among successor theories (see, for example, Ladyman (1998) and French and Ladyman (2011)).

Dana Tulodziecki, *Theoretical Continuity, Approximate Truth, and the Pessimistic Meta- Induction: Revisiting the Miasma Theory* In: *Contemporary Scientific Realism*. Edited by: Timothy D. Lyons and Peter Vickers, Oxford University Press. © Oxford University Press 2021. DOI: 10.1093/oso/9780190946814.003.0002 suggested by realists as possible candidates for preservation. I will make my case through discussing an example of such a theory: the 19th century miasma theory of disease. Specifically, I show that this theory made a number of important and successful use-novel predictions, despite the fact that its central theoretical element—miasma—turned out not to exist. After showing that miasma was crucially involved in virtually every successful prediction the miasma theory made, I argue that not just is there no ontological continuity between the miasma theory and its successor, but neither can a case be made for any other kind of continuity, be it in terms of structure, laws, mechanisms, or kind-constitutive properties. I conclude by arguing that realists face problems regardless of whether the miasma theory is approximately true: if it is, the prospects for any kind of substantive realism are dim; if it is not, the miasma case constitutes a strike against the "convergent" part of convergent realism.

I will proceed as follows: In section 2.2, I briefly outline the PMI and the responses by Worrall, Kitcher, and Psillos, including the more stringent notion of success that Psillos argues is required in order for theories to be genuinely successful. In section 2.3, I outline the most sophisticated version of the miasma theory of disease and show that it ought to be considered a genuinely successful theory, even on Psillos's own terms. Section 2.4 is concerned with showing that none of the extant realist responses to the PMI can account for the case of the miasma theory. After discussing some objections in section 2.5, I conclude in section 2.6.

2.2 The PMI and Realist Responses

The PMI received its most sophisticated and explicit formulation in Laudan (1981, 1984). The argument's main target is the Explanationist Defense of Realism, according to which the best explanation for the success of science is the (approximate) truth of our scientific theories (see, for example, Boyd (1981, 1984, 1990)). Anti-realists employ the PMI to undercut this alleged connection between truth and success by pointing to the (in their view) large number of scientific theories that were discarded as false, yet regarded as highly successful. Laudan, for example, provides a list of such theories that includes, among others, the phlogiston theory of chemistry, the caloric theory of heat, and the theory of circular inertia (1984: 121). However, since "our own scientific theories are held to be as much subject to radical conceptual change as past theories" (Hesse 1976: 264), it follows, the anti-realist argues, that the success of our current theories cannot legitimize belief in unobservable entities and mechanisms: just as past theories ended up wrong about their postulates, we might well be wrong about ours.[2]

Realist responses to the PMI typically come in several stages: first, realists try to winnow down Laudan's list by including only those theories that are genuinely successful (Psillos 1999: chapter 5). While, according to Laudan, a theory is successful "as long as it has worked reasonably well, that is, so long as it has functioned in a variety of explanatory contexts, has led to several confirmed predictions, and has been of broad explanatory scope" (1984: 110), Psillos argues that "the notion of empirical success should be *more* rigorous than simply getting the facts right, or telling a story that fits the facts. For any theory (and for that matter, any wild speculation) can be made to fit the facts—and hence to be successful—by simply "writing" the right kind of empirical consequences into it. The notion of empirical success that realists are happy with is such that it includes the generation of novel predictions which are in principle testable" (1999: 100). The specific type of novel prediction that Psillos has in mind is so-called use-novel prediction: "the prediction *P* of a known fact is use-novel *relative to a theory T,* if no information about this phenomenon was used in the construction of the theory which predicted it" (101). Once success is understood in these more stringent terms, Psillos claims, Laudan's list is significantly reduced.

With the list so shortened, the next step of the realist maneuver is to argue that those theories that remain on Laudan's list ought to be regarded as (approximately) true, since they don't, in fact, involve the radical discontinuity with later theories that Laudan suggests. Rather than being discarded wholesale during theory-change, it is argued that important elements of discarded theories get retained: according to Worrall (1989, 1994), theories' mathematical structures are preserved, according to Kitcher (1993) and Psillos (1996, 1999), those parts of past theories that were responsible for their success are. Because these components are preserved in later theories, the argument goes, we ought to regard as approximately true those parts of our current theories that are essentially involved in generating their successes, since those are the parts that will carry over to future theories, just as essential elements from earlier theories were carried over to our own. As I will show in section 2.4, however, none of these strategies will work for the miasma theory: despite the fact that the miasma theory made a number of use-novel predictions, and so counts as genuinely successful by realist standards, none of the elements or structures that were involved in those successes were retained by its successor.[3]

---

[2] Laudan's argument is usually construed as a reductio (see Psillos (1996, 1999)). For some recent discussions about how to properly interpret the argument, see Lange (2002), Lewis (2001), and Saatsi (2005).
[3] The miasma case is especially significant in view of the markedly different analysis that Saatsi and Vickers (2011) provide of Kirchhoff's theory of diffraction. Saatsi and Vickers diagnose a specific kind of underdetermination at work in the Kirchhoff case and argue that "it should not be implausible

2.3 The Miasma Theory and Its Successes

The most sophisticated version of the miasma theory saw its heyday in the mid-1800s. The situation with respect to the various accounts of disease at that time was fairly complicated, however, and so it is in some sense misleading to speak of *the* miasma theory of disease, since there was a whole cluster of related views that went under this label, rather than one easily identifiable position (see, for example, Baldwin (1999), Eyler (2001), Hamlin (2009), Pelling (1978), and Worboys (2000)). However, since all members of that cluster shared basic assumptions about the existence and nature of miasma, I will disregard this complication here, and treat them as one. According to this basic miasma view, diseases were caused and transmitted by a variety of toxic odors ("miasmas") that themselves were the result of rotting organic matter produced by putrefaction or decomposition. The resulting bad air would then act on individual constitutions to cause one of several diseases (cholera, yellow fever, typhus, etc.), depending on a number of more specific factors. Some of these were thought to be extraneous, such as weather, climate, and humidity, and would affect the nature of the miasmas themselves; others were related directly to the potential victims and thought to render them more or less susceptible to disease, such as their general constitution, moral sense, age, and so on. Lastly, there were a variety of local conditions that could exacerbate the course and severity of the disease, such as overcrowding and bad ventilation.

Although the miasma theory is sometimes contrasted with so-called "contagionist" views of disease (the view that diseases could be transmitted directly from individual to individual), this opposition is also misleading, for two reasons: first, because both contagionists and anti-contagionists subscribed to the basic miasmatic assumptions just described, with the debate not centering on the existence or nature of miasmas, but, rather, on whether people themselves were capable of producing additional miasma-like effects and through these "exhalations" directly give the disease to others. Second, although some diseases were generally accepted as contagious (smallpox, for example), and some were generally held to be non-contagious (malaria), most diseases (yellow fever, cholera, typhoid fever, typhus, and so on) fell somewhere in between these

to anyone that given the enormous variation in the nature and methods of scientific theories across the whole spectrum of 'successful science', some domains of enquiry can be more prone to this kind of underdetermination than others" (44). In fact, they believe that "witnessing Kirchhoff's case, there is every reason to expect that from a realist stance we can grasp the features of physics and mathematics that contribute to such differences" (44). As a result, they take themselves to have "argued for the prima facie plausibility" of a realist response that focuses on "showing how the field of theorizing in question is idiosyncratic in relevant respects, so that Kirchhoff's curious case remains isolated and doesn't provide the anti-realist with grounds for projectable pessimism" (44). The miasma case, however, is not prone to any of the idiosyncrasies Saatsi and Vickers identify (or any other idiosyncrasies, as far as I can tell).

two extremes. Pelling (1978: 9) appropriately terms these diseases the "doubtful diseases": instead of being straightforwardly contagionist or anti-contagionist about them, people espoused so called "contingent contagionism" with respect to them, holding that they could manifest as either contagious or non-contagious, depending on the exact circumstances and, sometimes, even transform from one into the other (see also Hamlin (2009)).

This version of the miasma theory was extremely successful with respect to a number of different types of phenomena. Most famous are probably the sanitary successes that it ultimately inspired, but it also managed to provide explanations of disease phenomena that any theory of disease at the time had to accommodate. These included the fact that diseases were known to be highly seasonal, that particular regions (especially marshy ones) were affected particularly harshly, that specific locations (prisons, workhouses, etc.) were often known to suffer worse than their immediate surroundings, that sickness and mortality rates in urban centers were much worse than those in rural areas, and why particular geographical regions/countries were struck much worse by disease than others. The miasma theory managed to explain all of these through its claims that decomposition and putrefaction of organic material was responsible for producing miasmas. Diseases peaked when conditions for putrefaction were particularly favorable: this was the reason why certain diseases were particularly bad during periods of high temperature and in certain geographical regions (for example, the many fevers in Africa), why urban centers were much more affected than rural areas, and why even specific locations in otherwise more or less healthy areas could be struck (sewage, refuse, and general "filth" would sit around in badly ventilated areas). It should also be noted that miasma theorists were not just making vague or simple-minded pronouncements about stenches producing toxic odors, but embraced very specific and often highly complex accounts of how various materials and conditions gave rise to miasmas—Farr, for example, drew in some detail on Liebig's chemical explanations (1852, lxxx-lxxxiii; see also Pelling (1978: chapters 3 and 4) and Tulodziecki (2016)). Moreover, there were debates about exactly what sorts of materials were prime for potential miasmas, such as debates about various sources of vegetable vs. animal matter, and so on. Since, however, I cannot do justice to the details of these accounts and their corresponding successes here, I will focus on a somewhat simpler, yet particularly striking, example of use-novel success, while merely noting that others could be given.

The example in question is that of William Farr's (then famous) elevation law (1852). Farr (1807–83), although not himself a physician, was viewed as an authority on infectious diseases in mid-1800s Britain. Among the positions he held were that of Statistical Superintendent of the General Register Office and member of the Committee of Scientific Inquiries. Through his various writings, especially those on the various big British cholera epidemics, he established his credentials as a medical authority.[4]

As we have seen, according to the miasma theory, any decomposing organic material could in principle give rise to miasmas. However, it was thought that the soil at low elevations, especially around riverbanks, was a particularly good source for producing highly concentrated miasma, since such soil held plenty of organic material and the conditions for putrefaction were particularly favorable. It thus followed directly from the miasma theory that, if miasmas were really produced in the manner described and responsible for disease, the concentration of noxious odors ought to be higher closer to such sources and dissipate with increasing distance. Correspondingly, it was to be expected that mortality and sickness rates would be higher in close proximity to sources of miasma, declining as one moved away. Farr confirmed that this was the case, in a number of different ways.

First, he found that "nearly 80 percent of the 53,000 registered cholera deaths in 1849 occurred among four-tenth of the population living on one-seventh of the land area" (Eyler 1979: 114), and, moreover, that, "cholera was three times more fatal on the coast than in the interior of the country" (Farr 1852: lii). Furthermore, he found that those deaths that did occur inland were in either seaport districts or close to rivers, noting that "[c]holera reigned wherever it found a dense population on the low alluvial soils of rivers" and that it "was almost invariably most fatal in the port or district lying lowest down the river" (lii). Concerning coastal deaths, he found that the "low-lying towns on the coast were all attacked by cholera," while the high-lying coast towns "enjoyed as much immunity as the inland towns" (liv). Further, he noted that mortality increased and decreased relative to the size of the port, with smaller ports having lower mortality. The Welsh town of Merthyr-Tydfil constituted an exception to this, having a "naturally" favorable location, yet relatively high mortality. However, Farr also noted that Merthyr-Tydfil had a reputation, with the Health of Towns' Commissioners' Report noting that "[f]rom the poorer class of the inhabitants, who constitute the mass of the population, throwing all slops and refuse into the nearest open gutter before their houses, from the impeded courses of such channels, and the scarcity of privies, some parts of town *are complete networks of filth emitting noxious exhalations*" (liv). In addition, much of the refuse was being carried to the local riverbeds, with the result that "the stench is almost intolerable in many places" (lv). In one area, close to the river, an "open, stinking, and nearly stagnant gutter, into which the house refuse is as usual flung, moves slowly before the doors" (lvi).

---

[4] For more on Farr's life and ideas, see Eyler (1979).

All of these phenomena were exactly what was to be expected on the miasma theory: wherever there was disease, one ought to be able to trace it back to miasmatic conditions and, similarly, wherever conditions particularly favorable to decomposition were to be found, disease ought to be rampant. If there were exceptions to the general rule, towns such as Merthyr-Tydfil that were not naturally vulnerable, one ought to be able to find alternative sources of miasma without difficulty. Farr also proceeded to check these results against a variety of data from different countries and concerning different diseases, and further confirmed what he had found. Moreover, the miasma theory did not merely accommodate these findings but, rather, all of the phenomena followed naturally from the account.

In addition, while the miasma theory made predictions about what areas ought to be affected by cholera and to what degrees, for example, none of these predictions were confirmed until Farr analyzed the data from the General Board of Health in the late 1840s (indeed, due to the fact that much of this data was not collected until shortly before that time, it would have been impossible to confirm these predictions until then). Thus, since it was not even clearly known to what extent these predictions were borne out, they could not have played a role in formulating the miasma theory in the first place and, so, ought to count as use-novel. One might object that, use-novel or not, these predictions were simply too vague to qualify a theory as genuinely successful in the realist sense. I will note, however, that (i) the predictions were as specific as a theory of this type would allow, (ii) the predictions were no more vague than Snow's later predictions about what ought to be expected on the (correct) assumption that cholera was waterborne (see Snow 1855a, 1855b) and, so, if one regards Snow's predictions as successful, one ought to also regard the miasmatic predictions as successful, and (iii) that miasmatists actually did a lot better than this (and, indeed, better than Snow ever did) by providing some detailed and quantifiable results.

Farr's elevation law is a particularly striking example of this and it is to this that I will turn now. Farr's law related cholera mortality to the elevation of the soil. However, Farr did not just predict that there ought to be a relationship between these two variables, but upon analyzing more than 300 pages of data, he found that the "mortality from cholera is in the inverse ratio of the elevation" (Farr 1852: lxi). Farr grouped the various London districts by their altitude above the Thames high water mark, in brackets of 20 feet (0–20, 20–40, and so on) and was able to capture the exact relation between the decline of cholera and increased soil elevation in the form of an equation. Specifically, he found that:

The mortality from cholera on the ground under 20 feet high being represented by 1, the relative mortality in each successive terrace [i.e. the terraces numbered

(x)			
Elevation of Districts, in feet.	Number of Terrace from bottom.	Deaths from Cholera in 10000 Inhabitants.	Calculated Series (1.)
Feet.			
20—	1	102	$\frac{102}{1} = 102$
20—40	2	65	$\frac{102}{2} = 51$
40—60	3	34	$\frac{102}{3} = 34$
60—80	4	27	$\frac{102}{4} = 26$
80—100	5	22	$\frac{102}{5} = 20$
100—120	6	17	$\frac{102}{6} = 17$
340—360	18	7	$\frac{102}{18} = 6$

Figure 2.1   Farr, William. (1852). "Report on the Mortality of Cholera in England, 1848–49." London: Printed by W. Clowes, for H.M.S.O., 1852, p. lxii.

"2," "3," etc.] is represented by ½ [for terrace 2], ⅓ [for terrace 3], ¼, ⅕, ⅙: or the mortality on each successive elevation is ½, ⅔, ¾, ⅘, ⅚, &c. of the mortality on the terrace immediately below it. (ibid.: lxiii; see Figure 2.1)

He then generalized this result: "Let e be any elevation within the observed limits 0 and 350, and c be the average rate of mortality from cholera at that elevation; also let e' be any *higher* elevation, and c' the mortality at that higher elevation" (lxiii). Then, adding a as a constant, Farr found that the formula

$$c = c' \times (e' + a)/(e + a)$$

represented the decreasing series he obtained when considering the mortality from cholera for districts with specific mean elevations. Farr then calculated the expected series according to the formula, compared it to the actual series recorded in London, and found remarkable agreement (see Figure 2.2, lxiii).

Seeking further confirmation, Farr immediately proceeded to "submit the principle to another test, by comparing the elevation and the mortality from cholera of *each sub-district*," and found that this "entirely confirms the announced law" (xv–xvi).[5] Trying to illustrate just how good Farr's numbers were and how

---

[5] The tables appear on pp. clxvi–ix of Farr's *Report*. (1852).

Mean Elevation of the ground above the High-water Mark.	Mean Mortality from Cholera.	Calculated Series.
o	177	174
10	102	99
30	65	53
50	34	34
70	27	27
90	22	22
100	17	20
350	7	6

Figure 2.2   Farr, William. (1852). "Report on the Mortality of Cholera in England, 1848–49." London: Printed by W. Clowes, for H.M.S.O., 1852, p. lxiii

convincing they must have seemed, Langmuir (1961: 174) plotted Farr's result about cholera mortality in the various London subdistricts, grouped by elevation (Figure 2.3). According to Langmuir, Farr had "found a confirmation that I believe would be impressive to any scientist at any time" (173).

Even more remarkably, it turned out that Farr's predictions did not just hold for the (sub-)districts of London, but were also confirmed by others in different regions. For example, "William Duncan, Medical Officer of Health for Liverpool, wrote that when he grouped the districts of his city by elevation as Farr had done, that cholera mortality in the last epidemic obeyed Farr's elevation law for Liverpool as well" (Eyler 1979: 228).

Now, these predictions of Farr's were certainly use-novel: Farr was predicting new phenomena that were hitherto unknown and that were later borne out by a variety of data from different regions, in different contexts, and from different times. Moreover, Farr's law clearly could not have played a role in the construction of the miasma theory since, first, it was obviously not even formulated by then, but, second, even the data on which the law was based (the statistics from the General Register Office, collected on Farr's initiative) did not exist and, indeed, in the case of Duncan in Liverpool, no one had even thought about collecting the relevant information. In short: it followed from Farr's law that cholera mortality and soil elevation ought to exhibit a specific relation that was then found to occur in the various sub-districts of London and various other parts of the country.

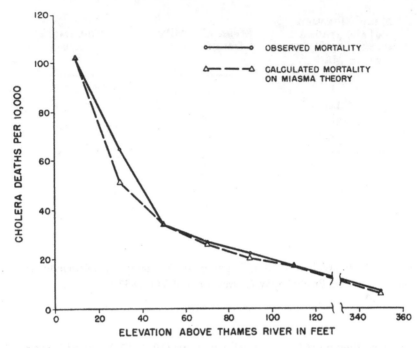

Figure 2.3   Correlation of cholera mortality and elevation above the Thames River, London, 1849. Langmuir, Alexander D. (1961). "Epidemiology of Airborne Infection." *Bacteriological Reviews*, 25(3), p. 174.

2.4 Miasmas and Realists

As we have seen, the miasma theory made a number of use-novel predictions and so, by the realist's own standard, ought to count as genuinely successful.[6] However, the existence of such cases—successful, yet false theories—is not enough, by itself, to spell trouble for realists. After all, realists themselves admit that there are such examples. They also argue that we can explain these cases by showing that crucial elements of those theories—those that were essentially involved in the theory's success—are preserved by later

---

[6] Doppelt (2007) argues that requiring novel predictions is not sufficient for genuine success; instead, we ought to also look for various explanatory virtues, such as consilience, simplicity, or unifying power. To make a detailed case for this would take us too far afield here, but it can be shown that the miasma theory also exhibits some of these more traditional virtues. Certainly one might argue that independent strands of evidence (epidemiological and pathological, for example) produced a degree of consilience and that the miasma theory unified a number of different phenomena (regarding the various diseases) in a simple and elegant way by providing essentially the same kind of explanation for a number of different afflictions.

theories. Kitcher (1993: 149) distinguishes working from presuppositional posits and argues that cases support the PMI only if it is found that a theory's working posits (those elements of a theory essential for making predictions) do not refer. Psillos resorts to what he calls the *divide et impera* move, and claims that "it is enough to show that the theoretical laws and mechanisms which generated the successes of past theories have been retained in our current scientific image" (1999: 108). Worrall (1989: 11) also argues that parts of past theories get retained, but instead of focusing on theoretical content, Worrall's suggestion is that there is retention at the structural level, specifically in the form of mathematical content. This view does not commit Worrall to the existence of unobservable entities and mechanisms, the way the approaches of Kitcher and Psillos do, but it also goes beyond the merely empirical.

Despite the fact that we have several options for what sorts of elements might be retained in order to support some version of realism or another, it turns out that the miasma theory does not exhibit any of the above realist continuities. First and foremost, miasmas were crucially involved in virtually every prediction the miasma theory made—thus, some kind of miasmatic continuity is essential to both Kitcher and Psillos: miasmas are working, not presuppositional posits on Kitcher's terms, and essential, not idle, components on Psillos's. Because miasmas were thought to be the cause of diseases and also the mechanism for transmission, all of the predictions—such as where the incidence of disease ought to be particularly high or low, what populations would be particular targets for diseases, what regions ought to suffer to what degrees, and correlations between distance from concentrated sources of miasma and mortality rates—essentially depended on miasmas themselves: without reliance on the concept of miasma and its rotting sources, there simply would not have been any predictions whatsoever.[7]

Now, while it's clear that there is no straightforward ontological continuity between the miasma theory and the germ theory, perhaps some sort of referential stability can be salvaged in other ways. But there are no other viable candidates: the mechanisms of disease transmission were not retained—bad

---

[7] Vickers' discussion of possible refinements of the *divide et impera* move is worth mentioning here, in particular his suggestion that, instead of the usual focus on working posits "it remains possible that we might develop a recipe for identifying certain idle posits" (2013: 209). Vickers develops an account of how this might have gone in the case of Kirchhoff's theory of diffraction. However, it is unclear how to extend Vickers' conclusions from the Kirchhoff case to the miasma theory, and so I will not discuss his views in any detail here. Note, however, that Vickers mentions further potential cases (see also Lyons (2006), especially his discussion of Kepler's *Mysterium Cosmographicum*).

smells don't transmit disease—and neither were any of the laws.[8] Specifically, not only is there no analogue of Farr's law in any of the modern disease theories, but not even the phenomenon associated with it—the connection between cholera mortality and soil elevation—was retained.

Incidentally, this last example is also the best bet as a candidate for Worrall's structure. Since Worrall emphasizes mathematical components, and there is virtually no mathematical structure to be had in the miasma theory, the only candidates for this view were of the kind put forward by Farr. One might object that there were other kinds of statistics that were retained: claims, for example, about the peaks, courses, and durations of epidemics. Moreover, these didn't involve or rely on miasma or its properties, and so one might think that these are prime candidates for preservation. The problem with these, however, is that while these are good candidates for preservation they are not good candidates for realism, since they are all observational. True, they don't depend on any theoretical components of the miasma theory (or any other disease theory, for that matter), but they don't depend on any other theoretical account either. Rather, this is merely empirical data—a constraint with which any viable theory of disease has to work. Realists and anti-realists agree on these data sets, and there's absolutely no debate here about approximate truth or realism about unobservables—and realism about observables was never the issue.

So, to sum up: the miasma theory made several use-novel predictions and so counts as genuinely successful on the realist's own terms. Further, it does not exhibit any kind of realist continuity, neither theoretical, nor referential, nor structural. There are no miasmas, the laws the theory gave rise to have been abandoned, its mechanisms of transmission turned out not to exist, and the properties that were ascribed to miasmas are not now ascribed to any of its etiological successors.

2.5   Objections

One objection realists might raise against Farr is that his prediction was not really use-novel, since it was based on empirical data and hence insufficiently grounded in theoretical assumptions. Thus, the realist might argue, as Vickers has done with respect to Meckel, that Farr "really reached his conclusion not via his (false) theoretical ideas, but rather via his empirical knowledge" (Vickers 2015).

---

[8] One also cannot make a case for Psillos's kind-constitutive properties (see Psillos 1999: chapter 12): miasma is akin in this respect to phlogiston, not the luminiferous ether, and an analogous version of the argument that Psillos makes against phlogiston will also work for miasma.

First, note that it is true that once all the data is in, in some sense the original theory that prompted the prediction can be discarded. However, just because this is possible does not mean that there was no theoretical basis for the prediction in the first place. In Farr's case, there were strong theoretical underpinnings: both his predictions and the empirical data he used were dependent on his disease theory. This data was not just readily available for him; instead, he needed to collect and generate it, which, given the extremely limited contemporary means at his disposal, amounted to a gargantuan effort. Moreover, he did not just have to gather the raw data, which was already extremely work-intensive but, in order to make well-formed predictions, he also needed to organize and interpret this data appropriately. Even here, his work was not straightforwardly empirical but based on a number of theoretically generated assumptions, for example about how to construct appropriate mortality rates, and so on. None of these efforts made sense unless one subscribes to a disease theory according to which such relationships are to be expected. Without his disease theory, there would never have been any reason for Farr to gather any of the data he did and, indeed, on other disease theories—such as Snow's, say—there would not have been any point in collecting it. Without Farr, quite possibly the relation between elevation and cholera mortality would have never been discovered; certainly, without the accompanying theory, it would have been meaningless.

One might still object that Farr's prediction amounts to mere curve-fitting and his data to "more of the same."[9] However, there is nothing "mere" about curve-fitting in science. Curve-fitting is widespread, standard practice, and it is a mistake to think that it involves nothing besides empirical data. On the contrary, most scientific curve-fitting has important theoretical elements: most obviously, it has to be decided what the relevant parameters of the curve are, what family of curves is appropriate, and so on. Further, picking a curve comes with predictions about future data points, and here different curves will of course make different predictions. As a result, finding that future data fits one's curve confirms the idea that one has hit on the (or at least a) right relation. Moreover, this data is "more of the same" only if one has, in fact, hit upon a curve that works.[10] Importantly for this discussion, all sorts of scientific laws involve curve-fitting, so if realists object to Farr's law on these grounds, they ought to also object to other laws.[11]

---

[9] An anonymous referee voiced this objection.

[10] One might make this argument for other laws, too: once one has hit on a working law, any data it predicts will be "more of the same," but this is so only if the law already works.

[11] Interestingly, curve-fitting "was first proposed by Adrian Marie Legendre (1752–1833) and Carl Friedrich Gauss (1777–1855) in the early 19th century as a way of inferring planetary trajectories from noisy data" (Forster 1999: 197). For some examples of other scientific theories that involve curve-fitting and for more details on the theoretical work that goes into successful curve-fitting, see Forster (2003).

Regardless of whether there is curve-fitting, however, not just do seemingly "merely" empirical laws come with theoretical components, it is also hardly ever possible (if ever at all) to give entirely theoretical deductions of laws. Most, if not all, scientific laws are based at least partially on data, and so usually at least some data is necessary in order to come up with the right form of a law in the first place. The fact that empirical data is used in the formulation of a law should not count against it—far from it, in fact. Science is, after all, empirical, and an overwhelmingly large number of scientific laws seek to describe empirical phenomena. It is a non-surprising consequence of this that, once we have a certain law, we are free to kick away the theoretical apparatus that generated it. As a result, the fact that the theoretical assumptions that go into a prediction can at some point be discarded is, once again, not an objection to Farr in particular, but to any theory involving laws that seek to describe empirical phenomena. Thus, realists should say about Farr's law exactly what they say about other laws. And I take it that most realists do, in fact, want to keep other laws since, if laws are taken away as a source of confirmation or sign of a theory's success, this would unduly limit the class of theories that are even potential candidates for "genuine" success and, hence, realist warrant.[12]

Lastly, it is important to note that whether Farr's prediction was based on empirical data is, at any rate, irrelevant in assessing use-novelty: the only criterion that matters for use-novelty is whether the prediction's content was used in the construction of the miasma theory, which it was not. Even though the miasma theory only received sophisticated treatment in the 19th century, some version or other of it, but in particular those parts of the theory from which the elevation prediction logically followed, had been around for a long time. The data, moreover, as we have seen, did not exist until Farr generated it; yet the theory had been around since long before his lifetime, regardless of whether one wants to trace the basics back to Hippocrates, Galen, Sydenham, or someone else.

Another avenue of realist criticism is to grant that Farr's prediction was use-novel but to argue that we should not on that basis infer that his theory was successful in a way that supports the success-to-truth inference. As we have already seen, realists place great importance on the kind of success involved in this inference. The standard line is that use-novel predictive success is what is required for genuine success but it is, of course, open to realists to argue that this notion needs to be refined further. On this view, what the Farr case shows is that use-novel predictive success is insufficient for genuine success and that

---

[12] Note that this is compatible with the view that in certain domains the theoretical plays a heavier role than in others, where the empirical is a more common starting point. In fact, this does not strike me as implausible. The general point about the interplay between the theoretical and the empirical in the generation of scientific laws is a question of degree, not kind, and so I do not see how this could be an objection to Farr without also being an objection to other laws.

further work needs to be done in order to capture what truly realism-warranting success looks like. In other words, Farr's law is seen as a counterexample to the view that use-novel predictions are signs of genuine success. But, what might this more refined notion of genuine success look like? One suggestion comes from Vickers' discussion of Meckel's prediction of gill slits in human embryos (Vickers 2015). In 1811, the German anatomist J. F. Meckel predicted that the human embryo ought to have gill slits, a prediction that was found to be correct by fellow anatomist M. H. Rathke in 1827. Since this prediction was temporally (and use-) novel, it counts as an instance of novel predictive success for Meckel's theory. For the realist to preserve the success-to-truth inference, it now needs to be the case that those parts of Meckel's theory that were responsible for this success were approximately true and were retained in successor theories. However, Vickers argues, it looks like this is not the case here: those of Meckel's assumptions responsible for the gill slit prediction cannot be regarded as approximately true and so it seems realists have a problem. The case of Farr, they might suggest, falls into the same category. What Vickers points out is that while Meckel might have made a novel prediction, "the prediction isn't 'risky' in the way the prediction of the Poisson white spot was for Fresnel's theory of light" and that this lack of riskiness "reduces the significance of the prediction" (2015). Further, Vickers invokes Bayes' theorem to argue that in cases of "remarkable novel predictive success," the prior probability of the evidence is small and, thus, such a prediction—if found to be true—ought to raise our degree of belief in the theory much more than a true prediction whose prior probability was high (2015). In Meckel's case, Vickers suggests, the prediction was not surprising and had a high prior probability, since "with a little bit of imagination" Meckel could have come to the gill slit conclusion given the other empirical knowledge of the day. The same response might be given to Farr.[13]

Let's start with this last point: Could Farr have come to his conclusions about cholera mortality and elevation without the miasma theory, given the existing empirical knowledge and with a little bit of imagination? Presumably we don't want to speculate about Farr's counterfactual psychology, so I take it what is at issue is whether it is plausible to think that someone would or could have come to the same conclusions as Farr, absent the miasma theory, and given the empirical knowledge of the day. It is not. The given empirical knowledge concerned data about the seasonality of cholera, the fact that it was worse in certain localities than others, such as swamps and areas rife with sewage, and so on. Making an inference from this sort of data to a relationship between mortality and elevation only makes sense on the assumption that there is a connection between disease

---

[13] In personal communication, Vickers has stated that this is, in fact, his preferred response to the Farr case.

and decomposition and that disease can be transmitted through the air, both central assumptions of the miasma theory.[14] Without these assumptions, the prediction about elevation and cholera mortality just disappears. Further, since without these core theoretical assumptions, the prediction did not even make sense, it is hard to see how Farr could somehow have made the elevation prediction merely from other empirical knowledge. Both assumptions do theoretical work for Farr and on taking out either one—the connection to decomposition or beliefs about aerial transmission—there is no reason to expect the elevation relation. As a result, it is extremely hard to see how someone could have predicted it without either of these components.[15]

It is perhaps noteworthy here that John Snow, famous for his view that cholera was water-, not air-borne and on whose view there wasn't a central connection between cholera and putrefaction, did not predict the elevation relation. In fact, not just did Snow not predict it, he did not even accommodate it, which was one of the biggest criticisms voiced by his contemporaries (see, for example, Parkes (1855)).[16] Thus, not everybody came to Farr's conclusion; notably, those who had a different disease theory did not. Further, the other well-known empirical knowledge had been around for a long time, yet it took Farr to predict a form of the elevation hypothesis. Presumably, if the elevation relation had somehow followed straightforwardly from this existing empirical knowledge, other people would have predicted it long before Farr.[17] So, to sum up: it is implausible to think that Farr could have come to his conclusions about elevation and cholera without the central theoretical assumptions of the miasma theory.

What about the other part of the objection, that Farr's prediction was not as risky as that of Poisson and, therefore, less significant? The first thing to note is

---

[14] There are, of course, as with any other theory, some auxiliary assumptions involved, such as assumptions about air dilution (that the air is cleaner the further away it is from miasmatic sources and so on). An anonymous referee has suggested that different auxiliaries are needed for different diseases and that the "miasma theory of cholera" is different from, say, the "miasma theory of typhoid fever." This, however, is to misunderstand the context of mid-19th century disease discussions. Diseases, at the time, were not thought of as entities that would attack the body from the outside and make it sick; instead, the prevalent view was a more physiological notion of disease, according to which, in some sense, diseases originated in the victim. The ontological conception became influential later in the century, with the idea that different diseases were different species and with the advent of early germ theories. For more detail, see Hamlin (1998: chapter 2).

[15] I have previously argued that both assumptions ought to count as essential for realists. For more detail, see Tulodziecki (2017).

[16] We can now partially explain the elevation law in terms of the water-hypothesis. Snow, however, did not do so.

[17] On the flipside, we might also have the following worry: assume we could somehow make an argument for the claim that Farr could have come to the elevation conclusion just from existing empirical knowledge, without miasmatic assumptions. Given the assumption that he somehow could have done so, even if it is hard to see how, we might now worry that we are able to make this same argument in many other instances and, in particular, in instances that realists want to keep as examples of genuine successes. Given enough ingenuity, perhaps we can always construct such a hypothetical argument (see also Greg Frost-Arnold's comment in response to Vickers (Frost-Arnold 2015)).

that even granting that Farr's prediction was not as risky or significant as Poisson's does not mean that Farr's prediction was not good. I have no qualms with the claim that Poisson's prediction was superior to Farr's; however, it is not clear how typical or frequent predictions like Poisson's actually are and what percentage of scientific predictions are Poisson-like.[18] Indeed, it is entirely possible, if not plausible, that such predictions, while exemplars of excellent predictions, are somewhat atypical. Just as most scientists—even very good ones—are not Curies or Einsteins, most predictions—even very good ones—are not like Poisson's. Thus, Farr's prediction might have been very good, even if it was less good than Poisson's. And, thus, the mere fact that his prediction was less significant than Poisson's is not a reason to think that Farr's prediction should not be viewed as an instance of genuine success. But what about the claim that Farr's prediction was less risky? Even without a specific notion of riskiness it is clear that it would have been bad for Farr if it had turned out that there was no relationship between elevation and cholera mortality since this clearly followed from the miasma theory. If no relation had been found, there would have been a need for another miasmatic explanation (such as more sewage at high altitudes, for example) and, absent any such explanation, the absence of a relationship would have constituted a strong argument against the theory.[19,20]

Still, a realist might argue, what matters is that what followed deductively from the miasma theory was only the general fact that there ought to be some relation between elevation and cholera mortality, but not the exact form of this relation or even the fact that it was possible to capture it through a mathematical law. This is true; however, once again, the fact that what followed logically from the theory was only a general, not a specific prediction should not be held against it, since this is all that could follow from such a theory, even under the very best circumstances. In physical theories, laws and predictions, even if heavily based on data, eventually usually have to be integrated into the theory mathematically,

---

[18] It actually strikes me as an interesting project to look into the different types of prediction one might find in different domains, what categories they might fall into, how frequent they are, and so on. I would not find it surprising if different kinds of predictions were prevalent in diffcrent fields.

[19] The same was true for other localities where high mortality was expected but not found, and for localities where low mortality was expected, but it turned out to be high. Indeed, many of the contemporary discussions of the miasma theory and its competitors centered around just such issues.

[20] What about the thought that Farr's prediction did not have a low prior probability, but Poisson's did? I honestly do not know how to assign the priors in this case. It makes sense to assign Poisson's prediction a low prior probability but, in this case, the prediction was against the prevalent theory. In Farr's case, however, the prediction followed from a theory that was already the dominant disease theory. It thus seems reasonable to think that its predictions, especially plausible ones like the elevation relation, would not have been surprising to anyone. But what would the prior probability have been if the miasma theory had been new? Certainly, given the humoral theory, according to which diseases were an imbalance of individual humors, the probability of Farr's prediction would have been extremely low. What would the prior have been on Snow's view, or some of the other alternatives? Since we do not want to make the priors contingent on historical circumstances, it is unclear to me how to determine what Vickers considers the crucial term, $p(E/-T)$.

in a way that gives rise to various theoretical deductive relationships. It is true that this is not the case here, but this is not surprising, given that the miasma theory, and indeed theories of disease etiology in general, are not mathematised in the same way in which physical theories are. The point here is that Farr is making the best kind of prediction that can be expected from this kind of theory. Predictions about causal factors involved in a given disease are not of the mathematical kind, and the best predictions possible involve likely disease incidence, mortality, and so on. However, these predictions are, by their nature, statistical at best (if that) and, as a result, their exact numerical form will never be logically entailed by a particular disease etiology. The most that can ever follow deductively from such a theory are general population-level claims.

It is also notable in this context that our current theories do not, in kind, do much better on this front than Farr did with respect to elevation, despite the fact that we have vastly improved resources. Indeed, it is precisely because theories about disease etiology do not usually make numerically precise predictions in the form of laws that Farr's prediction seems so remarkable and was seen by many as such a triumph. Lastly, it is noteworthy that Snow also did not make better predictions than Farr; in fact, Snow's predictions were worse, since they never approached any sort of numerical precision. What followed deductively from Snow's claim that cholera was waterborne were predictions of the same type as Farr's: population-level predictions about disease incidence and mortality, such as "disease incidence ought to be higher among people who drink from contaminated water sources" (since, of course, not everyone who drank contaminated water became sick). Thus, no theory of this kind could make predictions which were both mathematically precise and logically entailed by the theory in question.

This, in turn, gives rise to the following problem for realists: If Farr made the best kind of prediction that this type of theory (especially given the limited means available) allowed for, and realists do not consider this good enough for genuine success, they end up with a view according to which no theory of this kind can, even in principle, make predictions that would render it genuinely successful. As a result, we would never, in any such case, be licensed to make the success-to-truth inference, and thus never, in any such case, be warranted in being realists about such theories, at least not on the basis of their predictions, no matter how good.[21] Thus, if the elevation prediction is not good enough for genuine success,

---

[21] This is not to say, of course, that Farr's prediction, on its own, made the miasma theory successful, or that a single prediction of this type could warrant realist commitment in the (approximate) truth of a theory. Obviously, a theory's success depends on many factors. The point is, rather, that the best type of prediction a theory can make should play a large role in assessing that theory's success and that, if such predictions are discounted as a source of success for some class of theories, it is hard to see how theories in that class could ever count as genuinely successful, even in principle.

realists will have to pay a cost that goes much beyond the specific example of Farr and that now also includes, as we have seen, Snow's water-hypothesis and contemporary disease etiologies. In fact, the cost might be even greater, since there are plenty of theories whose predictions are not predominantly mathematical in nature or concern population-levels. Realists might be willing to bite this bullet, but the result would be a much narrower, perhaps domain-specific realism.

2.6 Conclusion

It follows from earlier sections that the case of the miasma theory supports the PMI: it does so by supporting the view that there is no connection between success and approximate truth.[22] Regardless, realists need some kind of account of what is going on in this case. Perhaps the most obvious option is to identify some other kind of continuity at work in the miasma case and add it to the possibilities presented earlier. In that event, perhaps the realist intuition that the (working parts of the) predecessors to our current theories are approximately true can be preserved.

However, this strategy is problematic: first, because it is not clear what possible candidates for retention are left after we have excluded theoretical entities, mechanisms, laws, and structures. Second, even if a candidate could be found, it's unlikely that the resulting proposal could be generalized to the cases in the already existing literature, and which already have other realist accounts. One might not be worried by this and instead be tempted by a kind of pluralism about retention: different types of preservation apply to different types of theories under different circumstances. I am not unsympathetic to this proposal, but it comes with one big worry for the realist, namely whether the resulting position has any realism left in it. After all, if the proposal is that theories get better over time, and increasing success brings with it increasing approximate truth, yet this approximate truth consists of such different things in different cases, it's not clear what we should be realist about. We might still be committed to some vague idea of progress, but we won't be able to tell what that progress involves. In particular, we cannot be realist about any particular element: we cannot be realist about the parts that were involved in predictions, since, sometimes, those are not retained; we cannot be realists about structures, since, sometimes, those are not retained; we cannot be realists about theoretical entities or mechanisms, since, sometimes, they are not retained. The same goes for a theory's laws and, indeed, if realists want to argue that the miasma theory is approximately true, they'll have to keep

---

[22] Note that of course it is not enough to establish the PMI. In order to achieve the latter, it also needs to be shown that these kinds of cases are pervasive (see also Kitcher (1993: 138–139)).

adding to this list. But this means that, from the perspective of our current theories, we simply have no idea which of *their* parts we ought to be realists about, beyond a vague commitment to the view that something about them is probably right, although we have not the slightest inkling what.

The alternative, of course, is to bite the bullet in the miasma case and accept that this theory, although genuinely successful, was simply not one that was approximately true. However, this is also a somewhat uncomfortable position for the realist, quite independently of the fact that this case now supports the PMI. The source of this discomfort is that we now have a case in which we went from a not even approximately true theory directly to a successor (the germ theory) that we do take be true—and not just approximately true, but true simpliciter. This is uncomfortable, because the whole realist idea is that there is a relation between increasing success over time and increasing truth-content, and if it now turns out that discontinuities between theories might sometimes be completely radical, then this does shed some doubt on the "converging" part of convergent realism. Of course just because approximate truth might not be necessary for truth simpliciter does not mean that it is not sufficient, and it doesn't follow from the fact that (some) true theories had no approximately true predecessors that approximate truth is not a reliable guide to truth as such.[23] The point is rather that, if the convergent realist's story is right—if the right way to view the history and development of science is as a succession of increasingly true theories that get better and better over time—then we ought to expect the predecessors of theories that we now know to be true to be approximately true. Finding true theories that do not conform to this expectation does not invalidate the story—one case does not make a pattern—but it is a strike against the plausibility of the general story that convergent realists tell.

Acknowledgments

For helpful comments, many thanks to Tim Lyons, Marshall Porterfield, and Jan Sprenger. Special thanks are due to Peter Vickers for numerous exchanges and conversations about this case, and for generous written comments on a different but related paper.

---

[23] For that, we would need cases of approximately true predecessors that led to false successors, and it is not clear how we would find these, since we also need true successors in order to assess these claims in the first place.

References

Baldwin, P. (1999). *Contagion and the State in Europe, 1830-1930*. Cambridge: Cambridge University Press.

Boyd, R. (1981). "Scientific Realism and Naturalistic Epistemology," in P.D. Asquith and T. Nickles (eds) *PSA 1980, Vol. 2*, East Lansing, MI: Philosophy of Science Association, pp. 613–662.

Boyd, R. (1984). "The Current Status of the Realism Debate," in J. Leplin (ed.) *Scientific Realism*, Berkeley: University of California Press, pp. 41–82.

Boyd, R. (1990). "Realism, Approximate Truth and Philosophical Method," in C.W. Savage (ed.) *Scientific Theories, Minnesota Studies in the Philosophy of Science, Vol. 14*, Minneapolis: University of Minnesota Press, pp. 355–391.

Doppelt, G. (2007). "Reconstructing Scientific Realism to Rebut the Pessimistic Meta-Induction." *Philosophy of Science* 74 (1): 96–118.

Eyler, J. M. (1979). *Victorian Social Medicine: The Ideas and Methods of William Farr*. Baltimore/London: Johns Hopkins University Press.

Eyler, J. M. (2001). "The Changing Assessments of John Snow's and William Farr's Cholera Studies." *Sozial und Präventivmedizin* 46: 225–232.

Farr, W. (1852). *Report on the Mortality of Cholera in England, 1848–49*. London: W. Clowes.

French, S., and Ladyman, J. (2011). "In Defence of Ontic Structural Realism," in A. Bokulich and P. Bokulich (eds.) *Scientific Structuralism (Boston Studies in the Philosophy of Science: Volume 281)*, Dordrecht: Springer, pp. 25–42.

Forster, M. R. (1999). "Curve-fitting problem." In R. Audi (ed.) *The Cambridge Dictionary of Philosophy, Second Edition*, Cambridge: Cambridge University Press, pp. 197–198.

Forster, M.R. (2003). "Philosophy of the Quantitative Sciences: Unification, Curve Fitting, and Cross Validation." Manuscript. Available at http://philosophy.wisc.edu/forster/papers/Part1.pdf

Frost-Arnold, G. (2015). "Comment on Vickers (2015)." https://thebjps.typepad.com/my-blog/2015/06/srpetervickers.html, last accessed 24 June 2018.

Hamlin, C. (1998). *Public Health and Social Justice in the Age of Chadwick: Britain, 1800–1854.* Cambridge: Cambridge University Press.

Hamlin, C. (2009). *Cholera: The Biography*. New York: Oxford University Press.

Hesse, M. (1976). "Truth and the Growth of Scientific Knowledge." *Proceedings of the Biennial Meeting of the Philosophy of Science Association* 2: 261–280.

Kitcher, P. (1993). *The Advancement of Science*. New York: Oxford University Press

Ladyman, J. (1998). "What is Structural Realism?" *Studies in History and Philosophy of Science Part A* 29 (3): 409–424.

Lange, M. (2002). "Baseball, Pessimistic Inductions and the Turnover Fallacy." *Analysis* 62: 281–285.

Langmuir, A.D. (1961). "Epidemiology of airborne infection." *Bacteriological Reviews* 25 (3): 173–181.

Langmuir, Alexander D. (1961). "Epidemiology of Airborne Infection." *Bacteriological Reviews*, 25(3): 174

Laudan, L. (1981). "A Confutation of Convergent Realism." *Philosophy of Science* 48 (1):19–49

Laudan, L. (1984). *Science and Values.* Berkeley and Los Angeles: University of California Press

Lewis, P. (2001). "Why the Pessimistic Induction Is a Fallacy." *Synthese* 129: 371–380.

Lyons, T.D. (2006). "Scientific Realism and the Stratagema de Divide et Impera." *British Journal for the Philosophy of Science* 57 (3): 537–560.

Parkes E.A. (1855). "Mode of Communication of Cholera. By John Snow, M.D. Second Edition." *British and Foreign Medico-Chirurgical Review* 15: 456.

Pelling, M. (1978). *Cholera, Fever and English Medicine: 1825–1865.* Oxford: Oxford University Press.

Psillos, S. (1996). "Scientific Realism and the 'Pessimistic Induction.'" *Philosophy of Science* 63 (3): 314.

Psillos, S. (1999). *Scientific Realism: How Science Tracks Truth.* London: Routledge

Saatsi, J. (2005). "On the Pessimistic Induction and Two Fallacies." *Philosophy of Science* 72 (5): 1088–1098.

Saatsi, J., Vickers, P. (2011). "Miraculous Success? Inconsistency and Untruth in Kirchhoff's Diffraction Theory." *British Journal for the Philosophy of Science* 62 (1): 29–46.

Snow, J. (1855a). "On the comparative mortality of large towns and rural districts, and the causes by which it is influenced." *Journal of Public Health, and Sanitary Review* 1: 16–24.

Snow, J. (1855b). *On the Mode of Communication of Cholera* (second much enlarged ed.). London: J. Churchill.

Tulodziecki, D. (2016). "From Zymes to Germs: Discarding the Realist/Anti-Realist Framework," in T. Sauer and R. Scholl (eds.) *The Philosophy of Historical Case-Studies,* Boston Studies in Philosophy of Science, Switzerland: Springer, 265–284.

Tulodziecki, D. (2017). "Against Selective Realism(s)." *Philosophy of Science* 84 (5): 996–1007.

Vickers, P. (2013). "A Confrontation of Convergent Realism." *Philosophy of Science* 80 (2): 189–211.

Vickers, P. (2015). "Contemporary Scientific Realism and the 1811 Gill Slit Prediction," *Auxiliary Hypotheses: Blog of the British Journal for the Philosophy of Science,* http://thebjps.typepad.com/my-blog/2015/06/srpetervickers.html, last accessed 24 June 2018.

Worboys, M. (2000). *Spreading Germs: Diseases, Theories, and Medical Practice in Britain, 1865–1900.* Cambridge: Cambridge University Press.

Worrall, J. (1989). "Structural realism: The best of both worlds?" *Dialectica* 43 (1–2): 99–124.

Worrall, J. (1994). "How to Remain (Reasonably) Optimistic: Scientific Realism and the" Luminiferous Ether." *PSA: Proceedings of the Biennial Meeting of the Philosophy of Science Association 1994,* 1: 334–342.

3

What Can the Discovery of Boron Tell Us About the Scientific Realism Debate?

Jonathon Hricko

Education Center for Humanities and Social Sciences
National Yang-Ming University
jhricko@ym.edu.tw

3.1 Introduction

In much of the recent work on the scientific realism debate,[1] realists and their anti-realist opponents have focused on theories with two significant features. First of all, those theories purport to describe some kind of unobservable reality, for example, unobservable entities like electrons or genes. Second, those theories exhibit novel predictive success, i.e., they generate true predictions that scientists did not use when constructing those theories. The clearest cases of such success involve temporal novelty, where a prediction of some phenomenon is temporally novel if that phenomenon was not known at the time the theory was constructed. Other cases involve only use-novelty, where a prediction is use-novel, but not temporally novel, if it predicts a phenomenon that was known at the time the theory was constructed, but knowledge of that phenomenon was not used in the construction of the theory.[2] In this chapter, I'll focus on an example of a *temporally* novel predictive success.

One of the central issues in the realism debate concerns whether the novel predictive success of a theory constitutes a sufficient reason for thinking that (at least some of) that theory's claims regarding unobservables are (at least approximately) true. Realists of various stripes argue that some form of realism or other provides the best explanation of why theories exhibit novel predictive success (Musgrave 1988; Psillos 1999). This argument is the so-called no-miracles argument, the name of which comes from the realist's contention that, without a

---

[1] For a comprehensive, up-to-date survey of the debate, see Alai (2017).
[2] See Psillos (1999, 105–107) for a good discussion of novelty.

Jonathon Hricko, *What Can the Discovery of Boron Tell Us about the Scientific Realism Debate?* In: *Contemporary Scientific Realism*. Edited by: Timothy D. Lyons and Peter Vickers, Oxford University Press. © Oxford University Press 2021. DOI: 10.1093/oso/9780190946814.003.0003 realist explanation, novel predictive success would be miraculous. Anti-realists argue that the success of a theory does not warrant or require any sort of realist attitude toward theoretical claims regarding unobservables (van Fraassen 1980; Stanford 2006). Although realists and anti-realists disagree about this issue, it's worth emphasizing that they actually have much else in common. Their assessment of a theory's claims regarding observables need not differ at all. And when it comes to theories that have not exhibited novel predictive success, today's realists and anti-realists may be equally skeptical of such a theory's claims regarding unobservables.

Theories that, by present lights, are not even approximately true and yet exhibited novel predictive success pose one of the strongest challenges to realism.[3] Recent work has uncovered many examples of such false-but-successful theories (Lyons 2002, 2006, 2016, 2017; Vickers 2013). The challenge for the realist is to clearly articulate what sort of realism is warranted in such cases. Many realists have argued that the most viable option is some form of selective realism, which involves a commitment, not to the (approximate) truth of entire theories, but only to certain parts of theories. Selective realism comes in a number of different varieties.[4] To take one example, deployment realists (Kitcher 1993; Psillos 1999) argue that we ought to commit only to the parts of theories deployed in deriving successful novel predictions. Their strategy is to explain the novel predictive success of false theories in terms of the (approximately) true parts of those theories.

Insofar as this strategy of providing such a realist explanation is successful, selective realists can accommodate these examples of false-but-successful theories. However, even if there were just one example of novel predictive success that realists failed to explain (i.e., one miracle), that example would pose a challenge to realism insofar as it would undermine the claim that realism is *required* to explain such success.[5] After all, if a non-realist explanation is required in one case, why stop there? Nothing compels us to endorse a realist explanation once we see that a non-realist explanation is going to be required regardless of what realists have to say about all of the other examples. Realists might retreat to the claim that miracles are permitted provided that they are rare. Additional examples of novel predictive success that don't yield to realist explanations cast doubt on this

---

[3] Throughout the chapter, when I claim that past theories are/are not approximately true, these claims should be understood as abbreviations for the conditional claim that, if our current theories are approximately true, then these past theories are/are not approximately true. Otherwise, I might be accused of presupposing realism about our current theories, and of thereby begging some central questions in the realism debate.

[4] See Lyons (2017, 3215) for a useful list of selective realist positions.

[5] This point is especially clear within the context of Lyons's pessimistic meta-*modus tollens* (2002, 67).

further claim. The many examples of false-but-successful theories thus pose a challenge to the viability of such a realist retreat.

My goal in this chapter is to present an example of a false-but-successful theory that has not yet received sufficient attention within the realism debate and use that example to pose a challenge to selective realism. The false-but-successful theory is the oxygen theory of acidity due to Antoine-Laurent Lavoisier (1743–94). The temporally novel prediction that it made was that boracic acid consists of oxygen combined with a hypothetical, combustible substance that Lavoisier called the boracic radical. And that prediction was subsequently confirmed. In 1808, the British chemist Sir Humphry Davy (1778–1829) used potassium to extract oxygen from boracic acid and thereby discovered that boron is the boracic radical, while the French chemists Joseph Louis Gay-Lussac (1778–1850) and Louis Jacques Thénard (1777–1857), working together but independently of Davy, did the same. This example of novel predictive success poses a strong challenge to selective realism because the parts of Lavoisier's theory responsible for its success are not even approximately true.

I proceed as follows. In section 3.2, I present Lavoisier's oxygen theory of acidity and show why it is, from our present perspective, not even approximately true. In section 3.3, I identify four novel predictions that Lavoisier made regarding boracic acid. In section 3.4, I demonstrate the success of these predictions by examining the work by Davy, Gay-Lussac, and Thénard that constituted the discovery of boron. In section 3.5, I show how the derivation of these predictions made use of claims from Lavoisier's theory that are not even approximately true. I go on to consider some ways in which selective realists might try to accommodate this example of novel predictive success, and I argue that these attempts are unsuccessful. Sections 3.2 through 3.5 present the main argument of the chapter in a relatively straightforward way, without anticipating and responding to some relevant objections and without exploring aspects of the history of this episode that, while relevant, are not essential to understanding the main argument. I flag these issues in footnotes as they arise and direct the reader to the relevant subsections of section 3.6, in which I address these issues more fully.

3.2 The False Theory

The false theory at the center of this example of novel predictive success is Lavoisier's oxygen theory of acidity. According to that theory, all acids contain oxygen and a non-oxygen component that Lavoisier called the "acidifiable base" or "radical" of the acid, two terms that he regarded as synonyms (1790 [1965], 65). Lavoisier was led to this theory by the fact that the acids that were most well understood at the time were all shown to contain oxygen. Lavoisier holds that different kinds of acids differ from one another in one of two ways. They may contain different radicals, as phosphoric acid and sulfuric acid do—the radicals in these acids are phosphorus and sulfur, respectively (66). And even if they contain the same radical, they may still differ from one another by virtue of the fact that they contain different amounts of oxygen, as sulfuric acid and sulfurous acid do (66–68). Moreover, Lavoisier emphasizes that the base or radical of an acid can be either a simple substance or a compound (115–116, 176–177). For example, he claims that many vegetable acids have compound radicals composed of carbon and hydrogen.

Oxygen is the substance that plays the most central role in Lavoisier's theory of acidity. It is the element that all acids share, and which "constitutes their acidity" (65). Oxygen, for Lavoisier, is what gives acids their acidic properties, for example, their sour taste. Lavoisier labels oxygen "the acidifiing [sic] principle" (65) and in fact the very name "oxygen" that Lavoisier proposed for this substance comes from the Greek words for "acid-generator" (51; Chang 2012b, 9). Though to be sure, as Le Grand (1972, 11–12) has emphasized, Lavoisier admitted that many compounds that contain oxygen are not acids, and his point was that oxygen is a necessary condition for acidity, not a sufficient one.

Lavoisier's theory is also a theory of the formation of acids. After recounting a series of experiments in which acids are formed by the combustion of the acidifiable bases phosphorus, sulfur, and carbon, Lavoisier writes: "I might multiply these experiments, and show by a numerous succession of facts, that all acids are formed by the combustion of certain substances" (1790 [1965], 64). Lavoisier's view, then, is that an acidifiable base must be a combustible substance. In order to understand Lavoisier's view regarding the formation of acids by means of combustion, we must start with his claim that oxygen gas is a compound of caloric and the base of oxygen gas ("oxygen base" for short, following Chang (2011, 415)); and it is oxygen base which is the true acidifying principle (Lavoisier 1790 [1965], 51–52). Lavoisier illustrates the combustion of an acidifiable base and the subsequent formation of an acid in terms of the example of phosphorus, the acidifiable base of phosphoric acid:

> at a certain degree of temperature, oxygen possesses a stronger elective attraction, or affinity, for phosphorus than for caloric; . . . in consequence of this, the phosphorus attracts the base of oxygen gas from the caloric. (57)

In other words, phosphorus effects the decomposition of oxygen gas into caloric and oxygen base; oxygen base then combines with phosphorus to form an acid. More generally, "[b]efore combustion can take place, it is necessary that the base of oxygen gas should have greater affinity to the combustible body than it has to caloric" (414–415). Hence, Lavoisier's theory of acidity makes use of his theories of combustion and caloric as well. Lavoisier goes on to propose the term "oxygenation" to name the process by which combustible substances combine with oxygen, in which case acidifiable bases are converted to acids by oxygenating them (61–62).

As a whole, Lavoisier's theory of acidity isn't even approximately true.[6] By the early years of the 19th century, chemists already knew of two counterexamples to Lavoisier's theory, namely, prussic acid (hydrocyanic acid, HCN) and muriatic acid (hydrochloric acid, HCl), neither of which contains oxygen, Lavoisier's acidifying principle. Additionally, by positing the muriatic radical as the non-oxygen component of muriatic acid, Lavoisier posited a non-existent entity.[7] Moreover, if one or both of our current conceptions of acidity are approximately true, then Lavoisier's theory cannot be. According to the Brønsted-Lowry concept, acids are proton donors. And according to the Lewis concept, acids are electron pair acceptors. While these concepts diverge from one another in interesting ways,[8] neither gives any sense to the claim that oxygen is the acidifying principle, which is the central claim of Lavoisier's theory.

It's worth pressing this point regarding the falsity of Lavoisier's theory a bit further by considering the theories of combustion and caloric that Lavoisier made use of in the context of his theory of acidity. One might think that Lavoisier's theory of combustion, at least, is approximately true. But it's not obvious that it is. Lavoisier's theory requires the presence of oxygen gas for combustion. But by the early 19th century, chemists knew of cases of combustion that occurred without the presence of oxygen gas. Moreover, combustion, for Lavoisier, requires the decomposition of oxygen gas into caloric and oxygen base (1790 [1965], 414–415).

The caloric theory of heat has been a problem case for realists at least since it appeared on Laudan's well-known list of false-but-successful theories (1981, 33). Even if we confine our attention to the period in which Lavoisier was working, this theory exhibited a number of successful explanations. As Chang makes clear, these included explanations of:

the flow of heat toward equilibrium, the expansion of matter by heating, latent heat in changes of state, the elasticity of gases and the fluidity of liquids, the heat released and absorbed in chemical reactions, [and] combustion. (2003, 907)

Moreover, these explanations made use of a number of assumptions about the nature of caloric which are not even approximately true. These included the

[6] See Chang (2012b, 8–10) for a good discussion of the problems with Lavoisier's theories of acidity, combustion, and caloric, which has informed the discussion in the remainder of this section.
[7] See Hricko (2018) for a detailed discussion of the muriatic radical within the context of the realism debate.
[8] See Chang (2012a) for a good discussion of the ways in which these concepts diverge.

assumption "that heat was a 'self-repulsive' (or 'elastic,' or 'expansive') substance, while it was attracted to ordinary matter"; "the postulation that caloric existed in two different states: sensible and latent"; and the identifications of caloric as a chemical element, of sensible heat with free or uncombined caloric, and of latent heat with caloric that was chemically combined with matter (907–908). The appearance of caloric in Lavoisier's theory of acidity thus provides some additional support for the claim that this theory was not even approximately true.

3.3 The Novel Predictions

The novel predictions on which I'll focus concern the composition of what Lavoisier called boracic acid. Lavoisier writes:

> The boracic radical is hitherto unknown; no experiments having as yet been able to decompose the [boracic] acid; we conclude, from analogy with the other acids, that oxygen exists in its composition as the acidifying principle. (1790 [1965], 245)

As we'll see shortly, there is a bit more to the predictions than what Lavoisier says here.

Lavoisier's novel predictions regarding boracic acid belong to a larger group of similar novel predictions that he made on the basis of his theory of acidity. At the time, there were a number of acids that chemists could neither produce from simple substances nor decompose into simple substances. These acids included boracic acid, muriatic acid, and fluoric acid (hydrofluoric acid, HF). Lavoisier hypothesized that these acids are composed of oxygen and the boracic, muriatic, and fluoric radicals, respectively. Regarding these radicals, Lavoisier writes:

> the combinations of these substances [the radicals], either with each other, or with the other combustible bodies, are hitherto entirely unknown . . . We only know that these radicals are susceptible of oxygenation, and of forming the muriatic, fluoric, and boracic acids. (209–210)

None of these radicals had been isolated, and in this passage, Lavoisier predicts that they are combustible substances that form acids by means of oxygenation. This prediction is a straightforward consequence of his theory of acidity, according to which acids contain oxygen and a combustible radical. It's also worth emphasizing that Lavoisier's theory predicts that the only components of these three acids are oxygen and their respective radicals. The radical, for Lavoisier, is the acid's non-oxygen component, and cases in which an acid has more than one non-oxygen component are cases in which the radical is a compound as opposed to a simple substance.

In order to gain a better understanding of these predictions, it's worth going into a bit more detail regarding how Lavoisier understood the hypothetical substances that he referred to as the boracic, muriatic, and fluoric radicals. The terms boracic radical, etc., are best understood as descriptions of the role that these substances were hypothesized to play within their respective acids, i.e., the role of the combustible component substance that forms an acid by oxygenation. These terms should not be understood as names of chemical substances, because they were temporary placeholders for the names of whatever might be discovered to play this role. Lavoisier included all three radicals in his table of simple substances (175).[9] By doing so, he did not mean to imply that they would be discovered to be previously unknown substances, elementary or otherwise. Regarding the muriatic radical, Lavoisier recognized that it might be "discovered to be a known substance, though now unknown in that capacity" (72), and this point applies to the other two radicals as well. If, say, the boracic radical turned out to be a known substance or a compound of known substances, then Lavoisier would have simply eliminated the term boracic radical from his table of simple substances. If, however, the boracic radical turned out to be a previously undiscovered simple substance, then Lavoisier would have replaced the term boracic radical with a name for this new substance. The important point here is that Lavoisier's theory of acidity did not entail that the three radicals are three hitherto undiscovered chemical substances.

At this point, we can distinguish four predictions that Lavoisier made regarding the composition and formation of boracic acid:

- Boracic acid contains oxygen.
- Boracic acid contains a combustible substance (the boracic radical).
- The only components of boracic acid are oxygen and the boracic radical (though the boracic radical itself might be a compound radical).
- Boracic acid is formed by the combustion of the boracic radical with oxygen.

Importantly, these predictions are novel predictions. In fact, the novelty of these predictions is temporal novelty, which arises from the fact that chemists at the time had not yet succeeded in decomposing boracic acid into simpler substances or producing it by means of simpler substances. Hence, the composition of

---

[9] He thereby indicated his view that they are elements. In section 3.6.2, I discuss the prediction that the boracic radical is an element and conclude that it is not relevant to the issue concerning whether selective realists can accommodate false-but-successful theories.

boracic acid and the way in which it is formed were unknown and so couldn't have been written into the oxygen theory of acidity at the time Lavoisier constructed it.

3.4 The Predictive Success

In order to determine whether Lavoisier's predictions were successful, we must first identify what substance chemists at the time referred to when they used the term boracic acid. In fact, they primarily had in mind the substance that we know as boron trioxide (B_2O_3).[10] Boron trioxide is composed of oxygen and a combustible substance, namely, boron; and it can be formed by the combustion of boron with oxygen. Moreover, it turns out that, unlike the muriatic radical, the boracic radical exists—it is what we now call boron.[11] Hence, Lavoisier's oxygen theory of acidity, although not even approximately true, exhibited novel predictive success. I'll now examine the work in the early 19th century that confirmed Lavoisier's predictions regarding boracic acid.

Three chemists are credited with decomposing boracic acid and isolating boron: Davy in England, and Gay-Lussac and Thénard in France (Lowry 1915, 288; Fontani, Costa, and Orna 2015, 30). Because they suspected boracic acid to contain oxygen, they attempted to decompose it with potassium, a substance known to have a strong affinity for oxygen. Gay-Lussac and Thénard isolated boron by heating boracic acid and potassium in a copper tube and washing the mixture that they obtained with water to separate boron from the other substances (1808, 169–171). They report that they thereby produced a greenish-brown substance (171). Davy first attempted the decomposition by acting on moistened boracic acid with a battery, but without much success (1808, 43; 1809, 75–76). Davy (1809, 76–78) eventually succeeded using more or less the same methods as Gay-Lussac and Thénard. He reports that he produced a dark, olive-colored powder (1809, 78). Today, we would say that these chemists produced relatively impure samples of amorphous boron.[12]

By producing these impure samples of boron, these chemists thought that they had isolated the base or radical of boracic acid. Gay-Lussac and Thénard label the substance that they had isolated the "radical boracique," for which they propose a new name: "bore" (1808, 171, 173). Davy, however, initially

---

[10] I discuss the details regarding the identification of boracic acid with boron trioxide in section 3.6.1.

[11] In section 3.6.4, I consider and respond to an objection to the identification of boron with the boracic radical.

[12] Given the impurity of the samples, one might question whether these chemists really confirmed Lavoisier's predictions. I consider this issue in section 3.6.3.

suspected that the substance he had isolated was a compound of oxygen and "the true basis of the boracic acid," which he conjectured was a metal and for which he proposed the name "boracium" (1809, 84–85). Several years later, he concluded that this conjecture had not been vindicated, and in light of the analogy between the substance that he had isolated and carbon, he proposed another name for what he calls "the basis of the boracic acid": "boron" (1812, 178). The terminology that these chemists used does not necessarily imply a commitment to Lavoisier's theory of acidity. In fact, at the time they discovered boron, Davy was skeptical of Lavoisier's theory of acidity while Gay-Lussac and Thénard were committed to it.[13] And by 1812, although Davy was still using the term "basis" to refer to the non-oxygen component of boracic acid, he had more decisive reasons for rejecting Lavoisier's theory since, in 1810, he had argued that the components of muriatic acid are hydrogen and chlorine (Davy 1811). However, Davy did previously claim that "the combustible matter obtained from boracic acid, bears the same relation to that substance, as sulphur and phosphorus do to the sulphuric and phosphoric acids" (1809, 82), which basically amounts to an identification of boron with the boracic radical. Importantly, regardless of their attitude toward Lavoisier's theory of acidity, these three chemists isolated the substance that Lavoisier had hypothesized years earlier; in other words, boron is the boracic radical.

Once they had isolated boron, Davy, Gay-Lussac, and Thénard also determined some of its properties, most notably, that it is a combustible substance that combines with oxygen to form boracic acid. Both Davy (1809, 79, 82; 1812, 179) and Gay-Lussac and Thénard (1808, 172–173) describe the experiments by which they converted the newly discovered substance into boracic acid by means of combustion with oxygen. And by doing so, they also showed that the components of boracic acid are boron and oxygen.

At this point, we can conclude that Lavoisier's four novel predictions, which I listed in section 3.3, were successful. The work of Davy, Gay-Lussac, and Thénard showed that boracic acid, which we know as boron trioxide, contains both oxygen and a combustible substance, namely, boron. Moreover, these chemists confirmed Lavoisier's prediction regarding the formation of boracic acid, namely, that it forms by the combustion of the boracic radical (boron) with oxygen. And by both producing and decomposing boracic acid, they provided a convincing demonstration that its sole components are boron and oxygen.

---

[13] See Brooke (1980, 124, 158) for Davy's skeptical attitude toward Lavoisier's theory of acidity and Crosland (1978, 45) for Gay-Lussac's acceptance of that theory.

3.5 The Challenge for Selective Realism

At this point, the upshot is that we can add Lavoisier's oxygen theory of acidity to the increasingly long list of theories that, although not even approximately true, exhibited novel predictive success. As I discussed in section 3.1, such theories pose a challenge to realism. My goal in this section is to consider two ways in which a selective realist might respond to this challenge and argue that neither of these ways is successful.

Before doing so, it's worth going into a bit more detail regarding the challenge that the selective realist faces. The derivation of Lavoisier's four successful predictions involved a number of claims from his oxygen theory of acidity that, by present lights, are not even approximately true. These claims include:

(1) Oxygen gas is a compound of caloric and oxygen base.
(2) All acids contain oxygen base, which is the acidifying principle, i.e., the substance in virtue of which acids have their acidic properties.
(3) All acids contain an acidifiable base or radical, which is the non-oxygen component of the acid, and which may be a simple substance or a compound.
(4) Acids form by the combustion of an acidifiable base, which requires that oxygen base has a stronger affinity for the acidifiable base than it does for caloric; oxygen base then combines with the acidifiable base to form an acid.

Lavoisier's prediction that boracic acid contains oxygen made use of (1), (2), and (4). His prediction that it contains a combustible substance made use of (1), (3), and (4). His prediction that it contains only these two substances made use of all of these claims, as did his prediction regarding the formation of boracic acid in terms of the combustion of an acidifiable base with oxygen. If our current theories are correct, then none of these four claims is even approximately true. (1) is false in light of the rejection of the caloric theory of heat. Moreover, we now consider oxygen gas to be an elementary substance, not a compound. (2) is false, not just because there are acids that do not contain oxygen, but also because we've rejected the idea that a chemical substance can be an acidifying principle in Lavoisier's sense. (3) is false because Lavoisier characterized the radical of an acid as its non-oxygen component, and so acids that lack oxygen also lack radicals in Lavoisier's sense. And (4) is false, not just because it involves the caloric theory of heat, but also because there are acids that do not form by combustion with oxygen because they do not contain oxygen.

The first selective realist response I'll consider is to show, contrary to first appearances, that the theoretical claims required for deriving Lavoisier's successful predictions are in fact approximately true. One way the selective realist might attempt to do so makes use of Vickers's (2013, 199) distinction between derivation external posits (DEPs) and derivation internal posits (DIPs), along with his notion of the working part of a DIP.[14] Realist commitment does not extend to DEPs, which merely inspire scientists to consider ideas; and it may not even extend to all parts of a DIP. Realist commitment extends only to the working part of a DIP, i.e., the part that "actually contributes to a given derivational step," and not to the other parts, which Vickers labels "surplus content" (201).

In order to evaluate this response, I'll focus on one of Lavoisier's predictions, namely, the prediction that boracic acid contains oxygen. While this prediction was in fact motivated by (1), (2), and (4), for the sake of argument, let's grant that (1) and (4) are DEPs, and only (2) is a DIP. It's plausible that (2) has some surplus content and that the working part of (2) is a more minimal claim that suffices for deriving the prediction without mentioning oxygen base or an acidifying principle. What might this more minimal claim be? The claim that all acids contain oxygen is more minimal, and it suffices for deriving the prediction; but it is not even approximately true since not all acids contain oxygen. The claim that boracic acid contains oxygen is even more minimal, and also true; but it doesn't provide an explanation of this instance of novel predictive success since it is identical to the prediction itself, which makes the derivation trivial. The claim that many acids contain oxygen is more minimal, and also true; but it doesn't entail the prediction, and on its own, it doesn't even make the prediction likely since it's consistent with the claim that many acids do not contain oxygen. The challenge for the selective realist is to identify a working part that has two properties: it must be at least approximately true, and it must be sufficient for deriving the prediction in a nontrivial way. The problem for the selective realist is that the claims that have one property lack the other property.

Another consideration that makes this kind of selective realist response problematic concerns Lavoisier's predictions regarding the other two undecomposed acids: muriatic (hydrochloric) acid and fluoric (hydrofluoric) acid. The selective realist must explain why Lavoisier's prediction regarding the presence of oxygen in boracic acid succeeded while his predictions regarding the presence of oxygen in the muriatic and fluoric acids failed. And this explanation must be consistent with the fact that Lavoisier's predictions regarding the presence of oxygen in these three acids were derived in exactly the same way, as I discussed in section 3.3. These predictions were motivated solely by Lavoisier's theory of acidity and by the fact that these three substances are acids. Explaining why one prediction succeeded while the other two failed requires adopting a view that is historically

---

[14] Vickers uses the term posit "as essentially synonymous with 'proposition' or 'hypothesis' (broadly construed)" (2013, 198, fn 7).

inaccurate. This is the view that Lavoisier's derivations of these three predictions differ in some significant way, say, because he identified some property of boracic acid that muriatic acid and fluoric acid lack and that truly indicates the presence of oxygen in a compound. In short, for any candidate for the working part of a DIP, the selective realist will have to explain, in a historically plausible way, why it works when it comes to boracic acid and doesn't work when it comes to the other two acids. It's worth emphasizing that the issue here is not simply about failed predictions. It's about whether, in light of those failed predictions (which were derived in exactly the same way as the successful predictions), the working parts in the derivation of the successful predictions can be regarded as approximately true. My claim is that they cannot.

The second selective realist response is that the novel predictive success of Lavoisier's theory is not impressive enough to pose a challenge to realism. Vickers (2013, 195–198) argues that successful novel predictions need to be sufficiently impressive in order to pose a challenge to realism. For Vickers, the distinction between impressive and unimpressive novel predictive successes is a matter of degree. One way in which they differ in degree concerns "the degree to which the prediction could be true just by luck (perhaps corresponding to the Bayesian's 'prior probability')" (196). If the probability that Lavoisier's predictions could be true just by luck is sufficiently high, then we have a group of lucky guesses that the realist can dismiss, as opposed to "miracles" that require a realist explanation.

Given the state of knowledge regarding acids in the years leading up to Davy's and Gay-Lussac and Thénard's work on boracic acid, what can we say about the prior probability of Lavoisier's predictions being true? Many acids (e.g., sulfuric acid and phosphoric acid) were known to contain oxygen and a combustible substance and to form via the combustion of that substance with oxygen. However, there were problem cases. In 1787, Claude Louis Berthollet (1748–1822) demonstrated that prussic acid (hydrocyanic acid, HCN) contains hydrogen, carbon, and nitrogen, but not oxygen (1789, 38). Thomas Thomson (1773–1852) mentions another problem case: "Sulphurated hydrogen [i.e., hydrogen sulfide, H_2S], for instance, possesses all the characters of an acid, yet it contains no oxygen" (1802, 4). In light of these problem cases, assigning a very high prior probability to Lavoisier's predictions regarding boracic acid would have been unwarranted. In that case, Lavoisier's predictions were, at least to some degree, impressive.

Vickers (2013, 196) discusses two other senses in which predictions can be unimpressive; but neither of them applies to Lavoisier's predictions regarding boracic acid. First of all, Vickers argues that vague predictions are unimpressive. His example is the prediction that the planet Venus is hot, which is successful so long as Venus has a temperature greater than, say, 50°C. The problem here seems to relate, not just to the vagueness of the term hot, but also to its imprecision, since there are many ways in which Venus could be hot. Lavoisier's predictions regarding boracic acid are, however, both less vague and more precise than this prediction. Secondly, Vickers argues that predictions may be unimpressive if the match between the prediction and the experimental results is not sufficiently close. But Davy's and Gay-Lussac and Thénard's results match Lavoisier's predictions quite well. It's also worth mentioning that, for Vickers, qualitative predictions (e.g., the Poisson white spot) can be very impressive, and so the qualitative nature of Lavoisier's predictions is no reason to dismiss them as unimpressive.

I conclude that the novel predictive success of Lavoisier's oxygen theory of acidity poses a strong challenge to selective realism. The working parts of Lavoisier's theory are not even approximately true. And the predictive success of that theory is sufficiently impressive that the selective realist cannot dismiss Lavoisier's predictions as likely to be true just by luck.

3.6 Complications, Clarifications, Objections, and Replies

At this point, I've presented the main argument of the chapter. In the course of doing so, I've also flagged and so far ignored a number of relevant issues, to which I now turn.

3.6.1 What Did Boracic Acid Refer To?

In section 3.4, I claimed that when chemists working in the late 18th and early 19th centuries used the term boracic acid, they primarily had in mind the substance that we call boron trioxide (B_2O_3). Boron trioxide is the anhydride of boric acid (H_3BO_3), and one might wonder why boracic acid didn't refer solely to the substance that we call boric acid. If it did, then some of Lavoisier's predictions (e.g., the prediction that oxygen and a combustible substance are the only components of boracic acid) are false. Moreover, in order to conclude that the work of Davy, Gay-Lussac, and Thénard demonstrated the success of Lavoisier's predictions, we need to be sure that these three chemists were experimenting with the substance that was the subject of Lavoisier's predictions. Since my argument depends on the claim that chemists used the term boracic acid to refer primarily to boron trioxide, it's important for me to defend this claim.

The use of the term boracic acid to refer primarily to boron trioxide was a particular instance of a more general trend. A number of commentators have observed that chemists at the time used the term acid to refer to two kinds of substances that we distinguish today, namely, acids and their anhydrides (Laurent

1855, 44–45; Miller 1871, 314; Lowry 1915, 250). Knight (1992, 83) explains that when chemists at the time used the term acid, they had in mind primarily the anhydride, though they recognized that the presence of water was required for anhydrides to manifest their acidic properties. Crosland (1973, 307) points out that this water "was considered as no more an essential part of the acid than, say, water of crystallization is a part of certain salts." To take an example, Lavoisier (1790 [1965], 69) uses the term carbonic acid to refer to what we could call carbon dioxide (CO_2), the anhydride of carbonic acid. And when he claims that the components of carbonic acid are oxygen and carbon, he's not neglecting hydrogen as a component, since he had in mind the anhydride, and since he recognized that carbonic acid mixed with water does contain hydrogen.

Chemists at the time took the same attitude toward boric acid and its anhydride, boron trioxide. For instance, in his *Dictionary of Chemistry*, William Nicholson (1753–1815) writes:

> In a moderate heat this concrete acid melts with less intumescence than borax itself, and runs into a clear glass, which is not volatile unless water be present. This glass does not differ from the original acid, except in having lost its water of crystallization. (1795, 18)

The concrete acid is what we would now call boric acid, while the clear glass is its anhydride, which we would call boron trioxide. Lavoisier (1790 [1965] 244) distinguishes the combinations of boracic acid with other substances "in the humid way" and "via sicca," i.e., in the dry way, which corresponds to the distinction we draw between boric acid and boron trioxide today. Years later, Davy treats the acid and the anhydride as two forms of the same substance:

> Boracic acid, in its common form, is in combination with water; it then appears as a series of thin white hexagonal scales; ... By a long continuous white heat the water is driven off from it, and a part of the acid sublimes; the remaining acid is a transparent fixed glass, which rapidly attracts moisture from the air. (1812, 180)

The white hexagonal scales are what we would call boric acid, while the transparent fixed glass is what we would call boron trioxide.

Importantly, Lavoisier's predictions regarding boracic acid are primarily predictions regarding what we would call boron trioxide. Lavoisier was primarily concerned with the composition of boracic acid without water (i.e., boron trioxide); and he predicted that it is composed of oxygen and a combustible substance, namely, the boracic radical. He was also perfectly aware that boracic acid with water (i.e., boric acid) contains hydrogen since it contains water.

It is also the case that Davy, Gay-Lussac, and Thénard decomposed boron tri-oxide as opposed to boric acid. To be sure, there is some disagreement about this point in the literature, with some commentators claiming that it was boric acid (e.g., Fontani, Costa, and Orna 2015, 30) and others claiming that it was boron trioxide (e.g., Crosland 1978, 78). There are two good reasons for con-cluding that these chemists decomposed boron trioxide. First of all, Gay-Lussac and Thénard (1808, 170) describe the boracic acid they were working with as "very pure and vitreous," which looks like a description of boron trioxide as op-posed to boric acid. Second, decomposing boron trioxide by means of alkaline metals like potassium is a recognized method of producing impure samples of boron. Chemists in the mid-20th century did attempt to substitute boric acid for boron trioxide in such reactions, but these attempts were eventually abandoned because of the risk of producing an explosive mixture (Zhigach and Stasinevich 1977, 216–217).

Since boracic acid referred primarily to boron trioxide, we can conclude that Lavoisier's predictions were successful and that the work of Davy, Gay-Lussac, and Thénard confirmed those predictions.

3.6.2 Lavoisier's Prediction That the Boracic Radical Is an Element

In section 3.3, I briefly noted another prediction regarding the boracic, muri-atic, and fluoric radicals that one finds in Lavoisier's work, namely, that they are elements. Lavoisier included the three radicals in his table of simple substances, thereby indicating his view that they should be considered elements (1790 [1965], 175).

For Lavoisier, the terms element and simple substance are synonyms, and he understands elements as substances that chemists cannot decompose into sim-pler component substances (xxiv). As Scerri (2005, 129) puts it, elements, for Lavoisier, are "observable simple substances that can be isolated," i.e., from the other substances that they combine with to form compounds, which requires chemical decomposition. Lavoisier (1790 [1965], xxiv, 177) emphasizes that "simple" should be understood relative to our present state of knowledge, so that a substance that we regard as simple today may be shown to be a compound to-morrow if chemists succeed in decomposing it. Until chemists are able to decom-pose a substance, Lavoisier states that we should regard it as an element.

However, there's a puzzle regarding the inclusion of the three radicals on the table of simple substances. As Hendry (2005, 41–42) has noted, Lavoisier applies his notion of element in an inconsistent way—since the boracic, muri-atic, and fluoric acids had not yet been decomposed, they should appear on the table of simple substances. So why do their radicals appear instead? Lavoisier's theory of acidity and his notion of element give us the resources to solve this puzzle. The undecomposed acids do not appear on the table because, according to Lavoisier's theory of acidity, acids in general are compounds, not simple substances. Lavoisier doesn't explicitly justify the inclusion of the radicals on his table of simple substances; and since his theory allows for compound radicals, it's not immediately obvious why he includes them. The most plausible reason is that, since the radicals had not yet been isolated, they had not been shown to be compounds, in which case they belong on the table of simple substances. To be sure, Hendry is correct about Lavoisier's inconsistent application of his notion of element. My point is just that he had his reasons for being inconsistent.

The upshot of this way of solving the puzzle is that, although it turned out to be true that the boracic radical is an element, namely, boron, the predictive success in this case is not the success of one of Lavoisier's theories. Lavoisier's theory of acidity is what led him to omit the three undecomposed acids from his table of simple substances. It did not lead him to include the three radicals on that table since his theory allowed for the possibility of compound radicals. It was his notion of element that led him to include the three radicals, and this notion does not amount to a theory of the nature of elements in any deep sense. It basically amounts to a norm regarding the application of the term element: Until a substance is demonstrated to be a compound, we should call it an element. As Lavoisier (1790 [1965], xxiv) emphasizes, this notion does not contain within it a view regarding "those simple and indivisible atoms of which matter is composed." It was precisely this kind of metaphysical view that Lavoisier attempted to avoid by proposing his operational notion of element. If the boracic radical were isolated and subsequently shown to be a compound, that result would be consistent with both Lavoisier's theory of acidity and his notion of element. It would merely necessitate an update to the table of simple substances. Hence, although this instance of novel predictive success may be interesting for other reasons, it is not relevant to the issue of whether selective realists can accommodate cases of false-but-successful theories because it is not a success of a theory.

3.6.3 The Purity of the Samples

In section 3.4, I noted that Davy, Gay-Lussac, and Thénard isolated relatively impure samples of amorphous boron.[15] Given the impurity of the samples that these three chemists isolated, did they really confirm Lavoisier's predictions?

---

[15] Methods for producing pure samples of boron were not developed until the 20th century (Naslain 1977, 167).

In short, the samples that they isolated were pure enough to confirm the novel predictions from section 3.3, because those predictions don't require that the boracic radical is an absolutely pure chemical substance. Confirming the prediction that the boracic radical is an element might require isolating a sample with a greater degree of purity. But as I argued in section 3.6.2, this prediction is not relevant to the issue at hand. That said, it's worth going into a bit more detail regarding the ways in which Davy, Gay-Lussac, and Thénard isolated as pure a sample as they could, and how they thereby confirmed Lavoisier's predictions. In order to do so, I'll make use of Chen's (2016) account of experimental individuation and go into a bit more detail regarding the experimental work on boron in the early 19th century.

According to Chen, scientists experimentally individuate an entity or a sample of an entity when their experiments satisfy three conditions: (1) they separate the entity or sample from its surrounding environment, (2) they maintain the structural unity of the entity or sample, and (3) they manipulate the entity or sample so as to investigate other aspects of nature.

Conditions (1) and (2) are the conditions most relevant to the isolation of samples of chemical substances, and therefore to the confirmation of Lavoisier's prediction that boracic acid contains a combustible substance. According to condition (1), chemists must separate the substances that they produce "from their environments" (Chen 2016, 348), and "from the experimental instruments that may have helped produce [them]" (365). And according to condition (2), chemists must maintain the structural unity of these substances in the process. In general, structural unity is the idea that "the components of an individual are structured into a whole in some specific manner" (358). When it comes to chemical substances, maintaining structural unity includes such things as ensuring that a highly reactive substance that has been isolated doesn't react with other substances in its environment. These two conditions collectively yield an understanding of what it means to isolate a sample of a chemical substance. Moreover, both of these conditions relate to the quality of the sample that chemists manage to produce. We must ask whether chemists managed to produce and preserve a sample that is both large enough and pure enough for them to determine the properties of the substance.

To a large degree, Davy's and Gay-Lussac and Thénard's experiments satisfied these two conditions. Davy's initial attempts to obtain the substance by means of decomposing boracic acid with a battery were unsatisfactory precisely because they didn't satisfy these conditions. Davy (1809, 76) writes that, via this method, he "was never able to obtain it, except in very thin films," and so "[i]t was not possible to examine its properties minutely, or to determine its precise nature, or whether it was the pure boracic basis." By heating boron trioxide in a metal tube with potassium, these chemists were able to produce more of the substance in question. While Gay-Lussac and Thénard report using a copper tube for the experiment (1808, 170), Davy reports repeating the experiment with a number of different metal tubes, including tubes made of copper, brass, gold, and iron (1809, 76–77). According to Davy, "[i]n all cases, the acid was decomposed, and the products were scarcely different" (77). By showing that the results were the same regardless of what kind of metal was used, Davy goes some length toward showing that the substance had been separated from the instruments used to produce it. Both Davy (1809, 78) and Gay-Lussac and Thénard (1808, 171) also used water and muriatic acid to wash the mixture that resulted from this process and thereby separate the boron they had produced from other substances like potash and borate of potash. By doing so, they go some length toward showing that the substance had been separated from its surrounding environment. In sum, they were able to produce samples of the substance that were large enough and pure enough for them to determine some of its properties (e.g., combustibility) and conclude that it is a component of boracic acid.

Condition (3) is relevant to the confirmation of Lavoisier's prediction regarding the formation of boracic acid. This condition concerns the "instrumental use" of an entity or sample of an entity "to investigate other phenomena of nature" (Chen 2016, 358). Davy (1809, 79, 82; 1812, 179) and Gay-Lussac and Thénard (1808, 172–173) used the samples of boron they had isolated to form boracic acid by means of combustion with oxygen. Importantly, their samples of boron were pure enough to confirm Lavoisier's prediction. And in the course of confirming this prediction, these three chemists used boron as an instrument. However, it's not clear that they were using boron to investigate other phenomena of nature as opposed to continuing an investigation into the same phenomena.

Experimental individuation is a matter of degree. Samples of chemical substances can be more or less separated from their environments and from the instruments that produced them, they can be more or less pure, and their structural unity can be more or less maintained. The important point is that the samples of boron that chemists produced in the early 19th century were pure enough to confirm Lavoisier's predictions.

3.6.4 Is Boron the Boracic Radical?

In section 3.4, I claimed that boron is the boracic radical. This claim is crucial to my argument since, if boron is not the boracic radical, we can't conclude that Lavoisier's predictions regarding boracic acid are successful. Hence, this claim is in need of some defense, especially since Chang (2012b, 54) has labeled all three of Lavoisier's hypothetical radicals (the muriatic, fluoric, and boracic radicals) as "non-existent." If the boracic radical doesn't exist, then there is no way to identify it with boron.

Chang's attitude toward the three hypothetical radicals is presumably motivated by some observations he makes regarding Lavoisier's system of chemistry. For Lavoisier, the element oxygen is, among other things, what one obtains by taking an acid and removing its radical, or by taking oxygen gas and removing caloric. More schematically, "oxygen = acid – radical" and "oxygen = oxygen gas – caloric." Chang holds that these are "meaningless formulations in modern chemistry, with the empty set as the extension in each case" (2011, 417). For Chang, it makes no sense, from our current standpoint, to ask what the radical of an acid is, since the radical, for Lavoisier, is the acid's non-oxygen component, and since we've long ago given up the idea that all acids contain oxygen. Moreover, to say that the radical is the acid's non-oxygen component is to say that it is the component of the acid that is not oxygen *base*, i.e., the substance that results when oxygen gas loses its caloric. Chang's claim that the three hypothetical radicals don't exist is presumably motivated by the fact that the very notion of an acid's radical is tied so closely to Lavoisier's theory of acidity, which is not even approximately true if our current theories are correct.

Chang's view is quite reasonable when it comes to the muriatic and fluoric radicals. Since the muriatic and fluoric acids contain no oxygen, there is no way to identify the non-oxygen component. In section 3.3, I discussed how the terms muriatic radical and fluoric radical are best understood as temporary placeholders for the names of whatever substances play the role of the non-oxygen components within the muriatic and fluoric acids, respectively. Once chemists understood the composition of these two acids, this role no longer made sense, and so the most reasonable conclusion to draw is that these two radicals do not exist.

While Chang's view only concerns Lavoisier's three hypothetical radicals, it's worth emphasizing that it would be implausible to claim that all of Lavoisier's radicals are non-existent. Consider, for example, the sulfuric, phosphoric, and carbonic radicals, i.e., the radicals of sulfuric, phosphoric, and carbonic acid, respectively. If we were to deny that these three radicals exist, that would amount to denying that they should be identified with sulfur, phosphorus, and carbon, respectively. But in the case of these three acids, the role descriptions of the radicals make some sense. The role descriptions are approximately true in these cases, though it's fair to note that sulfur, etc., do not combine with oxygen *base* to form acids. Moreover, in order to understand Lavoisier's theory of acidity, one has to be able to see certain substances as radicals of acids. Without entering into the mindset that sulfur is the radical of sulfuric acid, it's difficult to comprehend Lavoisier's theory. So while we have good reasons to conclude that the muriatic and fluoric radicals do not exist, these reasons do not extend to the sulfuric, phosphoric, and carbonic radicals, and we have good reason to identify them with sulfur, phosphorus, and carbon, respectively. Importantly, these identifications don't require a commitment to Lavoisier's theory of acidity. We can of course admit that his theory as a whole was not even approximately true. These identifications merely require an attempt to see the world through Lavoisier's theory insofar as it's possible to do so.

The boracic radical is much more like the sulfuric, phosphoric, and carbonic radicals than the muriatic and fluoric radicals, and so it's better to conclude that it exists, and to identify it with boron. The role description in this case is approximately true—the boracic radical is what combines with oxygen to form boracic acid, and that is exactly what boron does. Moreover, identifying the boracic radical with boron helps us to understand the history of the discovery of boron. For example, it helps us to understand why two of the chemists credited with discovering boron, namely, Gay-Lussac and Thénard, label the substance they isolated the "radical boracique" (1808, 171). Importantly, one can label something as a base or radical while rejecting Lavoisier's theory, and this seems to be exactly what Davy himself did. As I noted in section 3.4, Davy (1812, 178) was still referring to boron as "the basis of the boracic acid" two years after he had argued that muriatic acid contains no oxygen. The primary reason for concluding that the muriatic and fluoric radicals do not exist is that neither muriatic acid nor fluoric acid contains oxygen; and that reason does not extend to the boracic radical. In short, insofar as there are good reasons for identifying, say, sulfur with the sulfuric radical, there are good reasons for identifying boron with the boracic radical.

3.7 Conclusion

In this chapter, I've attempted to gain a better understanding of the work in chemistry that led to the discovery of boron and the significance of this work for the scientific realism debate. I've added Lavoisier's oxygen theory of acidity to the list of theories that, although not even approximately true, exhibited novel predictive success. I've considered some ways in which the selective realist may attempt to accommodate this instance of novel predictive success. And I've argued that these attempts are unsuccessful. In that case, Lavoisier's oxygen theory of acidity poses a strong challenge to selective realism.

Acknowledgments

This chapter branched off from a co-authored project with Ruey-Lin Chen, and I'd like to single him out for special thanks. This chapter also benefited from very helpful suggestions from Timothy Lyons. Thanks to the organizers and participants of the Quo Vadis Selective Scientific Realism? conference at Durham University in August 2017, especially Hasok Chang, Ludwig Fahrbach, Amanda J. Nichols, Myron Penner, Jan Potters, Juha Saatsi, Yafeng Shan, and Peter Vickers. Thanks to three anonymous referees for their helpful comments and suggestions. Finally, thanks to the Ministry of Science and Technology in Taiwan for supporting this work (MOST 106-2410-H-010-001).

References

Alai, Mario. 2017. "The Debates on Scientific Realism Today: Knowledge and Objectivity in Science." In *Varieties of Scientific Realism: Objectivity and Truth in Science*, edited by Evandro Agazzi, 19–47. Cham: Springer.

Berthollet, Claude Louis. 1789. "Extrait d'un mémoire sur l'acide prussique." *Annales de Chimie* 1: 30–39.

Brooke, John Hedley. 1980. "Davy's Chemical Outlook: The Acid Test." In *Science and the Sons of Genius: Studies on Humphry Davy*, edited by Sophie Forgan, 121–175. London: Science Reviews.

Chang, Hasok. 2003. "Preservative Realism and Its Discontents: Revisiting Caloric." *Philosophy of Science* 70 (5): 902–912.

Chang, Hasok. 2011. "The Persistence of Epistemic Objects through Scientific Change." *Erkenntnis* 75 (3): 413–429.

Chang, Hasok. 2012a. "Acidity: The Persistence of the Everyday in the Scientific." *Philosophy of Science* 79 (5): 690–700.

Chang, Hasok. 2012b. *Is Water H_2O? Evidence, Realism and Pluralism*. Dordrecht: Springer.

Chen, Ruey-Lin. 2016. "Experimental Realization of Individuality." In *Individuals across the Sciences*, edited by Alexandre Guay and Thomas Pradeu, 348–370. New York: Oxford University Press.

Crosland, Maurice P. 1973. "Lavoisier's Theory of Acidity." *Isis* 64 (3): 306–325.

Crosland, Maurice P. 1978. *Gay-Lussac: Scientist and Bourgeois*. Cambridge: Cambridge University Press.

Davy, Humphry. 1808. "The Bakerian Lecture [for 1807]: On Some New Phenomena of Chemical Changes Produced by Electricity, Particularly the Decomposition of the Fixed Alkalies, and the Exhibition of the New Substances Which Constitute Their Bases; and on the General Nature of Alkaline Bodies." *Philosophical Transactions of the Royal Society of London* 98: 1–44.

Davy, Humphry. 1809. "The Bakerian Lecture [for 1808]: An Account of Some New Analytical Researches on the Nature of Certain Bodies, Particularly the Alkalies, Phosphorus, Sulphur, Carbonaceous Matter, and the Acids Hitherto Undecompounded; with Some General Observations on Chemical Theory." *Philosophical Transactions of the Royal Society of London* 99: 39–104.

54 HISTORICAL CASES FOR THE DEBATE

Davy, Humphry. 1811. "The Bakerian Lecture [for 1810]: On Some of the Combinations of Oxymuriatic Gas and Oxygene, and on the Chemical Relations of These Principles, to Inflammable Bodies." *Philosophical Transactions of the Royal Society of London* 101: 1–35.

Davy, Humphry. 1812. *Elements of Chemical Philosophy.* Philadelphia: Bradford and Inskeep.

Fontani, Marco, Mariagrazia Costa, and Mary Virginia Orna. 2015. *The Lost Elements: The Periodic Table's Shadow Side.* Oxford: Oxford University Press.

Gay-Lussac, Joseph Louis, and Louis Jacques Thénard. 1808. "Sur la décomposition et la recomposition de l'acide boracique." *Annales de Chimie* 68: 169–174.

Hendry, Robin Findlay. 2005. "Lavoisier and Mendeleev on the Elements." *Foundations of Chemistry* 7 (1): 31–48.

Hricko, Jonathon. 2018. "Retail Realism, the Individuation of Theoretical Entities, and the Case of the Muriatic Radical. In *Individuation, Process, and Scientific Practices*, edited by Otávio Bueno, Ruey-Lin Chen, and Melinda B. Fagan, 259–278. New York: Oxford University Press.

Kitcher, Philip. 1993. *The Advancement of Science: Science without Legend, Objectivity without Illusions.* New York: Oxford University Press.

Knight, David M. 1992. *Humphry Davy: Science and Power.* Cambridge, MA: Blackwell.

Laudan, Larry. 1981. "A Confutation of Convergent Realism." *Philosophy of Science* 48 (1): 19–49.

Laurent, Auguste. 1855. *Chemical Method, Notation, Classification, and Nomenclature.* London: Harrison and Sons, St. Martin's Lane.

Lavoisier, Antoine Laurent. 1790 [1965]. *Elements of Chemistry.* 1st ed. Translated by Robert Kerr. New York: Dover. Translation of *Traité élémentaire de chimie*, 1789.

Le Grand, H. E. 1972. "Lavoisier's Oxygen Theory of Acidity." *Annals of Science* 29 (1): 1–18.

Lowry, Thomas Martin. 1915. *Historical Introduction to Chemistry.* London: Macmillan.

Lyons, Timothy D. 2002. "Scientific Realism and the Pessimistic Meta-*Modus Tollens*." In *Recent Themes in the Philosophy of Science: Scientific Realism and Common Sense*, edited by Steve Clarke and Timothy D. Lyons, 63–90. Dordecht: Kluwer.

Lyons, Timothy D. 2006. "Scientific Realism and the *Stratagema de Divide et Impera*." The *British Journal for the Philosophy of Science* 57 (3): 537–560.

Lyons, Timothy D. 2016. "Structural Realism versus Deployment Realism: A Comparative Evaluation." *Studies in History and Philosophy of Science* 59: 95–105.

Lyons, Timothy D. 2017. "Epistemic Selectivity, Historical Threats, and the Non-Epistemic Tenets of Scientific Realism." *Synthese* 194 (9): 3203–3219.

Miller, William Allen. 1871. *Elements of Chemistry: Theoretical and Practical. Part II. Inorganic Chemistry.* 3rd London ed. New York: John Wiley & Son.

Musgrave, Alan. 1988. "The Ultimate Argument for Scientific Realism." In *Relativism and Realism in Science*, edited by Robert Nola, 229–252. Dordrecht: Kluwer.

Naslain, R. 1977. "Crystal Chemistry of Boron and of Some Boron-Rich Phases; Preparation of Boron Modifications. In *Boron and Refractory Borides*, edited by Vlado I. Matkovich, 139–202. Berlin: Springer-Verlag.

Nicholson, William. 1795. *A Dictionary of Chemistry.* Vol. 1. London: G. G. and J. Robinson, Paternoster Row.

Psillos, Stathis. 1999. *Scientific Realism: How Science Tracks Truth.* London: Routledge.

Scerri, Eric R. 2005. "Some Aspects of the Metaphysics of Chemistry and the Nature of the Elements." *HYLE—International Journal for Philosophy of Chemistry* 11: 127–145.
Stanford, P. Kyle. 2006. *Exceeding Our Grasp: Science, History, and the Problem of Unconceived Alternatives.* New York: Oxford University Press.
Thomson, Thomas. 1802. *A System of Chemistry.* Vol. 2. Edinburgh: Bell & Bradfute, and E. Balfour.
van Fraassen, Bas C. 1980. *The Scientific Image.* Oxford: Clarendon Press.
Vickers, Peter. 2013. "A Confrontation of Convergent Realism." *Philosophy of Science* 80 (2): 189–211.
Zhigach, A. F., and D. C. Stasinevich. 1977. "Methods of Preparation of Amorphous Boron." In *Boron and Refractory Borides*, edited by Vlado I. Matkovich, 214–226. Berlin: Springer-Verlag.

4

No Miracle after All

The Thomson Brothers' Novel Prediction that Pressure Lowers the Freezing Point of Water

Keith Hutchison

School of Historical and Philosophical Studies
University of Melbourne
keithrh@gmail.com

4.1 A Problem and Its Context

Defenders of scientific realism sometimes suggest that it would be "miraculous" for false theories to generate successful predictions. But it is fairly common for theories to make good predictions. So it must be rare for such theories to be false. The predictive success of science, then, seems good evidence for its truth.

The early history of thermodynamics, however, poses a couple of pressing challenges to simplistic understandings of this no-miracle argument. For it seems to provide clear examples of false theories producing successful predictions. Rankine's thermodynamics of around 1850, for example, anticipated a string of experimental effects, even though it was based on a model of heat (as a non-random vortex motion in the interior of atoms) that is now thoroughly ignored. Indeed it was the surprising success of one of these predictions that prompted Rankine to start publishing his investigations.[1] About the same time he predicted that saturated steam would have a *negative* specific heat, a phenomenon that soon seemed to be equally confirmed by observation, albeit via ambiguous data.[2] More momentous however was Rankine's use of the same model of heat to predict (around 1854) the existence of the entropy function, a function that

---

[1] RANKINE, 'On an equation' (1849) = *Misc. Sci. Pap.*, pp.1–12. The prediction here was of the *form* of the relationship between the temperature and saturated vapour pressure of a liquid/vapour mixture, without a specification of precise numerical values, and this may well place it outside the intended scope of the no-miracle argument.

[2] For an overview, see: HUTCHISON, 'Miracle or mystery', pp.106–7 and 'Mayer's hypothesis', esp. pp.288–294.

Keith Hutchison, *No Miracle after All* In: *Contemporary Scientific Realism*. Edited by: Timothy D. Lyons and Peter Vickers, Oxford University Press. © Oxford University Press 2021. DOI: 10.1093/oso/9780190946814.003.0004 consequently became familiar to physicists a decade before Clausius renamed it in his famous 1865 declaration of the connection between entropy and the arrow of time.[3] I surveyed these predictions in a paper published some 20 years ago, though I then left unresolved the ultimate impact of Rankine's successes on the defence of realism, for I underestimated the obstacles to showing that his predictions really did depend on the discredited theory.[4] Even now, that task remains beyond me, if only because of the logical maze that I glimpse within it.

So I wish to analyse here a quite different example of a successful prediction from the early history of thermodynamics, one that was (beyond any doubt) made using another abandoned theory, the 1824 proto-thermodynamics of Sadi Carnot.[5] This magnificent analysis of the efficiency of steam engines was balanced upon a postulate that its author personally doubted, the conservation of heat. Accordingly, it allowed no inter-conversions between heat and other forms of what would today be called energy, and for this reason was unequivocally abandoned around 1850, as the first formulations of modern thermodynamics emerged, with its so-called first law replacing Carnot's postulate.[6]

But just before this happened, William Thomson (the later Lord Kelvin) was able to confirm the magnitude of an "entirely novel physical phaenomenon" predicted "in anticipation of any direct experiments on the subject" by his brother James on the basis of Carnot's theory: a lowering of the freezing point of water, produced by an increase in the external pressure.[7] The Thomson brothers rightly saw the success of this prediction as good evidence for a theory that was attractive for a lot of other reasons. But at the same time, James Joule was producing strong evidence that Carnot was wrong, and William was much puzzled, acknowledging the persuasiveness of Joule's investigations. He concluded that a resolution of the dilemma required further empirical data—but this was wrong. For in 1850, Clausius managed to harmonise the two understandings by purely intellectual (and relatively harmless) modifications to the Carnot theory.

---

[3] Rankine called it the "thermodynamic function", and denied (e.g. *Misc. Sci. Pap.*, p.233) that it had any asymmetric tendency to increase. For its first explicit identification, see pp.351 2 of his *Misc. Sci. Pap.* For Clausius's renaming, see the English trans. on pp.364–5 of *Mechanical theory*. A prediction like this, that a particular state-function exists, is definitely subject to empirical verification, but it is certainly atypical, so could perhaps be outside the scope of the no-miracle argument.

[4] HUTCHISON, 'Miracle or Mystery' (2002). In that study, I did consider the dependence question, but only briefly (on pp.109–11), but that discussion now feels too thin. As I recall, those remarks were inserted at the request of a referee.

[5] CARNOT, *Réflexions sur la puissance motrice du feu*.

[6] For a survey, see e.g., CARDWELL, *Watt to Clausius*, esp. pp.186–260.

[7] The whole process took place between 1847 and 1850. Though the prediction came out under James's name, William was heavily involved and may even have been the principal investigator. For the chronology etc., see: SMITH & WISE, *Energy and empire*, pp.294–99. For the original prediction, see: J. Thomson, 'Theoretical considerations' and W. Thomson 'The effect of pressure', reprinted together on pp.156–69 of W. Thomson, *Math. Phys. Pap.*, v.1. My quotations are from pp.165–6 of William's paper. Given that the sensitivity of the boiling point of water to pressure had long been known, it is odd that the sensitivity of the freezing point was (apparently) quite unsuspected in 1848.

To preserve Carnot's central claim, however, Clausius did need to invoke a new premise—a second law—and with these decisive steps, modern thermodynamics emerged. Thomson soon accepted Clausius's understandings.

My concern here is with that slightly earlier prediction however. What was the thinking that led James to conclude that external pressure lowered the freezing point? How did he defend this conclusion? How did he estimate the magnitude of the lowering? Did he really use the theory that was about to be discarded?

By answering these historical questions, I shall demystify the prediction, by showing that though James's argument uses suppositions rejected in modern thermodynamics, and applies those falsehoods to a situation where the two theories differ, the quantitative difference between the two accounts of James's phenomenon is only marginal. One says that heat is conserved here, the other says it is not, but agrees that in this particular situation heat is nearly conserved. James's argument furthermore needed a principle that was central to both theories, but at the time he wrote that principle was only supplied by the Carnot theory. So his prediction did indeed need the precarious theory. Yet the critical inference did not itself directly use those portions of the Carnot theory that were soon abandoned. The false components did however give credence to the true portions—so mattered. But there is (seemingly) little miracle in using the truish sections of a false theory to make a prediction. To demolish the no-miracle argument, a bigger miracle is needed, one where a successful prediction hinges on bigger deviations from the truth.

In discussing this episode, I deploy a very naive distinction between "true" and "false" statements. This greatly simplifies my narrative, for it avoids complicated hedging, circumlocutions, etc. The associated over-simplifications seem to me to do no harm, because the nature of "scientific truth" is not the issue here and does not need to be confronted. But my language might at times seem annoying to the fastidious reader.

4.2 The Carnot Theory of 1824

The half-century beginning with 1775 saw a ten-fold increase in the efficiency of steam engines, an improvement so extraordinary that the question began to circulate whether there was any limit to this process.[8] Attempts to answer this question were generally of limited scope, focusing too closely on familiar devices, but one was quite exceptional. In a masterpiece of elegant analysis, Sadi Carnot introduced (in 1824) the notion of a "heat-engine," one characterised by the single fact that it produces mechanical work by

---

[8] See, e.g. FOX, 'Introduction etc.', pp.2–12 and CARDWELL, Watt to Clausius, pp.153–81.

exchanging heat with its environment. The recent prevalence of the notion that heat was a special fluid, caloric, suggested that there was, accordingly, some sort of analogy between heat-engines and water-engines, given that the latter operated through an exchange of *water* with their environment. This analogy lies at the heart of Carnot's discussion, but it also incorporates the flaw that eventually led to the analysis being abandoned, the presumption that heat (like water) was conserved. Yet it provided Carnot with his key insight, an insight that survived the later demise of the 1824 analysis: just as a water-engine requires two "reservoirs," a high one to supply water and a low one to receive "waste," a heat-engine requires two heat-reservoirs, one at a high temperature to *supply* heat, and one at a low temperature to *receive* heat. (Watt's separate condenser had made it abundantly clear that steam engines produced waste-heat, and that, accordingly, their coal needed to be supplemented by a source of cold, simply for the engine to function.)

The same analogy also provided Carnot with the idea of a heat-*pump*, and that of a *reversible* engine, and he was then able to make a simple adjustment to an argument that had already been applied to water-engines, to show that if it was accepted that perpetual motion was impossible, no heat-engine could be more efficient than one which was perfectly reversible (for given source and sink levels). This is Carnot's answer to the question challenging his contemporaries, and to enhance its bona fides, Carnot showed that his theory was very fertile indeed. He produced some rough calculations of the maximal efficiencies he had identified, rough only because too much of the requisite data was lacking; and he produced a series of novel lemmata about the properties of fluids, typical of those deductions which are a standard feature of later thermodynamics.[9]

4.3 Thomson Revives the Carnot Theory

But Carnot's bold theory attracted little attention— until the late 1840s, after a copy of his memoir had been discovered by William Thomson, then working in Regnault's laboratory in Paris (where the sort of data needed to complete Carnot's calculations was being systematically generated). Thomson was enthusiastic about the analysis. He soon published his own rehearsal of the theory, with some of the core calculations systematized (via Regnault's new data). But before this came out, Thomson also showed that Carnot's understanding provided a rationale for absolute thermometry, a temperature scale independent of

---

[9] For the efficiency estimates, see CARNOT, *Réflexions*, pp.94–101. For the lemmata, see: *ibid.*, pp.78–90; FOX, 'Introduction etc.,' pp.132–45; CARDWELL, *Watt to Clausius*, pp.201–4.

the idiosyncrasies of any particular chemical substance. And he noticed that the Stirling air-engine vividly illustrated Carnot's claims: it used two heat reservoirs and could be run in two different directions: as a heat-engine; or as a heat-pump—where the transfer of heat was palpable.[10]

4.4 The Rationale Behind James Thomson's Prediction

When a model Stirling engine was run as a pump, shifting heat between two reservoirs at the same temperature, virtually no work was required to operate the pump (just as Carnot's theory would predict).[11] It was this fact that led to James's prediction, because it meant that heat could easily be pumped from beneath the piston of a cylinder containing a water-ice mixture at 0°C to (say) an icy lake at the same temperature (where the heat received would melt some ice without changing the lake's temperature). Yet if this was done, some of the water in the cylinder would freeze, and in doing so it would expand (for water colder than 4°C expands when cooled). And it was well known that the forces generated by freezing water are very great, for they often caused serious damage. So quite a high force could be applied to the piston, without preventing the expansion, and combining this force with the movement of the piston would yield a supply of external work that exceeded (by far) the negligible quantity of work consumed in pumping the heat to the lake. So a perpetual motion would seem to be on offer. Or rather there had to be some flaw in the reasoning here: some phenomenon, hitherto overlooked, must be sabotaging the proposal.

Reflection quickly shows one such flaw: the argument that threatens a perpetual motion presumes the temperature in the high-pressure cylinder to be the same as that in the lake, i.e. that the temperatures of the two different ice-water mixtures are identical, despite the different pressures at which the phases change. (For it is this identity that enables the more or less work-free pumping of the heat from the cylinder.) So if the temperature at which water freezes changes with the pressure the argument collapses. Indeed, if the temperature of the high-pressure cylinder were lower than that of the lake (acted on by normal atmospheric pressure) work would be consumed in pumping the heat "uphill" into the lake, and this work could suffice to prevent a perpetual motion.

---

[10] W. THOMSON, 'Notice of Stirling's air engine.'
[11] So this minor episode might provide another example of a false theory successfully predicting an unknown effect.

4.5 A More Polished Analysis

Such were the thoughts that led to James's discovery, though his published announcement of the discovery includes a more systematic analysis, one extended to include a quantitative estimate of the magnitude of the newly appreciated effect.[12] Let us now sketch that argument.

We begin by observing that to understand the operation of any engine, we obviously need to make a clear distinction between the fuel (materials which take on permanently new forms in the course of the operation of the device) and the engine proper (that which suffers no permanent change, the "working substance" as it is often called, plus its mechanical container, etc.). Carnot accommodates this fundamental distinction by having his engines operate in cycles, their materials passing through a sequence of changes that bring them back to their original configuration. To avoid wastage, these changes need (as mentioned earlier) to be reversible; and in a thermal context, the simplest such engine would seem to be one with a single working substance undergoing just four changes, two to exchange heat with the two different reservoirs (each at the same temperature as the engine materials[13]), and two to switch the temperature of those materials with no exchange of heat. The materials need to be brought, back and forth, between the pair of temperatures asserted by Carnot to be essential to a heat-engine.[14]

Adapting this idea of a "Carnot cycle" to his context, James considers an engine composed of a cylinder containing a mixture of water and ice beneath a piston. As the engine operates, the quantities of water and ice here vary, as ice sometimes melts, and as water sometimes freezes, in association with the energetic exchanges between the engine and its environment.

To see the details, imagine the cylinder to be thoroughly insulated, with a relatively large weight on the piston, so that the water-ice mixture is at a highish pressure, and in internal equilibrium at whatever temperature happens to be the melting point of water for that pressure. Suppose now (as a preliminary) the

[12] For the more informal discussion, see J. THOMSON, 'Theoretical considerations,' pp.157–9; for the quantitative version, see pp.160–4.

[13] It might (very reasonably) be objected that no heat will flow into the reservoir if it is at the same temperature as the ice-water mixture. Like Carnot before him (e.g. *Réflexions*, p.68n), James recognized this problem, and avoided it. I, by contrast, create it here—by using a circumlocution that is quite common in thermodynamics, which serves to abbreviate the wording, and which I presume will be familiar to any concerned reader. Since he was writing at the very beginning of thermodynamic analysis, James could not presume such familiarity and explains the logic more fully, overtly imagining a small difference in temperature, but one that could be made arbitrarily small, perhaps infinitesimal (etc.) Some such artifice lies behind many references to isothermal change in thermodynamic argument, and is routinely deployed in descriptions of a Carnot cycle.

[14] Thermocouples however operate without undergoing any change themselves, but electrical engines were a rarity at the time of the Thomson brothers' discovery, so I just ignore any minor challenge they might pose to my narrative.

insulation is removed, and the cylinder brought into thermal contact with a heat reservoir at that same temperature, and that heat flows out of the cylinder and into the reservoir, so some of the water in the cylinder freezes, and the resulting expansion causes the piston to lift the weight, doing mechanical work on its environment.

After a period of expansion, one chosen simply for pragmatic reasons, the cylinder is insulated again, and the weight on the piston is slowly reduced. As the pressure is reduced, the mixture in the cylinder expands, and work is done on the environment, while water freezes, or ice melts, to bring the cylinder to a new temperature, that which is the melting point of water at the new, reduced, pressure. This is to be the first stage of our cycle. For stage 2, the insulation is removed, and the cylinder is brought into contact with a second heat-reservoir, one at the same temperature as the low-pressure cylinder. Heat is now injected into the cylinder, melting some of the ice, so that the piston descends (as the volume of its contents decreases) and work is done on the system. After a convenient time, the cylinder is insulated again, and stage 3 begins: the weight on the piston is slowly increased, further compressing the mixture in the cylinder, with yet more work being done by the environment. The stage terminates when the weight on the piston reaches its original value. Along the way, some water will freeze, or some ice will melt, and the cylinder will return to its original temperature. The final, fourth, stage then mimics the preliminary process described earlier, beginning as thermal contact with the first heat-reservoir is reestablished. Expansion resumes and heat flows to the reservoir, until the initial volume of stage 1 (= the final volume of the preliminary stage) is reached. The cycle is now complete, and can be repeated if desired etc.

This cycle constitutes a heat-engine. For it exchanges nothing but heat and work with its environment. And it produces mechanical work: for each change of volume takes place in two directions, and the compression that reverses an expansion takes place at a lower pressure than the corresponding expansion. So if Carnot is right, heat must be flowing from a high-temperature reservoir to a low-temperature one. But we know that in this case heat is flowing *into* the reservoir during stage 4 and *out of* the reservoir during stage 2 (while no heat is flowing during stages 1 and 3). So the high-pressure stage 4 must take place at a lower temperature than the low-pressure stage 2. In other words, an increase in pressure reduces the freezing point.

4.6 No Miracle Involved in This Prediction

There is no doubt that James's prediction here uses the Carnot theory, for it is this that tells James heat must flow from a high temperature source to a low temperature sink. But Carnot does more for James than simply assert this necessity, it embeds the requirement in a sustained analysis which enhances its plausibility. The fact that the Carnot theory requires such a thermal descent is not, however, what makes it false. The theory is false because of something quite different, the fact that it declares the quantity of heat released to the sink identical to that acquired from the source. Carnot's discussion made much use of the heat-conservation that this alleged identity illustrates,[15] and the rationale for insisting upon a descent of heat presumes conservation. But once that descent is accepted, the prediction requires no further appeal to conservation. Indeed, James's derivation works just as well in post-1850 thermodynamics where the heat released to the sink differs from that taken from the source. Cosmetic adjustments do, of course, need to be made to articulations of the story; and a different context (like that provided by Clausius), is needed to make the necessity of a low temperature sink plausible. So James may be thought of as taking a trustworthy derivation and placing it in bad company (via the alternative rationale for its central premise). But no-one is challenged by a trustworthy argument making a successful prediction. In other words, James's prediction requires no miracle, and the episode shows that a false theory really can—sometimes—make a successful prediction. But it is a weak counter example to the no-miracle argument, for the prediction proceeds via the true components of a false theory. To get a better counter-example, it would be necessary to show something stronger: that James's deduction would collapse if denied some false component of the Carnot theory, heat-conservation in particular. But it does not collapse. All that happens is that its critical premise is left without a rationale. The Rankine story is far more powerful, then, and does seem to be a real counter-example—if I got that story right.

4.7 Some Methodological Hesitation

Before moving on to James's quantitative discussion, where he estimates the magnitude of the effect predicted, thoroughness requires some navel-gazing, in which we briefly consider the relationship between my summary of James's argument and his own wording. Needless to say the two differ, though the discrepancies are no more severe than those standard in the history of science, where historians routinely paraphrase their sources. (Indeed, I find I cannot trust a historian whose articulation of an argument too closely mirrors that of the original version: it suggests the argument has not been mastered.) Accordingly, we need

---

[15] The argument given by Carnot (at p.69 of *Réflexions*) to show that no engine can be more efficient than a reversible uses the conservation of heat. See also: CARNOT, *Réflexions*, pp.65, 76n, 100–101; FOX, 'Introduction etc.', pp.131, 150.

to consider how much this matters in the context of the no-miracle argument—arguing that the differences are of no moment.

Like me, James gives two versions of the argument (see n. 12), one informal (and focused on that key insight supplied by the Stirling engine) and the other phrased, more systematically, in terms of a Carnot cycle. James however introduces the Carnot cycle as a means of making his quantitative estimate, and regards the effect itself as established by the more informal argument; while I use that cycle to perform both these tasks (with the quantitative evaluation delayed until later). Otherwise, our discussions of the cycle are very close, though that of James seems more verbose, aimed at an audience with little exposure to thermodynamic analysis (as, e.g. indicated in n. 13).

Our informal arguments however differ markedly, mainly because I cannot follow the original argument in detail. There may (accordingly) be some serious blemish in James's prediction, a blemish hidden by my rephrasing.[16] Our concern here, however, is with the plausibility of the prediction, whether the conclusions are reasonably justified by the premises (and their context), not whether a particular string of words adequately displays that connection. So a blemish in some wording of the argument is of no moment, so long as some minor redrafting sustains the conclusion—without introducing new premises. James may have written his argument down hastily, and not done full justice to his thoughts, or he may have gone to an opposite extreme, and penned numerous drafts, none of which is canonical. In either extreme, the published argument does not tell the full story, and the real argument for the prediction is that elaboration of the overt wording that each of us privately prepares, as we absorb the message conveyed to us by the written summary. So there are (I suggest) no grounds for concern in the mere existence of small discrepancies between James's words and my own phrasing of the argument that supplies the prediction.

4.8 The Quantitative Calculation

To estimate the magnitude of the effect he had discovered, i.e. how large a reduction ("$-t$") in the freezing point is produced by a given increase ("$[+]p$") in pressure, James examines a concrete example of the Carnot cycle we introduced to set out the qualitative argument.[17] To reveal the relationship between p and t,

---

[16] James does not, for example, evaluate the work done on the environment during the adiabatic stages of his cycle, just treating the work done in these stages as negligibly small. I, by contrast, expanded my description of the cycle slightly, in such a manner as would (hopefully) reduce any uncertainty here. James's brevity probably does not matter, but I am not sure. The issue hardly seems to matter, since it is so easy to rephrase James's argument.

[17] James THOMSON, *loc. cit.*, n.12 = 'Theoretical considerations,' pp.160–4.

he uses three "known" quantities: (1) how much expansion is produced by the freezing of water; (2) the latent heat required to melt ice at 0°C; and (3) the efficiency of a Carnot engine that operates by absorbing heat at temperature 0°C and releasing heat at temperature $-t$°C. At that time, there was no novelty in the first two of these data, while the third had recently been calculated by William, in his systematic revision of Carnot's fragmentary calculations.

The first of these items gives James the work done by his engine, as a multiple of p—viz. "0.087 × p." (James ignores the work done during the changes in temperature, since the adiabatic stages of the cycle are so small, "infinitesimal" so to speak.) The other two quantities give another, quite different, expression for this same work, now as a multiple of t—viz. "4925 × 4.97 × t," just the product of William's figure for the efficiency ("4.97"), the fuel consumed ("4925") and the magnitude of temperature drop ("t"). Equating these two figures, gives James his final estimate, "$t = 0.00000355p$," for the change in melting point produced by an additional pressure p. Without specifying James's units (viz. "cubic feet"; "foot-pounds"; "pounds on a square foot"; "the quantity of heat required to raise a pound of water from 0 to 1 degree centigrade") these results are verging on the meaningless, but our aim here is not to reveal the actual magnitude of the effect he discovered, just to check that James's calculation requires no appeal (direct or indirect) to the conservation of heat.

We have clearly not established such an independence yet, for though the first two items of data make no discernible appeal to conservation, the earlier calculation of efficiency was made in the context of the Carnot theory, and used data that had been prepared in a laboratory that did not embrace the possibility that heat was created or destroyed. So application of this calculation to James's cycle could well have required some identification, overt or hidden, of the heat supplied to some engine with that it released. And some such comparison had indeed been required in the very calculations used by James, those performed by William to indicate that his first "absolute" scale of temperatures was in good agreement with familiar laboratory scales. When the Carnot theory was abandoned, William realised his calculations were suspect, so checked the agreement he had claimed for the earlier scale, and found it no longer applied. He decided to rescue the convenience it had promised, by developing a rather different absolute scale, effectively the one in use today.[18]

We thus need to scrutinise James's use of William's earlier calculations. But before doing so, it is worth observing that James's method of calculating the magnitude of the effect he had discovered is effectively a standard component of modern thermodynamics textbooks, used to generate the so-called Clausius-Clapeyron equation, relating the variation in the boiling-point of a liquid to

[18] HUTCHISON, 'Mayer's hypothesis,' pp.294–300; CHANG, *Inventing temperature*, pp.182–6.

the external pressure. A version of this equation had been published by Émile Clapeyron, in his important 1834 injection of Carnot's discussion into a far more mathematical, and especially graphical, idiom.[19] The later survival of the calculation suggests that the result was not essentially tied to heat-conservation, but we need here to confirm that suggestion—for on a strictly literal reading of James's account, a dependence of heat-conservation is evident.

4.9 William Thomson's Figure for the Efficiency of James's Engine

James tells us that his estimate of the efficiency of his engine comes from "tables deduced . . . from the experiments of Regnault" and published in his brother's major paper of 1849—to which James's own discussion is attached as an appendix (in the 1882 reprint). Strictly speaking this is not correct, as the coldest engine listed in William's table is one that operates just *above* 0°C, while James's engine operates just below that temperature. But William's figure for the efficiency of a perfect engine taking heat from 1°C to 0°C is 4.96, and James has presumably got his 4.97 by extrapolation (something perfectly reasonable), or by access to data not used in the final version of William's table.

William had published those efficiency figures as the ones applicable to *all* Carnot engines, so James was entitled to apply the figure to his own ice/water device, and that is the logic used by James here.[20] Yet an appeal to heat-conservation is lurking beneath the surface here, for (as observed in n. 15) the argument that the figure at issue applied to *all* engines used the problematical presumption. But a validation of this figure does not require a *universal* conservation of heat. It is enough that heat be conserved in two particular engines, that used by William to compute the efficiency and that used by James to evaluate the change in the freezing point. And both these engines operate over a very small temperature range, where the difference between the fuel heat and the waste heat is tiny. So the claim that heat is conserved is effectively true in these particular circumstances, and the suspect evaluation of the "infinitesimal" engine's efficiency is in fact reliable, as William later checked.[21]

But there is also a risk, far less obvious, of a further appeal to heat-conservation in William's use of Regnault's experimental data. For Regnault could well have

---

[19] See the three equations on p.90 of CLAPEYRON, 'Motive power.' For modern versions, see PIPPARD, *Classical thermodynamics*, p.54 (eq. 5.12) and FERMI, *Thermodynamics*, pp.63–69. Clausius points out that James's prediction is easily transferred to the new theory: see his 1850 remarks on pp.80–3 of *Mech. Theory*.

[20] The critical passage is on p.163 of *Math. Phys Papers*, v.1. William's table is on p.140.

[21] See n. 22.

presumed that heat was conserved in making his measurements, and this background assumption could have tainted his data. When, soon after these events, William accepted that Joule had refuted the conservation of heat and endorsed Clausius's modification of the Carnot theory, William published a review of the data behind James's calculation and concluded that it had not been compromised by any such presumption by Regnault.[22]

I see no reason to doubt William's judgment here, all the more so because he was certainly happy to admit error. For at the very same time, William reviewed his earlier estimates of the efficiency of Carnot engines, concluding that these were seriously in error, and publishing revised estimates (effectively those in use today).[23] This meant that the table cited by James was vitiated by the presumption of heat-conservation made in its compilation, but it turned out that the errors got smaller and smaller as the range of temperatures over which the engine operated shrank. So the one figure used by James, the 4.97, was quite acceptable. Once again, there is no dependence on the false components of Carnot's account.

4.10   Concluding Review: What Can We Take Home?

At the risk of some repetition, let us summarise my conclusions. The no-miracle argument presumes it highly unlikely that a substantially false theory will make a precise, non-trivial, and correct observational prediction, if that prediction makes important use of the false components of the theory at issue. At first sight, the Thomson episode seems to provide one of these very rare events. It thus challenges the no-miracle argument by making it seem these predictions may not be so rare after all.

But closer scrutiny shows the damage is minor, and indeed gives us a glimpse of how a false theory will sometimes manage a surprising prediction that does not seem mysterious. The brothers' argument definitely uses a false theory, but what makes that theory false is its presumption that heat is conserved. This matters to the prediction for two quite different reasons. Firstly, because it gets involved in calculating the magnitude of the effect discovered. If we overlook these calculations for the moment, we see a second, quite different role for the Carnot falsehood. It provides the context for the core premise of the argument, the insistence that heat needs to descend for mechanical work to be produced.

---

[22] See W. THOMSON, *Math. Phys. Papers*, v.1, pp.193–5. CHANG, *Inventing temperature*, pp.76–77, 86–7, 99–100 notes that Regnault was extremely cautious about allowing theory to taint his measurements and (pp.174–5) that William Thomson found this caution a bit excessive. But Chang notes too (77, n.48) that some of Regnault's calorimetric measurements did require heat to be conserved. Thomson had, of course, worked with Regnault's laboratory, so his assessment here would be especially reliable.

[23] W. THOMSON, *Math. Phys. Papers*, v.1, pp.189–91, 197–8.

Without the context supplied by Carnot's false interpretation of heat-engines, the premise of the Thomson argument would be ill-supported. But "ill-supported" is not the same as "false." So heat-conservation was needed to make the prediction plausible, but it was not needed simply to make the prediction.

This last remark is perhaps in slight error though, for I made it under the tentative presumption that we could ignore the associated calculations. Heat-conservation was used in the calculations twice, in both cases to declare that the quantity of heat involved in one process was the same as that used in a different one. The Clausius theory insists these declarations are false, but also allows that in the particular processes at issue, the difference between the two quantities is tiny. So the errors in the quantitative estimation are of no moment. The Carnot theory has a big flaw in it, but that does not mean that every calculation that invokes that flaw will make big errors. A big error can sometimes have tiny consequences.

For the purpose of this discussion, it is granted that scientists sometimes come up with theories that are deemed true and make successful predictions. Given this, it is reasonable to acknowledge that some of the false theories devised by scientists are approximately true. And it is barely miraculous for an approximately true theory to make—sometimes—much the same successful prediction as a true theory.

This is not what happens with the Thomson prediction, though, for the Carnot theory is *not* approximately true. But there are certain restricted situations in which the two theories, Carnot and Clausius, do approximate each other closely. Here, they effectively overlap and make extremely similar predictions about observable phenomena. In these circumstances, a prediction made by the false theory is no more surprising than the same prediction being made by the true theory. James's prediction is an example of this.

It is not a case of a false theory making the very same predictions as a true one. The predictions differ, but get very close together in special circumstances, and indeed (in our example) they get too close to be separated observationally. If a false theory were to make exactly the same significant predictions as some genuinely different true theory, that would be far closer to a miracle.

References

CARDWELL, Donald S. L. *From Watt to Clausius: The rise of thermodynamics in the early industrial age.* (London: Heinemann, 1971).

CARNOT, Sadi. *Réflexions sur la puissance motrice du feu.* [1824]. Critical edition, ed. Robert FOX. (Paris: Vrin, 1978). My pages reference are however, to Fox's English translation, *Reflexions on the motive power of fire*, (Manchester: Manchester University Press, 1986).

CHANG, Hasok. *Inventing temperature: measurement and scientific progress*. (Oxford: University Press, 2004)

CLAPEYRON, Emile. "Memoir on the motive power of heat," [1834]. Trans. Eric MENDOZA on pp.71–105 of: MENDOZA, ed. *Reflections*.

CLAUSIUS, Rudolf J. E. *The mechanical theory of heat, with its applications to the steam-engine and to the physical properties of bodies*. Trans. J. HIRST. (From several German texts). (London: Van Voorst, 1867).

FERMI, Enrico. *Thermodynamics*. (New York: Dover, 1956). Originally 1936.

FOX, Robert. "Introduction" [on pp.1-57] and annotations etc. to CARNOT, *Reflexions*.

HUTCHISON, Keith. 'Mayer's hypothesis: a study of the early years of thermodynamics.' *Centaurus*. 20 (1976): 279–304.

HUTCHISON, Keith. "Miracle or mystery? Hypotheses and predictions in Rankine's thermodynamics." On pp. 91–120 of: CLARKE, Steve & Timothy LYONS, (ed). *Recent themes in the philosophy of science. Scientific realism and commonsense*. (Dordrecht: Kluwer, 2002).

MENDOZA, Eric, ed. etc. *Reflections on the motive power of fire*. English trans. of CARNOT, *Réflexions* and papers by CLAPEYRON and CLAUSIUS. (New York: Dover, 1960).

PIPPARD, Alfred B. *Elements of classical thermodynamics for advanced students of physics*. (Cambridge: Cambridge University Press, 1966).

RANKINE, William J. M. *Miscellaneous scientific papers*. With a memoir of the author by P. G. TAIT. Ed. W. J. MILLAR. (London, 1881).

RANKINE, William J. M. "On an equation between the temperature and the maximum elasticity of steam and other vapours." *Edinb. New Phil. J.* 47 (1949): 29–42. Reprinted on pp.1–12 of RANKINE, *Misc. Sci. Pap.*

SMITH, Crosbie & M. Norton WISE, *Energy and empire: a biographical study of Lord Kelvin*. (Cambridge: Cambridge University Press, 1989.

THOMSON, James. "Theoretical considerations on the effects of pressure in lowering the freezing point of water." Read 2 Jan 1849. *Trans. R. Soc. Edinb.* 16 (1849): 575–80. Reprinted on pp.156–64 of William THOMSON, *Math. Phys. Papers*, v.1. Page references are to this reprint.

THOMSON, William. "The effect of pressure in lowering the freezing point of water, experimentally demonstrated." Read 21 Jan 1850. *Proc. R. Soc. Edinb.* 2 (1851): 267–71. Reprinted on pp.165–9 of William THOMSON, *Math. Phys. Papers*, v.1.

THOMSON, William. "Notice of Stirling's air engine." Read 21 Apr. 1847. *Proc. Phil. Soc. Glasg.* 2 (1847): 169–70. Reprinted on pp.38–39 of William THOMSON, *Math. Phys. Papers*, v.5. Appears to be a summary, written by someone else.

THOMSON, William. *Mathematical and physical papers*. Vol. 1. (Cambridge: University Press, 1882). Vol. 5, ed. J. Larmor. (Cambridge: University Press, 1911).

5

From the Evidence of History to the History of Evidence

Descartes, Newton, and Beyond

Stathis Psillos

Department of History and Philosophy of Science
University of Athens
psillos@phs.uoa.gr

5.1 Introduction

The pessimistic induction (PI) takes it that the evidence of the history of science is an indictment for scientific realism: if past theories have been invariably false, then it will be very unlikely that current theories will turn out to be true, unless a sort of epistemic privilege is granted to current theories. But, the argument might go, what could possibly be the ground for this privilege? Assuming a substantial difference between current theories and past (and abandoned) ones would require that the methods used by current scientists are substantially different than those of the past scientists. But methodological shifts, the argument would conclude, are rare, if they occur at all.

Yet, standards of evidence as well as conceptions of explanation have changed as science has grown. If we look at these changes, we may draw different conclusions regarding the bearing of the history of science on scientific realism. Theories are not (nor have they been) invariably worthless because false. Some have been outright falsehoods, while others hold in their domains under various conditions of approximation. More importantly, some are such that important parts of them have been retained in subsequent theories. Some, but not others, have been licensed by more rigorous standards of success. Some, but not others, have initiated research programs, which result in successive approximations to truth. Hence, the history of science carries a mixed message, which needs to be read with care.

Stathis Psillos, *From the Evidence of History to the History of Evidence* In: *Contemporary Scientific Realism.*
Edited by: Timothy D. Lyons and Peter Vickers, Oxford University Press. © Oxford University Press 2021.
DOI: 10.1093/oso/9780190946814.003.0005

In this chapter, I will look into two episodes in the history of science: the transition from the Cartesian natural philosophy to the Newtonian one, and then to the Einsteinian science. These are complex episodes, and I can only offer sketches here. But the point I want to drive home should still be visible: though the shift from Descartes's theory to Newtonian mechanics amounted to a wholesale rejection of Descartes's theory (including his account of vortices as well as his methodology), in the second shift, a great deal was retained; Newton's theory of universal gravitation gave rise to a research program which informed and constrained Einstein's theory. Newton's theory was a lot more supported by the evidence than Descartes's and this made it imperative for its successor theory to accommodate within it as much as possible of Newton's theory: evidence for Newton's theory became evidence for Einstein's. Besides, a point that these two successive cases of theory-change will highlight is that the very judgment that some theories are more supported by the evidence than others, and the concomitant judgement that some standards of evidence and explanation are better than others, need not be made in a post hoc way by current philosophers of science. Rather, they are made by the protagonists of these episodes themselves.

This double case study will motivate a rebranding of the divide et impera strategy against PI that I introduced in my (1996) and developed in my (1999). This rebranding urges shifting attention from the (crude) evidence of the history of science to the (refined) history of evidence for scientific theories. What's the connection between the two? As I will show in some detail before the conclusions, to paraphrase Clausewitz, "moving from the evidence of history to the history of evidence" is the continuation of divide et impera *by other means*.

The reader might wonder at the outset: are the lessons drawn from this double case generalizable? Similar studies in optics and the theories of heat that I have performed in the past (see my 1999, chapter 6) suggest they are. But all these are theories of physics and the reader might still wonder whether these lessons can be extended to other areas of science. I will leave it open whether they can or cannot. The key point is that it is sufficient to neutralize the force of the PI to show that some typical historical cases, which have been used to feed epistemic pessimism about current science, do not license this pessimism. On the contrary, they suggest that a more fine-tuned account of theory-change in science gives us reasons for epistemic optimism.

Here is the road-map. In section 5.2, I examine in some detail Descartes's account of method and explanation. I focus my attention on the role of hypotheses in Cartesian science and the circumstances under which they might be taken to be true. In section 5.3, I discuss Newton's more stringent account of scientific explanation and his critique of Descartes's vortex theory of planetary motions. I focus my attention on the fact that Newton's initiated a research program which aimed at the discovery of physical causes of the systematic discrepancies between the predictions of the theory and the actual orbits of the planets; a program which put Newton's theory to a series of rigorous tests. In section 5.4, I discuss the retentionist pattern that Newton's theory of gravity set in motion, as this was exemplified in Einstein's general theory of relativity (GTR) and in particular in the explanation of the anomalous perihelion of Mercury. In section 5.5, I compare the strategy suggested in this paper with the earlier divide et impera strategy and show their substantial similarity. Finally, I draw some general conclusions concerning the need to shift our attention from the evidence of the history of science to the history of evidence for scientific theories.

5.2 Descartes on Mechanical Hypotheses

Descartes was the first to put forward a comprehensive theory of the world based on entities invisible to the naked eye. All worldly phenomena, all macroscopic objects and their behavior, are accounted for by the motion(s), notably collisions, of invisible corpuscles, subject to general and universal laws of nature. Within this framework, to offer a scientific explanation of a worldly phenomenon X was to provide a particular configuration Y of matter in motion, subject to laws, such that Y could cause X. Elsewhere, I have explained the status and role of laws of nature in Cartesian natural philosophy (cf. 2018b; forthcoming).[1] The key relevant idea is that in Cartesian physics, the possible empirical models of the world are restricted from above by *a priori* principles, which capture the fundamental laws of motion, and from below by experience. Between these two levels there are various mechanical hypotheses, which refer to configurations of matter in motion. As Descartes explains in (III, 46) of the *Principia*, since it is a priori possible that there are countless configurations of matter in motion that can underlie the various natural phenomena, "unaided reason" is not able to figure out the right configuration. Mechanical hypotheses are necessary, but experience should be appealed to, in order to pick out the one in conformity with the phenomena:

> [W]e are now at liberty to assume anything we please [about the mechanical configuration], provided that everything we shall deduce from it is {entirely} in conformity with experience. (III, 46; 1982, 106)

---

[1] There has been considerable debate about the status and role of laws of nature in Descartes. Some notable contributions include Ott (2009), Hattab (2000), and Wilson (1999, chapter 28).

There are two sorts of hypotheses in Cartesian physics and his concomitant hypothetico-deductive method: vertical and horizontal. "Vertical" I call the hypotheses which posit invisible entities and their primary modes (shape and size, i.e., bulk, and motion). These kinds of entities, being invisible, have to be hypothesized. But there are "horizontal" hypotheses too. These concern macroscopic (that is visible) entities. Being visible, these entities do not have to be hypothesized; yet their structural relations have to be hypothesized because different structures yield the same appearances. The prime example is the "system of the world"—the solar system. Here, there are competing, but empirically equivalent, possible structures of the planetary motions.

In *Principia* (IV, 201; 1982, 283), Descartes states that sensible bodies are composed of insensible particles. So insensible particles (of various bulks and speeds) are everywhere: they fill (literally) space. A number of problems crop up, at this juncture. How can there be epistemic access to these invisible entities? How is knowledge of them possible? How can hypotheses be justified and accepted? How can one of them be more preferable than another, given that there are more than one which tally with the phenomena? All these are familiar questions. They are at the heart of the current scientific realism debate. Let us see how Descartes answers them.

5.2.1 The Bridge Principle

Descartes's ingenious attempt to answer them is based on a bridge principle, that is a principle which bridges the gap between the visible macrocosm and the invisible microcosm. According to this principle, the properties of the minute particles should be modeled on the properties of macro-bodies. Here is how Descartes put it:

> Nor do I think that anyone who is using his reason will be prepared to deny that it is far better to judge of things which occur in tiny bodies (which escape our senses solely because of their smallness) on the model of those which our senses perceive occurring in large bodies, than it is to devise I know not what new things, having no similarity with those things which are observed, in order to give an account of those things [in tiny bodies]. {E.g., prime matter, substantial forms, and all that great array of qualities which many are accustomed to assuming; each of which is more difficult to know than the things men claim to explain by their means}. (IV, 201; 1982, 284)

In this passage Descartes advances a *continuity thesis*: it is simpler and consonant with what our senses reveal to us to assume that the properties of micro-objects are the same as the properties of macro-objects. This continuity thesis licenses certain kinds of explanations: those that endow matter in general, and hence the unobservable parts of matter, with the properties of the perceived bits of matter. It therefore licenses as explanatory certain kinds of unobservable configurations of matter; viz., those that resemble perceived configurations of matter. The key point here is that the bridge principle makes "visibility" irrelevant ontically, though epistemically important. It allows transferring knowledge gained by experience (and the fundamental laws) to the invisible particles. There is not much justification for this principle offered by Descartes. A principle such as this is assumed as required for gaining knowledge of the invisible based on experience.

Let's see how this bridge principle works in practice in the (in)famous vortex hypothesis, according to which the planets are carried by vortices around the sun. A vortex is a specific configuration of matter in motion—subtle matter revolving around a center. The underlying mechanism of the planetary system then is a system of vortices:

> [T]he matter of the heaven, in which the Planets are situated, unceasingly revolves, like a vortex having the Sun as its center, and . . . those of its parts which are close to the Sun move more quickly than those further away; and . . . all the Planets (among which we {shall from now on} include the Earth) always remain suspended among the same parts of this heavenly matter. (III, 30; 1982, 196)

The very idea of this kind of configuration is suggested by experience, and by means of the bridge principle it is transferred to the subtle matter of the heavens. Hence, invisibility doesn't matter. The bridge principle transfers the explanatory mechanism from visible bodies to invisible ones. More specifically, the specific continuity thesis used is the motion of "some straws {or other light bodies}. . . floating in the eddy of a river where the water doubles back on itself and forms a vortex as it swirls." In this kind of motion, we can see that the vortex carries the straws "along and makes them move in circles with it." We also see that

> some of these straws rotate about their own centers, and that those which are closer to the center of the vortex which contains them complete their circle more rapidly than those which are further away from it. (III, 30; 1982, 196)

Given the continuity thesis, we may transfer this mechanical model to the motion of the planets and "imagine that all the same things happen to the Planets; and this is all we need to explain all their remaining phenomena" (III, 30; 1982, 196).

5.2.2 The System of the World

How about horizontal hypotheses? Here the chief rivalry was between the three systems of the world: the Ptolemaic, the Tychonic, and the Copernican. The bridge principle cannot be of help. But experience can help refute one of them, viz., the Ptolemaic. As Descartes notes it is ruled out because it does not conform with observations, notably the phases of Venus (cf. III, 16; 1982, 89–90). But when it comes to Copernicus vs. Tycho, Descartes notes that they "do not differ if considered only as hypotheses"—they "explain all phenomena equally well," though Copernicus's hypothesis is "simpler and clearer." Hence, qua potential (hypothetical) explanations these two systems do not differ.

Descartes advocated a version of the Copernican theory because it's "the simplest and most useful of all; both for understanding the phenomena and for inquiring into their natural causes" (III, 19; 1982, 91). In his own version, he denied the motion of the earth "more carefully than Copernicus and more truthfully than Tycho" (III, 19; 1982, 91). To understand what goes on, we need to make a short digression into Descartes's theory of motion.

Notoriously, Descartes defines motion twice over. On the one hand, there is the ordinary account of motion as "*the action by which some body travels from one place to another.*" On the other hand, there is the "true" account of motion: "*the transference of one part of matter or of one body, from the vicinity of those bodies immediately contiguous to it and considered as at rest, into the vicinity off some] others*" (*Principia* II, 24, 25; 1982, 50–51). According to this account, motion is change of position relative to contiguous bodies that are considered at rest. But the earth is carried over by a vortex on which it is, so to speak, pinned; hence, it does not move (it is at rest) relative to its contiguous and at-rest bodies. Nor do the other planets. The observed changes in their positions are wholly due to the motion of the vortex which contains them.[2]

It is noteworthy that Tycho's account is in conflict with Descartes's theory of motion: the Earth appears immobile, but it is not at rest. The reason for this is based on a) the relativity of motion; and b) on Descartes's claim that a body is in motion only if it moves *as a whole*. As he observes in *Principia* (III, 38; 1982, 101–102), the relativity of motion implies that it could be either that the earth rotates and the heavens with the stars are rest, or the other way around. But, on the Tychonic hypothesis, condition b) implies that it is the earth which is actually in motion, since the earth moves *as a whole body* relative to the heavens (its separation from the contiguous heavens taking place along the *whole* of the surface of the earth), whereas the heavens with the stars in them are at rest, since the separation takes place over a small fraction of their surface, viz., this part which

---

[2] For an excellent account of Descartes's theory of vortices, cf. Aiton (1957).

is contiguous with the earth. In fact, there is a deeper reason why Descartes dismissed Tycho's model of the System of the World. Given that the planets (apart from the earth) revolve around the sun, and the sun revolves around the earth, it is hard to see how this combination of motions could be modeled by vortices. Hence, Descartes concluded that Tycho "asserted only verbally that the Earth was at rest and in fact granted it more motion than had his predecessor" (III, 18; 1982, 91).[3]

Despite his heliocentrism, Descartes noted that he accepted the Copernican account not as true, "but only as a hypothesis," which may be false. Hypotheses need not be true to be useful. But *can* they be true? Or better, under what circumstances, if any, should they be taken to be true?

5.2.3 Was Descartes an Instrumentalist?

Descartes seems to endorse an instrumentalist account of hypotheses. He says quite explicitly in the third part of the *Principia* (article 44) that though his hypotheses may be false, it is enough that "all the things which are deduced from them are entirely in conformity with the phenomena" (III, 44; 1982, 105). He further says, (IV, 204; 1982, 286), that his causal story enables us to understand a possible way the world might have been (in its invisible structure), but that we "shouldn't conclude that" it was actually "made in that way." To illustrate the difference, he envisages a watchmaker who could have produced "two equally reliable clocks that looked completely alike from the outside but had utterly different mechanisms inside." God, being the "supreme maker of everything," could have produced all we see in "in many different ways." In light of this, Descartes says, it is enough that his hypothesis "corresponds accurately with all the phenomena of nature." For after all, this is all that is necessary for all the practical applications of science, which are directed "towards items that are sense-perceptible."

Yet, to read Descartes as an instrumentalist would be far too quick. He certainly thinks that there is an alternative theory concerning the creation of the Solar System (and of everything else), viz., the story told in the Bible (cf. IV, 1; 1982, 181). However, he is clear that his own corpuscularian hypothesis offers a *better* explanation than a) no explanation at all and b) the biblical story. The reason for this is that his own hypothesis identifies the natural causes of things and is based on simple and "easy to know" principles. Descartes occasionally uses the expression "as if": "my hypothesis will be as useful to life as if it were true" or "as if they were [the] causes [of these effects]." But the context suggests that this use is not the same as the instrumentalist's. The key difference is that

---

[3] For a more detailed account of Descartes's theory of motion, cf. Garber (1992).

for Descartes "as if true" seems to mean something like "true for all practical purposes" (or, morally certain, as Descartes would put it). In the French translation of the *Principia* he added right after the claim that his hypothesis "will be as useful to life as if it were true": "because we will be able to use it in the same way to dispose natural causes to produce the effects which we desire" (III, 44; 1982, 105). And again: "And I do not think it possible to devise any simpler, more intelligible or more probable principles than these" (III, 47; 1982, 107).

Simplicity, intelligibility, and probability are marks of truth. Descartes adds novel predictions, too; it is further support of the truth of a hypothesis, that is of its being right about the causes of the phenomena, when it explains "not only the effects that we were initially trying to explain but all these other phenomena that we hadn't even been thinking about" (III, 42; 1982, 104). On various occasions, Descartes stresses that unification is a mark of truth too. In (III, 43; 1982, 104), he notes emphatically that it "can scarcely be possible that the causes from which all phenomena are clearly deduced are false." He grounds this on the further claim that if the principles are evident and the deductions of observations from them "are in exact agreement with all natural phenomena," it would "be an injustice to God to believe that the causes of the effects which are in nature and which we have thus discovered are false."

The best explanation is known with moral certainty, i.e., certainty for all practical purposes. What is known with moral certainty could be false in the sense that it would be in the absolute power of God to make things otherwise than our best theory says they are. Descartes uses the metaphor of decoding a secret message to make an analogy with the explanatory power of theories. The more complex the message is the more unlikely it is that the code that was devised to successfully decode it is a matter of coincidence. Similarly, the more a theory reveals about the "fabric of the entire World," the more unlikely it is that it is false (cf. IV 205; 1982, 287).

5.2.4 The Fate of Vortices

The picture that emerges from this account of Descartes's methodology is nuanced and intricate. Empirical adequacy, so to speak, is a necessary condition for the truth of a hypothesis. But it being insufficient, what should possibly be added to render a hypothesis true? Descartes seems clear that the theoretical virtues of simplicity, intelligibility, unification, and novel predictive success are enough to render a hypothesis the best explanation of the evidence, and hence true, though known with moral certainty. Descartes clearly thought that his own vortex theory met these conditions. He was wrong about this. Descartes did try to show how planets move in non-circular orbits—by assuming that the solar vortex is somewhat distorted because the neighboring vortices exert unequal pressures on it and then arguing that subtle matter at the wider part of the vortex moves more slowly than at the narrower parts thus making the planet recede further from the sun (cf. *Principia* II, 141; 1982, 169). But he otherwise paid no attention to Kepler's laws; nor did he try to accommodate them within his theory. Nor did he account for the parallax of the equinoxes. In sum, his theory was empirically inadequate and shown to be false by Newton, whose theory superseded Descartes's.

As we are about to see, Newton's theory offered not only a better explanation of gravitational phenomena than Descartes's but also a better account of explanation as well as a better research strategy for devising such explanations. In doing all this, Newton changed the standards of evidence.

5.3 Newton's Deduction from the Phenomena

As is well known, Newton subjected Cartesian physics to trenchant criticism. In the unpublished *De Gravitatione*, he ventured to dispose of Descartes's "fictions" (2004, 14). One of his key points was that Descartes's theory of motion leads to absurdities since, if we take it seriously, it follows that a moving body has neither determinate velocity nor trajectory of motion; hence, it does not move![4] Newton was particularly opposed to Descartes's theory of vortices. He produced a number of arguments against it. Toward the end of Book II of the *Principia*, Newton demonstrated that "the hypothesis of vortices can in no way be reconciled with astronomical phenomena and serves less to clarify the celestial motions than to obscure them" (1729 [2016], 790). In particular, Newton showed that the vortex hypothesis could recover neither Kepler's area law nor the harmonic law. In one key argument, Newton constructed a simple model of the solar system in which planets are carried by vortices and showed that planets will move more swiftly in the aphelion and more slowly in the perihelion, contrary to Kepler's area law.[5] In the second edition of the *Principia* in 1713, Newton started his famous *General Scholium* by stressing that "hypothesis of vortices is beset with many difficulties" (1729 [2016], 939) and added that the second and third laws of Kepler cannot be mutually satisfied by the layers of vortex and that the motions of comets are incompatible with the existence of vortices.

[4] For a thorough account of Newton's criticism of Descartes's theory of motion and of body, see Katherine Brading (2012). Brading argues that both Descartes and Newton shared a law-constitutive account of bodies but Newton developed it more systematically. See also Jalobeanu (2013).

[5] For a detailed account of Newton's criticism of Cartesian vortices, see Snow (1924) and Aiton (1958a).

In fact, after Newton's accommodation of Kepler's laws within his theory, Cartesian and non-Cartesian advocates of the vortex theory aimed to show how their favorite versions of the vortex theory could accommodate Kepler's laws.[6] But all attempts resulted in failures (cf. Aiton 1996).

5.3.1 What was the Problem with Vortices?

But the problem with vortices appeared to be *epistemological* too: they were hypothetical entities. In the preface to the second edition of the *Principia*, Roger Cotes equated vortices with "occult causes," being "utterly fictitious and completely imperceptible to the senses" (1729 [2016], 392).

But is the problem *really* that vortices, if they exist, are imperceptible? It seems it is not. To see why it is not let us make a short digression to Huygens and Leibniz.

5.3.1.1 Huygens vs. Leibniz
Christiaan Huygens came to doubt the vortex theory "which formerly appeared very likely" to him (1690 [1997], 32). He didn't thereby abandon the key tenets of mechanical philosophy. For Huygens too the causal explanation of a natural phenomenon had to be mechanical. He said, referring to Descartes:

> Mr Descartes has recognized, better than those that preceded him, that nothing will be ever understood in physics except what can be made to depend on principles that do not exceed the reach of our spirit, such as those that depend on bodies, deprived of qualities, and their motions. (1–2)

Huygens posited a fluid matter that consists of very small parts in rapid motion in all directions, and which fills the spherical space that includes all heavenly bodies. Since there is no empty space, this fluid matter is more easily moved in circular motion around the center, but not all parts of it move in the same direction. As Huygens put it "it is not difficult now to explain how gravity is produced by this motion." When the parts of the fluid matter encounter some bigger bodies, like the planets: "these bodies [the planets] will necessarily be pushed towards the center of motion, since they do not follow the rapid motion of the aforementioned matter" (16). And he added:

> This then is in all likelihood what the gravity of bodies truly consists of: we can say that this is the endeavor that causes the fluid matter, which turns circularly

---

[6] As Aiton (1958b) argues, it was only after the publication of Newton's *Principia* that Cartesians started to try to account for Kepler's laws.

around the center of the Earth in all directions, to move away from the center and to push in its place bodies that do not follow this motion. (16)

In fact, Huygens devised an experiment with bits of beeswax to show how this movement toward the center can take place. At the same time, however, he had no difficulty in granting that Newton's law of gravity was essentially correct when it comes to accounting for the planetary system, and especially Kepler's laws. As he put it:

> I have nothing against *Vis Centripeta*, as Mr. Newton calls it, which causes the planets to weigh (or gravitate) toward the Sun, and the Moon toward the Earth, but here I remain in agreement without difficulty because not only do we know through experience that there is such a manner of attraction or impulse in nature, but also that it is explained by the laws of motion, as we have seen in what I wrote above on gravity. (31)

Explaining the fact that gravity depends on the masses and diminishes with distance "in inverse proportion to the squares of the distances from the center" was, for Huygens, a clear achievement of Newton's theory, despite the fact that the mechanical cause of gravity remained unidentified. His major complaint was that he did not agree with Newton that the law of gravity applies to the smallest parts of bodies, because he thought that "the cause of such an attraction is not explicable either by any principle of mechanics or by the laws of motion" (31).[7]

Leibniz in his *Tentamen de motuum coelestium causis* (1689), adopted a version of the vortex theory despite his overall disagreement with Cartesian physics and metaphysics. On this theory, a planet is carried by a "harmonic vortex," the layers of which moved with speeds inversely proportional to the distance from the center of circulation, thereby satisfying Kepler's area law. But the planet is also subject to a radial motion (*motus paracentricus*), which is due to two forces acting on the planet: the gravity exerted by the sun on the planet and the centrifugal force arising from the circulation of the planet with the vortex. Given all this, Leibniz was able to show that an elliptical orbit requires that the gravity (which was conceived by Leibniz along the lines of the magnetic action) is inversely proportional to the distance, and then to deduce from this Kepler's harmonic law. In effect, Leibniz had posited two independent but superimposed vortices for each planet, the vortex that causes gravity and the harmonic vortex (cf. Aiton 1996, 12).

---

[7] For further discussion of Huygen's theory of gravity and his reaction to Newton's, see Martins (1994).

Huygens, in correspondence, (Huygens to Leibniz 11 July 1692) challenged Leibniz on this, arguing that the harmonic vortex was not necessary since Newton had shown that gravity was enough for Kepler's laws to hold. Leibniz replied that he could not abandon this "deferent matter" (the harmonic vortex) because it explains why "all the planets [as well as the satellites of Jupiter and of Saturn] move somewhat in the same direction"; a thing that Newton's theory just takes for granted (Leibniz 1989, 415). But, as was pointed out by Newton's disciple David Gregory in his *Astronomiae physicae et geometricae elementa* (1702), Kepler's third law was inconsistent with the idea that planets are carried by harmonic vortices. Besides, Leibniz's theory cannot account for the motion of the comets: comets, on Leibniz's theory, should also be drawn by harmonic vortices, since their motions satisfy Kepler's area law, but their orbits are inclined at large angles to the plane of circulation of the planetary vortices. How can it be that the comets' vortices and the planetary vortices do not interact with each other and each of them remains intact with no modification of its state of motion?[8] Newton himself took issue with Leibniz's account arguing that Leibniz had the wrong measure of centrifugal force, which was for Newton equal and opposite to the force of attraction and that Leibniz's mathematical reasoning was unsound (cf. Aiton 1962).[9]

Leibniz wrote to Newton on March 7, 1692/3. In this letter, he praised Newton for having made "the astonishing discovery that Kepler's ellipses result simply from the conception of attraction or gravitation and passage in a planet" (cf. Newton 2004, 106). He added, however, that he thought that Newton's account is incomplete, since it leaves out the causes of attraction, which Leibniz took to be "the motion of a fluid medium." Leibniz, as we have seen, had developed his own version of the vortex theory. But the real purpose of the letter was to solicit Newton's opinion regarding Huygens's vortex theory.

Newton's reply (October 16, 1693) was very illuminating. He objected to vortex theory on the grounds that the vortices would disturb the motions of the planets and the comets. He made the further point that the laws of gravity he discovered were *enough* to account for the motion of the heavenly bodies. Moreover, as he put it, "since nature is very simple, I have myself concluded that all other causes are to be rejected and that the heavens are to be stripped as far as may be of all matter, lest the motions of planets and comets be hindered or rendered irregular" (2004, 109). The point here is double: on the one hand, Newton's explanation of planetary motions is *simpler* than the various accounts based on vortices; on the other hand, the explanations based on vortices are inadequate since they were at odds with the regular motions of the heavenly bodies. The

---

[8] For an insightful discussion, see Bussotti (2015, section 4.2).
[9] For a detailed account of Leibniz's theory, see Aiton (1958a; 1962; 1996) and Bussotti (2015).

inferential strategy is guided by a theoretical virtue (simplicity) plus the available evidence against the rival theories.

But Newton went on to add something seemingly puzzling: "But if, meanwhile, someone explains gravity along with all its laws by the action of some subtle matter, and shows that the motion of planets and comets will not be disturbed by this matter, I shall be far from objecting" (2004, 109). This suggests that Newton was not *in principle* against a "deeper" explanation of gravity by reference to the action of some subtle matter. Newton's problem with vortex theory was not that it posited the action of some subtle matter. Hence his problem was not that the vortex theory referred to insensible particles and their motion. Rather his problem was that the very vortex theory was inadequate as a potential explanation of the heavenly orbits. Invisibility does not matter; explanation does!

5.3.2 How to Do Natural Philosophy

This point, viz., that the problem with vortex theory was that it does not offer a good explanation of gravity, is echoed by Roger Cotes: "For even if these philosophers [the Cartesians] could account for the phenomena with the greatest exactness on the basis of their [vortices] hypotheses, still they cannot be said to have given us a true philosophy and to have found the true causes of the celestial motions until they have demonstrated either that these causes really do exist or at least that others do not exist" (Newton 1729 [2016], 393). But, as Newton has shown, not only does the vortex hypothesis fail to account for the gravitational phenomena; there is a rival theory—Newton's own—that accounts for them.

In fact, Newton stressed repeatedly that gravity is a non-mechanical force; hence it cannot be modeled by action within a mechanical medium. In the *General Scholium* he noted that gravity does not operate "in proportion to the quantity of the *surfaces* of the particles on which it acts (as mechanical causes are wont to do) but in proportion to the quantity of *solid* matter" (1729 [2016], 943).

It transpires then that Newton's theory is superior to Descartes's as an *explanation*. As Aiton (1958a, 1958b) has shown, most Cartesians succumbed to Newton's theory, and by the 1740s, the vortex theory had been fully abandoned. But even before this, most Cartesians who advocated the vortex theory tried to make it consistent with Newton's.[10] But there is more to it. Newton goes beyond

---

[10] One of the last attempts to defend Descartes's theory over Newton was by the Swiss mathematician Johann Bernoulli who won the 1730 *Académie Royale des Sciences* prize for his essay on the causes of elliptical orbits in a Cartesian vortex. He disagreed with two "bold suppositions" made by Newton, which "shock the spirits accustomed to receive in Physics only incontestable and obvious principles." The first of these suppositions, according to Bernoulli, was "to attribute to the body a virtue or attractive faculty" (1730, 6), while the second is the perfect void. But Bernoulli only gestured

the hypothetico-deductive account of explanation that was favored by Descartes and places more stringent demands on explanation. Let us pursue this point further.

What counts as a good explanation, according to Newton? As Cotes notes, it is an explanation that shows that the posited causes exist or "at least that others do not exist" (Newton 1729 [2016], 393). Hence for an explanation to be good it should be shown that other potential explanations are excluded; in effect, that it is the *unique* explanation of the phenomena. An explanation need not be full or complete to be good. Newton considered it enough that he identified the law that gravity obeys. The fact, if it is a fact, that there is a hitherto unknown "cause" of gravity does not detract from the fact that he identified gravity as "the force by which the moon is kept in its orbit." His agnosticism as to the "physical seat" of gravity, as Andrew Janiak (2008, 56–57) has nicely put it, neither renders the law a mere calculating device nor prevents Newton from stressing that gravity "really exists."

Newton's "hypotheses non fingo" has become emblematic. Here is the full quotation:

> I have not as yet been able to deduce from phenomena the reason for these properties of gravity, and I do not "feign" hypotheses. For whatever is not deduced from the phenomena must be called a hypothesis; and hypotheses, whether metaphysical or physical, or based on occult qualities, or mechanical, have no place in experimental philosophy. In this experimental philosophy, propositions are deduced from the phenomena and are made general by induction. The impenetrability, mobility, and impetus of bodies, and the laws of motion and the law of gravity have been found by this method. And it is enough that gravity really exists and acts according to the laws that we have set forth and is sufficient to explain all the motions of the heavenly bodies and of our sea. (1729 [2016], 943)

In this, hypotheses of any sort are contrasted to propositions which "are deduced from the phenomena and are made general by induction." We shall see in the sequel how this claim may be understood. For the time being, the relevant point is that Newton criticizes a certain account of explanation. But on what grounds? This is explained by Cotes in his preface to the second edition of the *Principia*. Cotes aims to highlight Newton's methodological novelty, and he contrasts it with both the Aristotelian way and the Cartesian way. He does not doubt that Cartesian mechanical philosophy is an improvement over the Aristotelian science. Nor does he

---

as to how Kepler's laws could be accommodated in his system of vortices. For a detailed discussion, see Aiton (1996, 17–18). See also Iltis (1973).

doubt that the Cartesians were right in aiming to explain the perceptible in terms of the imperceptible. Indeed, Cotes praises the Cartesians for trying to show that the "the variety of forms that is discerned in bodies all arises from certain very simple and easily comprehensible attributes of the component particles" (Newton 1729 [2016], 385). Where did they go wrong then? In relying on speculative and unfounded hypotheses about the properties of the invisible corpuscles.

If Cartesians are right on demonstration based on laws, but wrong to rely on speculative and unfounded hypotheses, what is the alternative? It is to use the phenomena as premises such that, together with the general laws of motion, further laws are derived on their basis. What is thereby derived is not speculative and unfounded. Indeed, that's how Cotes describes Newton's method:

> Although they [the Newtonians] too hold that the causes of all things are to be derived from the simplest possible principles, they assume nothing as a principle that has not yet been thoroughly proved from phenomena. They do not contrive hypotheses, nor do they admit them into natural science otherwise than as questions whose truth may be discussed. (1729 [2016], 385)

The *aim* of explanation then changes: it is to use the general laws of motion and phenomena in order to deduce further special laws, which have been established in such a way that they cannot be denied while affirming the general laws and the phenomena. In this sense, the special laws have been deduced from the phenomena. They are then at least as certain as the phenomena and general laws of motion. That's how the law of gravity has been found.

5.3.3 Newton's Account of Explanation

We can then say that Newton's experimental philosophy reverses the order of explanation in science. Schematically put:

Descartes
    General Laws & Hypotheses → phenomena

Newton
    General Laws & phenomena → special laws

What, of course, Newton called "phenomena" in book 3 of the *Principia* were not data or observations but empirical structures that exemplify a certain pattern, captured by an empirical law (Kepler's area law, Kepler's harmonic law). So, in a certain sense, they exhibit robustness and have been built inferentially from data and observations. Hence, they are amenable to mathematical analysis. For instance, phenomenon one is that the satellites of Jupiter (the circumjovial planets), by radii drawn to the center of Jupiter, satisfy Kepler's area law and Kepler's harmonic law. And phenomenon six is that the moon, by a radius drawn to the center of the earth, satisfies Kepler's area law.

It is then straightforward for Newton to apply Propositions 1, 2, and 4 (corollaries 6 and 7) of Book 1 of the *Principia*, which are proved for bodies qua mathematical points and forces qua mathematical functions, to the phenomena. As is well known, Newton demonstrates the following two equivalences:

A body A moving in an ellipse (a conic section) around another body B which is in a focus of the ellipse describes equal areas in equal times (that is, it satisfies Kepler's area law) if and only if it is urged by a centripetal force that is directed toward body B.

A body A moving in an ellipse (a conic section) around another body B which is in a focus of the ellipse has its periodic time T as the 3/2 power of its radius r (that is, it satisfies Kepler's harmonic law) if and only if the centripetal force that urges it toward B is inversely proportional to the square of the distance r between the two bodies.[11]

What is special about Newton's account of explanation, as William Harper and George Smith point out, is that he explains a phenomenon by relying on equivalences such as those just described. These equivalences "make the phenomenon measure a parameter of the theory which specifies its cause" (Harper and Smith 1995, 147).[12] In Book 3, the mathematically described centripetal force of the foregoing equivalences is identified with gravity. In the *scholium* to Proposition 5, Newton says: "Hitherto we have called 'centripetal' that force by which celestial bodies are kept in their orbits. It is now established that this force is gravity, and therefore we shall call it gravity from now on" (1729 [2016], 806).

Gravity is rendered a universal force by means of Rule III of Philosophy, which is a rule of induction.[13] This rule licenses the generalization of a quality, which is found to all bodies on the basis of experiments, to all bodies whatsoever. Or, as Newton put it:

---

[11] In fact, Newton added a further proof of the inverse square law, one that as he says proves its existence "with the greatest exactness" (1729 [2016], 802). The proof is based on the empirical fact that "the aphelia [of the planets] are at rest." As Newton argued, invoking proposition 45 of Book 1, the slightest departure from the inverse square of the distance would "necessarily result in a noticeable motion of the apsides in a single revolution and an immense such motion in many revolutions." This "systematic dependence," as William Harper has put it, between the inverse square ratio and the immobility of the aphelia is something that makes Newton's account of gravity robust. Besides, as Harper (2011, 42ff) has argued, looking for systematic dependencies between theoretical parameters and the phenomena renders Newton's method distinct from a simple hypothetico-deductive method.

[12] For an elaboration of the idea of theory-mediated measurement, see Harper (2011).

[13] For a useful discussion of Newton's account of induction vis-à-vis Locke's and Hume's, see de Pierris (2015, chapter 3).

Finally, if it is universally established by experiments and astronomical observations that all bodies on or near the earth gravitate [lit. are heavy] toward the earth, and do so in proportion to the quantity of matter in each body, and that the moon gravitates [is heavy] toward the earth in proportion to the quantity of its matter, and that our sea in turn gravitates [is heavy] toward the moon, and that all planets gravitate [are heavy] toward one another, and that there is a similar gravity [heaviness] of comets toward the sun, it will have to be concluded by this third rule that all bodies gravitate toward one another. (1729 [2016], 796)

Unlike Descartes's bridge principle, Newton's Rule III, is a more rigorous bridge principle: not all observable properties of matter are transferable to all bits of matter (no matter how small); but only those shown to be possessed by experiments. It should be noted that this bridge principle is both horizontal (it extends to all observable bodies) and vertical (it extends to all unobservable particles of matter). As such, it is a vehicle for unification. In justifying rule III, which yields the unification, Newton appeals to nature's being "always simple and ever consonant to itself." Newton's theory of gravity is offered as the best explanation of the gravitational phenomena and is licensed as such by the theoretical virtues and (rigorous) evidence.

5.3.4 *Quam Proxime*

We know that Newton's theory of gravity was superseded by Einstein's GTR. But even if this hadn't happened, Newton's theory would still have been, at best, an approximation to the phenomena to be accounted for. In what Smith (2008) has called the *Copernican Scholium* of the "De Motu" tracts that preceded the *Principia*, Newton stated the great complexity of the actual planetary motions:

By reason of the deviation of the Sun from the center of gravity, the centripetal force does not always tend to that immobile center, and hence the planets neither move exactly in ellipses nor revolve twice in the same orbit. Each time a planet revolves it traces a fresh orbit, as in the motion of the Moon, and each orbit depends on the combined motions of all the planets, not to mention the actions of all these on each other. . . . But to consider simultaneously all these causes of motion and to define these motions by exact laws admitting of easy calculation exceeds, if I am not mistaken, the force of any human mind. (Herival 1965, 301)

Kepler's laws hold only approximately of the planets and their satellites mainly because the actual motions are lot more complicated than the theory could capture. Was this something to give Newton pause? To cut a long story short, it was certainly not. Not only Kepler's laws hold *quam proxime*, but Newton himself took pains to show that the equivalences he proved in Book I of the *Principia* hold *quam proxime* too. This phrase (which occurs 139 times in the *Principia*) literally means "most nearly to the highest degree possible"; it is probably best translated "very, very nearly" or, as is more customary, "very nearly."[14]

After proving Propositions 1 and 2 of Book I (which, as we have seen, assert the equivalence: the centripetal force by which a body is drawn to an immobile center of force is directed to this center iff that body satisfies Kepler's area law), Newton went on, in corollaries 2 and 3 of Proposition 3, to show that the centripetal force is directed to the center of force *quam proxime* iff Kepler's area law is satisfied *quam proxime*. More generally, the theoretical equivalences that Newton proved in Book I hold *quam proxime* too: p *quam proxime* iff q *qua proxime*. Hence, given the laws of motion, Newton's law of gravity is true at least *quam proxime* (cf. Smith 2002, 156).

Still, Newton also showed the conditions under which the theoretical equivalences would hold, not simply *quam proxime*, but *exactly* of the phenomena. As he characteristically put it in Proposition 13 of Book III: "if the sun were at rest and the remaining planets did not act upon one another, their orbits would be elliptical, having the sun in their common focus, and they would describe areas proportional to the times" (1729 [2016], 817–818). It follows, as Smith put it, that "the true motions would be in exact accord with the phenomena were it not for specific complicating factors" (2002, 157).

This interplay between exact, idealized motions and *quam proxime* actual motions renders Newton's account of explanation substantially different from Descartes's. As Ducheyne put it, Newton's target is not merely to explain Keplerian motions but rather to show that, given the laws of motion, there is a unique explanation of Keplerian motions (cf. 2012, 104). Given that Keplerian motions are idealized motions, Newton's target is to explain the actual deviations from these motions.

As Smith has noted, Newton's theory makes possible the following: On the assumption that the theory is exact (that is on the basis of idealizations embodied in the theory) there are systematic discrepancies between the predictions of the theory and the observations. These discrepancies constitute "second-order phenomena" which the theory, in the first instance, should explain by attributing them, as far as possible, to physical causes. Hence, the theory makes possible

---

[14] Ducheyne (2012, 82) offers a more literal translation as "as most closely as possible" or "uttermost closely."

further explanations of the discrepancies between predicted values and observed values of various parameters. The theory is being further elaborated and tested by finding the causes of discrepancies. This is achieved by "asking in a sequence of successive approximations, what further forces or density variations are affecting the actual situation?" (2014, 277).

The discovery of physical causes of these systematic discrepancies is a rigorous test of the physical theory because it shows

a) the circumstances under which the theory holds exactly;
b) the sense in which it can be taken to be an approximation of real worldly phenomena; and
c) the ways the theory can be improved upon by taking into account further causal factors in the world.[15]

The discovery of Neptune by Le Verrier in 1846 is a case in point. The idea that the theory holds *quam proxime* of worldly phenomena is tied to the continuous development and further testing of the theory, which leaves open the possibility that some phenomena may well be genuine exceptions to the theory; that is, that some phenomena might not be such that they can be accommodated within the theory by the method of successive approximations. One reason why Smith has called this kind of phenomena "second-order" is the fact that their being acknowledged as phenomena worthy of accommodation *presupposes* the theory that makes them available. That is, it presupposes that the theory offers the best explanation of the first-order phenomena. In the case of the planetary motions, these second-order phenomena, which call for a physical explanation, presuppose Newton's theory of gravity and in particular that it holds exactly. They are made available qua phenomena as discrepancies between the idealized predictions of Newton's theory (which hold on the assumption that it is an exact theory) and observations. As Smith put it: "they are what you get by subtracting observations from idealized calculated results." And he adds: "They are second-order because they categorically presuppose the theory of gravity, taken as holding exactly. They are phenomena because they are systematic and hence constitute regularities that cannot initially be identified because more dominant regularities mask them" (2014, 278). The method of successive approximations turn Newton's theory into a research program for its own development and further testing (cf. Smith 2002, 159). Instead of relying on ad hoc forces to save the law of gravity, the systematic discrepancies between the predicted theoretical

[15] For a detailed discussion of the method of successive approximations and the differences it introduces vis-à-vis the simple HD method, see Ducheyne (2012, 161ff).

planetary motions and the actual ones call for an explanation in terms of specific gravitational forces.

This kind of attitude toward the theory is embodied in Newton's Rule IV.

In experimental philosophy, propositions gathered from phenomena by induction should be considered either exactly or very nearly [*quam proxime*] true notwithstanding any contrary hypotheses, until yet other phenomena make such propositions either more exact or liable to exceptions. (1729 [2016], 796).

Being already an idealized description of the phenomena under investigation, Newton's theory is false of the actual world. But this kind of falsity is cheap and fully consistent with the theory being, in some sense, a good approximation of worldly phenomena. What mandates theory change is not falsity *per se* but rather the fact that the theory systematically fails to find robust physical explanations (framed with the conceptual resources of the theory) of further (including some second-order) phenomena.

5.4 Einstein's Retentionism

Unlike the transition from Descartes's theory to Newton's, the transition to Einstein's GTR highlights a pattern in theory-change which capitalizes on the successes (and I would add, the truth-content) of the predecessor theory to advance a better explanation of the phenomena. In the case we are about to discuss, the cause of falsity, as it were, was that spacetime is curved and that this contributes substantially to the explanation of the discrepancies found between prediction and observation. By the same token, however, identifying the causes of falsity presupposed treating Newton's theory as *partially* correct.

In a survey article on gravitation in 1903, the US astronomer Jonathan Zenneck presented the situation regarding Newton's gravitational theory as follows:

Two independent fields insure that Newton's law, even if not absolutely accurate, represents real conditions with a far-reaching accuracy unmatched by hardly any other law. In the astronomical domain, this law yields planetary motions not only to the first approximation (Kepler's laws); but even to the second approximation, the deviations of planetary motion due to perturbations by other planets follow from Newton's law so accurately that the observed perturbations led to the prediction of the orbit and relative mass of a hitherto unknown planet (Neptune). (1903, 86)

Yet, there are "astronomical observations that show deviations compared to calculations based on Newton's law." The most significant one that was cited was the "ca. 40″ per century in the perihelion motion of Mercury." As it was noted by Zenneck, the case of Mercury, unlike the case of Neptune, presented a real difficulty for Newton's theory of gravity. In fact, already in 1894, there had been attempts to revise Newton's law of gravity in order to account for the perihelion of Mercury—by Asaph Hall. As Zenneck noted:

> Indeed, M. Hall proved that the law previously examined by G. Green, which replaces $1/r^2$ with $1/r^{2+\lambda}$ where λ stands for a small number, is sufficient to explain the anomalous perihelion motion of Mercury, if $\lambda = 16 \cdot 10^{-8}$. This figure for λ would also give the right result for the observed anomalous perihelion motion of Mars, though for Venus and Earth the consequence would be somewhat too large a perihelion motion. (1903, 92)

But this revision led to conflicts with other phenomena, such as the position of the lunar apogee. Fixing the exponent of the inverse square law in such a way that it accounted for the various phenomena was proved to be more difficult than anticipated. An adequate account of the anomalous perihelion of Mercury had to wait until a new theory emerged: Einstein's GTR.

In November 18, 1915, Einstein published a paper in which, as he put it, he found "an important confirmation of this most fundamental theory of relativity, showing that it explains qualitatively and quantitatively the secular rotation of the orbit of Mercury (in the sense of the orbital motion itself), which was discovered by Leverrier and which amounts to 45 sec of arc per century" (quoted in, Kennedy 2012, 170). He was, of course, referring to the GTR, which he had presented a week earlier, in November 1915, in the same journal.[16]

Einstein took for granted that Newton's theory (aided by standard perturbation theory) successfully accounted for 531 arc-seconds per century of the total 575.1 arc-seconds per century precession of the perihelion of Mercury. In fact, taking seriously the 43.4 arc-seconds per century as the observed discrepancy that constituted an anomaly for Newton's theory *presupposed* that the remaining 531 arc-seconds were accounted for by the gravitational effects on Mercury of all other planets, as explained by Newton's theory. This suggests that the evidence for Newton's theory was solid enough to allow taking *this* theory as the benchmark for the explanation of the 531 arc-seconds per century.

In fact, Einstein never doubted that he had to recover Newton's theory of gravity as a *limiting case* of his own theory. In his letter to Hilbert (November 18,

---

[16] For a thorough and detailed study of Einstein's account of the anomalous perihelion of Mercury see Earman and Janssen (1993).

1915) he emphasized the role of the Newtonian limit in his formulation of the theory of relativity. Stressing that he arrived at the field equations first, he noted that three years ago

> it was hard to recognize that these equations form a generalisation, and indeed a simple and natural generalisation, of Newton's law. It has just been in the last few weeks that I succeeded in this (I sent you my first communication), whereas 3 years ago with my friend Grossmann I had already taken into consideration the only possible generally covariant equations, which have now been shown to be the correct ones. We had only heavy-heartedly distanced ourselves from it, because it seemed to me that the physical discussion had shown their incompatibility with Newton's law. (quoted in Renn et al. 2007, 908).

In his 1916 systematic presentation of GTR, Einstein (1916, 145) made clear that starting from the field equations of gravitation "give us, in combination with the equations of motion . . . to a first approximation Newton's law of attraction, and to a second approximation the explanation of the motion of the perihelion of the planet Mercury . . . These facts must, in my opinion, be taken as convincing proof of the correctness of the theory".

How can this be, if Newton's law is characteristically false? In other words, how can recovering Newton's law of gravity as an approximation be "convincing proof" for *Einstein's* theory, if Newton's theory has had no substantial truth-content that could (and should) be recovered by (and retained in) Einstein's theory? What's really interesting in Einstein's case is that Einstein's chief conceptual innovation, the curvature of spacetime, provided evidence for Newton's account of gravity being *in some sense* a correct approximation to the actual gravitational phenomena. Conversely, as Smith has insisted, Newton's law of gravity provided evidence for Einstein's own theory (and hence Einstein's own account of gravity) being more correct than Newton's. This is because Einstein's account identified "non-Newtonian sources of discrepancies, like the curvature of space," which provide "evidence for the previously identified Newtonian sources presupposed by the discrepancies" (2014, 328). In other words, the evidence for Newton's account of gravity was solid enough to allow taking *this* account as the benchmark for the explanation of the 531 arc-seconds per century.

In sum, some key theoretical components of Newton's theory of gravity—the law of attraction and the claim that the gravitational effects from the planets on each other were a significant cause of the deviations from their predicted orbits—were taken by Einstein to be broadly correct and explanatory (of at least part) of the phenomena to be explained.

An objection to this retentionist argument could be that there is indeed retention in the transition from Newton's theory of gravity to Einstein's, but this is only at the mathematical level.[17] Indeed, Stanford has capitalized on this objection in order to point out that "Newtonian mechanics is just plain false, and radically so." For him, the claim that "Newtonian mechanics itself is 'approximately true,'" can "only mean that its empirical predictions approximate those of its successors across a wide range of contexts. It cannot mean that it is approximately correct as a fundamental description of the physical world" (2006, 9).

Falsity, we noted already, is cheap. Newton's theory is false, and so is Descartes's. But this judgement obscures an important difference: Newton's theory is still used to model various phenomena while Descartes's is not. Newton's account of gravity is not like phlogiston or vortices. And this is so because Newton's account of gravity, though false, is still a good approximation of the gravitational phenomena. This feature is not accidental (nor a matter of convenience only). It relies on the fact that Newton's account of gravity identified features of gravity that were retained in Einstein's theory and form part of the current scientific image of the world. Einstein himself stressed that his field equations, together with the laws of motion, yield "to a first approximation Newton's law of attraction." As is well known, Einstein's chief point was that in a quasi-static and weak gravitational field, he could recover Newton's law of attraction, in the form of the gravitational potential. Besides, Newton correctly identified masses as causally relevant factors in the stability of the solar system and he accounted for gravity on their basis. For a realist, it should be enough that Newton's account of gravity was world-involving in a way that Descartes's was not.[18]

Here is a characteristic way to put the relation between GTR and Newton's theory of gravity:

> Our modern acceptance of general relativity is not only based on experiments or observations related to some of its special predictions but also on the fact that it incorporates the entire Newtonian knowledge on gravitation, including its relation to other physical interactions, that has been accumulated over a long period of time in classical physics and in the special theory of relativity. This knowledge embraces, among other aspects, Newton's law of gravitation including its implications for the conservation of energy and momentum, the relation between gravitation and inertia, the understanding that no physical action can propagate with a speed greater than that of light, which was first achieved by the field theoretic tradition of classical physics and then finally established with the formulation of special relativity, and, more generally, the local properties of space and time, also formulated in special relativity. (Renn and Sauer, 303)

---

[17] Many thanks to Peter Vickers for pressing me on this point.
[18] For a more detailed account of the "world-involving" explanation, see my (2017).

5.5 Divide et Impera by Other Means

Before some conclusions are drawn, let me discuss the connection of the strategy followed in the foregoing double case study with the move I introduced many years ago (cf. 1996) and called the divide et impera strategy to address PI. I already noted in the introduction that the "moving from the evidence of history to the history of evidence" strategy is, to paraphrase Clausewitz, the continuation of divide et impera *by other means*. To see this let us recall the core motivation and the core idea behind the divide-and-conquer strategy.

The key motivation was to refute PI by showing that the pessimistic conclusion that current theories are likely to be false because most past theories were false was too crude to capture the dynamics of theory change in science. The problem with the conclusion of the PI is not that it is false. As noted already, falsity is cheap; current theories are indeed likely to be false, strictly speaking, if only because they employ idealizations and abstractions. The problem with the conclusion of PI is that it obscures important senses in which current theories are approximately true (better put: they have truth-like components). More specifically, the conclusion of PI obscures the substantial continuity there is in theory-change. Now, the way I read PI was as a *reductio* of the realist thesis that currently successful theories are approximately true. PI, I argued, capitalizes on the claim that current theories are inconsistent with past ones. So, according to PI, if we were to assume, for the sake of the argument, that current theories are true, we would have to assume that past theories were false. But given that those false past theories were empirically successful (recall Larry Laudan's list of past successful-yet-false theories), truth cannot be the best explanation of the empirical success of theories, as realists assume.

The key idea behind the divide et impera gambit was that the premise of PI that needed rebutting was that past theories cannot be deemed truth-like because they are inconsistent with current theories. And the novelty of divide et impera, if I may say so, was that it refocused the debate on individual past theories and their specific empirical successes. If we focus on specific past theory-led successes (e.g., the empirical successes of the caloric theory of heat, or of Newton's theory of gravity), the following questions suggest themselves: How were these particular successes brought about? In particular, which theoretical constituents of the theory essentially contributed to them? An important assumption of the divide et impera strategy was that it is not, generally, the case that no theoretical constituents contribute to a theory's successes. By the same token, it is not, generally, the case that all theoretical constituents contribute (or, contribute equally) to the empirical successes of a theory. Theoretical constituents that essentially contribute to successes are those that have an indispensable role in their generation. They are those which really fuel the derivation of the prediction. Hence, the theoretical task was two-fold:

(i) to identify the theoretical constituents of past genuine successful theories that essentially contributed to their successes; and

(ii) to show that these constituents, far from being characteristically false, have been retained in subsequent theories of the same domain.

If both aspects of this two-fold task were fulfilled, the conclusion would be that the fact that our current best theories may well be replaced by others does not, necessarily, undermine scientific realism. Clearly, we cannot get at the truth all at once. But this does not imply that, as science grows, there is no accumulation of significant truths about the deep structure of the world. The lesson of the divide et impera strategy was that judgements from empirical support to approximate truth should be more refined and cautious in that they should only commit us to the theoretical constituents that do enjoy evidential support and contribute to the empirical successes of the theory. Moreover, it was argued that these re-fined judgements are made by the historical protagonists of the various cases. Scientists themselves tend to identify the constituents that they think responsible for the successes of their theories, and this is reflected in their attitude toward their own theories.

I recently reviewed the debate that the divide et impera strategy has generated (cf. Psillos 2018a). Hence, there is no point in repeating it here too. But given the outline just presented, it transpires that the key thought behind the divide et impera strategy was precisely that the evidence from the history of science was too crude, undifferentiated, and misleading to indict realism. Besides, this evidence portrayed theory-change in science as an all or nothing approach. The divide et impera move had in it the thesis that the proper target was looking at the history of evidential relations between the various successes of the theory and the parts of the theory that fueled them. In this sense, the "moving from the evidence of history to the history of evidence" strategy is a continuation of the divide et impera strategy.

But what are the "other means" suggested by paraphrasing Clausewitz? These are suggested by the dynamics of the theory-change we have discussed in the double case study. What makes Newton's strategy distinctive was a) the idea that a law (or a theory, for that matter) holds *quam proxime*; and b) the idea that the various discrepancies between the predictions of the theory and the observed facts can become the source of a strategy for the development of theory and for testing it more rigorously, until such time that there is reason to abandon the theory. Hence, as for instance Harper and Smith (1995) have repeatedly claimed, Newton significantly altered the criteria of rigorous testing of a theory. Now, a rigorously successful theory is never *really* abandoned, as Einstein made sure to stress. Not only because it holds *quam proxime*; but also (and mainly) because it identifies and reveals some elements of reality which, precisely because they enjoy evidential support, are retained in subsequent theories. Descartes's theory, on the other hand, was really abandoned. His theory was not fully unsuccessful. (Actually, Descartes's theory, as well as Leibniz's, offered accounts of the fact that all planets rotate around the sun in the same direction). Still, Descartes's theory failed to identify elements of reality which could be retained in subsequent theories; hence it failed to transfer whatever evidence there was for it to Newton's theory. And it was for this reasons that it was discarded. (The whole account of vortices was one such important failure.)

Hence, the other means that the "history of evidence" strategy adds to the divide et impera strategy is retentionism: those components of a false theory are deemed retainable by the historical actors for which there is strong evidence that identify and capture elements of reality, and which can therefore become evidence for the successor theory (which retains parts of the predecessor). Whereas Newton found little retainable in Descartes's theory, Einstein found much retainable in Newton's.

5.6 Conclusions

In the preface to the *Treatise on Light* (1690), Huygens summed up the Cartesian approach to hypotheses as follows:

> One finds in this subject a kind of demonstration which does not carry with it so high a degree of certainty as that employed in geometry; and which differs distinctly from the method employed by geometers in that they prove their propositions by well-established and incontrovertible principles, while here principles are tested by the inferences which are derivable from them. The nature of the subject permits of no other treatment. It is possible, however, in this way to establish a probability which is little short of certainty. This is the case when the consequences of the assumed principles are in perfect accord with the observed phenomena, and especially when these verifications are numerous; but above all when one employs the hypothesis to predict new phenomena and finds his expectations realized. (quoted in Harper 2011, 373)

We have seen that Newton replaced this conception of method with a more elaborate account of explanation, which, without neglecting the role of theoretical virtues, took it that the explanation offered by the theory should be taken to be *quam proxime* and should be further put to the test, while the theory is developed to account for more first- and second-order phenomena. Both Descartes and Newton developed their theories as the best explanation of the evidence, though they differed as to what counts as the best explanation.

Both Descartes's and Newton's theories are false. Both can be used to feed the premises of the PI. Yet, there is a substantial difference between them: Newton's theory has given rise to a research program that led to, and put constraints on, the subsequent theory.[19] Some key elements of Newton's account of gravity have become part of the current scientific image of the world. Newton's account of gravity has latched onto causal features of the world in a way that Descartes's has not. Newton's account of gravity holds *quam proxime*, and because of this it led to a retentionist pattern in theory-change. The evidence for Newton's theory became evidence for its successor GTR.

Looking grossly at the evidence from the history of science obscures a more fine-tuned look at the history of evidence for the truth—*quam proxime*—of scientific theories (past and present). In thinking about the bearing of the evidence on theories we should shift our attention from the evidence that comes from history (of science) to the history of the evidence for theories and in particular for the key theoretical claims they make.

Acknowledgments

Many thanks to Timothy Lyons and Peter Vickers for their encouragement and useful comments. I am also grateful to an anonymous reader for insightful criticism. Earlier versions of this paper were presented at the History of Science and Contemporary Scientific Realism Conference, IUPUI, Indianapolis, in February 2016; the philosophy seminar of the University of Hannover, in May 2016; and at the inaugural conference of the East European Network for Philosophy of Science (EENPS), Sofia, June 2016. I would like to thank all those who asked questions and made comments on these occasions. This paper would not have been written without the generous intellectual help of Robert DiSalle, William Harper, and Stavros Ioannidis.

References

Aiton, E. J. 1957. "The Vortex Theory of the Planetary Motions—I." *Annals of Science* 13: 249–264.

---

[19] For an elaborate account of how Descartes's and Newton's theories differ in the ways they represent the world mathematically, see Domski (2013).

Aiton, E. J. 1958a. "The Vortex Theory of the Planetary Motions—II." *Annals of Science* 14: 132–147.

Aiton, E. J. 1958b. "The Vortex Theory of the Planetary Motions—III." *Annals of Science* 14: 157–172.

Aiton, E. J. 1962. "The Celestial Mechanics of Leibniz in the Light of Newtonian Criticism." *Annals of Science* 18: 31–41.

Aiton, E. J. 1996. "The Vortex Theory in Competition with Newtonian Celestial Dynamics." In R. Taton and C. Wilson (eds.) *The General History of Astronomy: Planetary Astronomy from the Renaissance to the Rise of Astrophysics*; vol 2B, The Eighteenth and Nineteenth Centuries. Cambridge: Cambridge University Press, pp. 3–21.

Bernoulli. J. 1730. *Nouvelles Pensees vers le Systeme de M. Descartes, et la Maniere d'en Deduire les Orbites et les Aphelies des Planetes*. Paris: Rue S. Jacques.

Brading, K. 2012. "Newton's Law-Constitutive Approach to Bodies: A Response to Descartes." In A. Janiak and E. Schliesser (eds.) *Interpreting Newton: Critical Essays*. Cambridge: Cambridge University Press, pp. 13–32.

Bussotti, P. 2015. *The Complex Itinerary of Leibniz's Planetary Theory*. Dordrecht: Springer.

De Pierris, G. 2015. *Ideas, Evidence, and Method: Hume's Skepticism and Naturalism concerning Knowledge and Causation*. Oxford: Oxford University Press.

Descartes, R. 1982/1644. *Principles of Philosophy*. Trans. Valentine Rodger Miller and Reese P. Miller. Dordrecht: D. Reidel Publishing Company.

Domski, M. 2013. "Mediating between Past and Present: Descartes, Newton, and Contemporary Structural Realism." In Mogens Laerke, Justin E. H. Smith, and Eric Schliesser (eds.) *Philosophy and Its History: Aims and Methods in the Study of Early Modern of Philosophy*. Oxford: Oxford University Press, pp. 278–300.

Ducheyne, S. 2012. *The Main Business of Natural Philosophy: Isaac Newton's Natural-Philosophical Methodology*. Dordrecht: Springer.

Earman, J., & Janssen M. 1993. "Einstein's Explanation of the Motion of Mercury's Perihelion." In John Earman, Michel Janssen, and John D. Norton (eds.) *The Attraction of Gravitation: New Studies in the History of General Relativity*. Boston: Birkhaeuser, pp. 129–172.

Einstein, A. 1916. "*The Foundation of the General Theory of Relativity*" reprinted in A. Einstein *The Principle of Relativity*. Dover 1932.

Garber, D. 1992. *Descartes' Metaphysical Physics*. Chicago: University of Chicago Press.

Harper, W. 2011. *Isaac Newton's Scientific Methodology*. New York: Oxford University Press.

Harper, W. & Smith, G. E.1995. "Newton's New Way of Inquiry." In Jarrett Leplin (ed.) *The Creation of Ideas in Physics*. Dordrecht: Kluwer, pp. 113–166.

Hattab, H. 2000. "The Problem of Secondary Causation in Descartes: A Response to DesChene." *Perspectives on Science* 8: 93–118.

Herival, J. 1965. *The Background to Newton's Principia*. Oxford: Clarendon Press.

Huygens, C. 1690/1997. *Discourse on the Cause of Gravity*. Karen Bailey (trans.). Mimeograph.

Iltis, C. 1973. "The Decline of Cartesianism in Mechanics: The Leibnizian-Cartesian Debates." *Isis* 64: 356–373.

Jalobeanu, D. 2013. "The Nature of Body." In Peter R. Anstey (ed.) *The Oxford Handbook of British Philosophy in the Seventeenth Century*. Oxford: Oxford University Press, pp. 213–239.

Janiak, A. 2008. *Newton as a Philosopher*. Cambridge: Cambridge University Press.

Kennedy, R. 2012. *A Student's Guide to Einstein's Major Papers*. Oxford: Oxford University Press.

Leibniz, G. W. 1989. *Philosophical Papers and Letters*. L. E. Loemker (ed.) 2nd Edition. Dordrecht: Springer.

Martins, Roberto De. 1994. "Huygens's Reaction to Newton's Gravitational Theory." In J. V. Field and Frank A. J. L. James (eds.) *Renaissance and Revolution*. Cambridge: Cambridge University Press, pp. 203–213.

Newton, I. 1729/2016. *Mathematical Principles of Natural Philosophy*. I. Bernard Cohen, Anne Whitman, Julia Budenz (trans.), Oakland, CA: University of California Press.

Newton, I. 2004. *Philosophical Writings*. Cambridge Texts in the History of Philosophy. Andrew Janiak (ed.) Cambridge: Cambridge University Press.

Ott, W. 2009. *Causation and Laws of Nature in Early Modern Philosophy*. Oxford: Clarendon Press.

Psillos, S. 1996. "Scientific Realism and the 'Pessimistic Induction.'" *Philosophy of Science* 63: S306–14.

Psillos, S. 1999. *Scientific Realism: How Science Tracks Truth*. London; New York: Routledge.

Psillos, S. 2017. "World-Involving Scientific Understanding." *Balkan Journal of Philosophy* 9: 5–18

Psillos, S. 2018a. "Realism and Theory Change in Science." In *The Stanford Encyclopedia of Philosophy*, Summer 2018 Edition, Edward N. Zalta (ed.). URL = <https://plato.stanford.edu/archives/sum2018/entries/realism-theory-change/>.

Psillos, S. 2018b. "Laws and Powers in the *Frame of Nature*." In Lydia Patton and Walter R. Ott (eds.) *Laws of Nature*. Oxford: Oxford University Press, pp. 80–107.

Psillos, S. Forthcoming. "From Natures to Laws of Nature."

Renn, J. et al (eds.). 2007. *The Genesis of General Relativity*, 4 volume-set. (Boston Studies in the Philosophy of Science, vol. 250.) Dordrecht: Springer.

Renn, J. & Suer, T. 2007. "Pathways out of Classical Physics". In Jürgen Renn et al. (eds.) *The Genesis of General Relativity*. Dordrecht: Springer.

Smith, G. E. 2002. "The Methodology of the *Principia*." In I. B. Cohen and G. E. Smith (eds.) *The Cambridge Companion to Newton*. Cambridge: Cambridge University Press, pp. 138–73.

Smith, G.E. 2008. "Newton's Philosophiae Naturalis Principia Mathematica." In Edward N. Zalta (ed.) *The Stanford Encyclopedia of Philosophy* (Winter 2008 Edition), https://plato.stanford.edu/entries/newton-principia/.

Smith, G. E. 2014. "Closing the Loop: Testing Newtonian Gravity, Then and Now." In Zvi Beiner and Eric Schliesser (eds.) *Newton and Empiricism*, Oxford: Oxford University Press, pp. 262–351.

Snow, A. J. 1924. "Newton's Objections to Descartes's Astronomy." *The Monist* 34: 543–557.

Stanford, P. K. 2006. *Exceeding our Grasp*. New York: Oxford University Press.

Wilson, M. D. 1999. *Ideas and Mechanism: Essays on Early Modern Philosophy*. Princeton: Princeton University Press.

Zenneck, J. 1903. "Gravitation." In Arnold Sommerfeld (ed.) *Encyklopädie der mathematischen Wissenschaften*, Vol. 5 (Physics). Leipzig: Teubner, 25–67. Reprinted in Renn, Jürgen et al (eds) 2007. The Genesis of General Relativity, 4 volume-set. (Boston Studies in the Philosophy of Science, vol. 250.)

6

How Was Nicholson's Proto-Element Theory Able to Yield Explanatory as well as Predictive Success?

Eric R. Scerri

Department of Chemistry & Biochemistry
UCLA
scerri@chem.ucla.edu

6.1 Introduction

Let me begin by saying how grateful I am to Peter Vickers and Tim Lyons for having invited me to two of their conferences.[1] I can honestly say that I have seldom participated in such interesting interdisciplinary settings and from which I learned so much. The following article is based on a lecture that I gave at one of these meetings on the work of the English mathematical physicist John Nicholson. This particular episode in the history of physics has received little attention in the scientific realism debate even though Nicholson's theory had considerable explanatory and predictive success. What makes the case all the more remarkable is that by most standards almost everything that Nicholson proposed was overturned.

My study is most closely connected with the research interests of Peter Vickers who has published on the subject of inconsistent theories.[2] To quote from the brief directive that was given to speakers by the conference organizers: "The following is among the key historical questions to be asked: To what extent did theoretical constituents that are now rejected lead to significant predictive or explanatory successes?" Many authors, some of whom are represented in this

---

[1] I presented papers at the History of Chemistry and Scientific Realism meeting held in Indianapolis between December 6 and 7, 2014, as well as the meeting Contemporary Scientific Realism and the Challenge from the History of Science that also took place in Indianapolis.

[2] P. Vickers, *Understanding Inconsistent Science*, Oxford, Oxford University Press, 2013.

Eric R. Scerri, *How Was Nicholson's Proto-Element Theory Able to Yield Explanatory as well as Predictive Success?*
In: *Contemporary Scientific Realism*. Edited by: Timothy D. Lyons and Peter Vickers, Oxford University Press. © Oxford University Press 2021. DOI: 10.1093/oso/9780190946814.003.0006

volume, appear to be somewhat puzzled by the fact that some inconsistent theories were able to attract a good deal of success. However, in the approach that I will be proposing, such features are seen to arise in a far more natural manner.

I will consider the work of Peter Vickers on quantum theory and quantum mechanics as an example of the current research that has focused on the scientific realism debate and the challenge from the history of science. In his recent book Vickers considers two well-trodden examples from the history of quantum theory. The first one consists of Bohr's calculation of the spectrum of He$^+$, a feat that brought admiration from no less a person than Albert Einstein when it was first published. Several authors have claimed that Bohr's theory was in several respects inconsistent to the point of containing internal contradictions.[3] And yet Bohr succeeded not only in calculating exactly the energy of the hydrogen atom but in also assigning the electronic configurations of many atoms in the periodic table. Even more dramatically, Bohr gave a highly accurate calculation for the energy of the helium +1 ion. The second episode began when Fowler criticized Bohr's initially published theory because it predicted an energy of precisely four times that of the hydrogen atom, or 4 Rydbergs, when in fact experiments revealed the energy of He$^+$ to be 4.00163 Rydbergs. In response, Bohr pointed out that he would take account of the reduced mass of the electron in order to carry out a more accurate treatment of the problem.[4] On doing so he obtained an energy of 4.00160 Rydbergs, which is what led Einstein to say, "This is an enormous achievement. The theory of Bohr must then be right."[5]

Vickers recounts a similar story concerning Arnold Sommerfeld.[6] It emerges that precisely the same formula is featured in Sommerfeld's semi-classical atomic theory as it is in the fully quantum mechanical theory that came later. This equivalence occurs in spite of the fact that the existence of electron spin and the wave nature of electrons, which play a prominent role in the fully quantum mechanical theory, were completely unknown when Sommerfeld published his account. Many commentators wonder how Sommerfeld could have arrived at the correct formula without knowing about these aspects of the more mature quantum mechanics, as opposed to the old quantum theory within which he worked. For example, while Sommerfeld assumed definite

[3] T. Bartelborth, "Is Bohr's Model of the Atom Inconsistent?" in P Weingartner and G. Schurz, eds., *Philosophy of the Natural Sciences, Proceedings of the 13th International Wittgenstein Symposium : H*, 1989, pp. 220–223.

[4] Bohr's revised calculation essentially consisted of considering the reduced mass of the hydrogen atom rather than assuming that the nucleus was infinitely heavier than the electron.

[5] http://galileo.phys.virginia.edu/classes/252/Bohr_to_Waves/Bohr_to_Waves. html#Mysterious%20Spectral%20Lines.

[6] Vickers's more recent views on the Sommerfeld question appear in P. Vickers, Disarming the Ultimate Historical Challenge to Scientific Realism, *British Journal for the Philosophy of Science*, 2018.

trajectories or orbits for electrons, the quantum mechanical account denies the existence of orbits and prefers to speak of orbitals in which definite particle trajectories are abandoned.

The response from somebody wishing to maintain a form of realism might be to claim that something about Sommerfeld's theory may have been correct, and it is that part of the theory which accounts for the fact that the two mathematical formulas in question were identical. Those wanting to defend such a position might claim that the core ideas were correct and were responsible for Sommerfeld success, in spite of his lack of knowledge of electron spin, the wave nature of electrons, and any other such later developments. Such a preservative realist would presumably claim that *some* aspect of Sommerfeld's theory was latching on to "the truth." There has been much discussion of this issue in the philosophy of science literature and especially in connection with the realism and anti-realism debate. Realists have tended to appeal to the slogan of divide and conquer (*divide et impera*) that was first proposed, as far as I am aware, by my old friend from graduate school in London, Stathis Psillos. Such an approach, although perhaps not precisely under the same banner, has also been promoted by Philip Kitcher and Larry Laudan.[7] Meanwhile Carrier, Chang, Chakravartty, and Lyons among others have expressed strong reservations about this strategy.[8] It should be noted that these debates have usually been waged over what were once very successful theories and scientific entities such as caloric, phlogiston, or the ether.

As I will endeavor to show, it becomes more difficult to argue for some form of "preservative realism" in cases like the one I am about to present, concerning the mathematical physicist John Nicholson who was a contemporary of Niels Bohr at the time of the old quantum theory. In fact I shall be arguing that it is rather difficult to find *anything* about Nicholson's view that has been preserved, except perhaps for the notion of the quantization of angular momentum, which ironically did not play a prominent role in any of the success that Nicholson's theories achieved at the time when they were first presented. What the defenders of preservative realism recommend in

---

[7] P. Kitcher, *The advancement of science*, Science without legend, objectivity without illusions, Oxford, Oxford University Press, 1993; L. Laudan, "A Confutation of Convergent Realism, *Philosophy of Science*, 48, 19–48, (1981).

[8] M. Carrier, "Experimental Success and the Revelation of Reality: The Miracle Argument or Scientific Realism." In Knowledge and the World: Challenges beyond the Science Wars, ed. Martin Carrier, Johannes Roggenhofer, Günter Küppers, and Philippe Blanchard, 137–61. Berlin: Springer (2004); A. Chakravartty, *A Metaphysics for Scientific Realism: Knowing the Unobservable*. Cambridge: Cambridge University Press (2007); H. Chang, "Preservative Realism and Its Discontents: Revisiting Caloric." *Philosophy of Science* 70, (5), 902–12 (2003); T.D. Lyons, "Scientific Realism and the Pessimistic Meta-modus Tollens." In *Recent Themes in the Philosophy of Science: Scientific Realism and Commonsense*, ed. Steve Clarke and Timothy D. Lyons, 63–90. Dordrecht: Kluwer, (2002).

the case of theories that featured the caloric or phlogiston, is that some parts of the theories tracked the truth while others did not. However, this strategy cannot be deployed in the case of Nicholson. Another strategy, that has been recommended by John Worrall, has been the notion of structural realism. Briefly put, this is the view that although the entities postulated by successive theories might change as history unfolds, the underlying mathematical structure is seen to persist and to display continuity. I believe that this form of strategy too will also fail in the case that I will be examining, all of which leads me to say that my historical case has a good deal to contribute to the general debate addressed in the present volume. I will now give a review of the scientific work in atomic physics of John Nicholson before turning back to the wider questions as to what one should make of the question of realism and the development of scientific theories.

6.2 Introduction to John Nicholson and His Early Work

John Nicholson was born in Darlington in County Durham in 1881. He attended Middlesbrough High School and then the University of Manchester, where he studied mathematics and physical sciences. He continued his education at Trinity College, Cambridge, where he took the mathematical tripos exams in 1904.[9] Nicholson won a number of prizes at Cambridge including the Isaac Newton Scholar Prize for 1906 and was a Smith Prizeman in 1907, as well as an Adams Prizeman in 1913 and again in 1917. His first position was as lecturer at the Cavendish Laboratory in Cambridge, followed by a similar position at Queen's University in Belfast. In 1912 Nicholson was appointed professor of mathematics at King's College, London. In 1921 he was named fellow and director of studies at Balliol College, Oxford, before retiring in 1930 due a recurring problem with alcoholism. He died in Oxford in 1955.

Nicholson proposed a planetary model of the atom in 1911 that had certain features in common with those of Jean Perrin, Hantaro Nagaoka, and Ernest Rutherford,[10] in that he placed the nucleus at the center of the atom. However, it

---

[9] The mathematical tripos was a distinctive written examination of undergraduate students of the University of Cambridge, consisting of a series of examination papers taken over a period of eight days in Nicholson's time. The examinations survive to this day although they have been reformed in various ways. A. Warwick, *Masters of Theory: Cambridge and the Rise of Mathematical Physics.* Chicago: The University of Chicago Press, 2003.

[10] E.R. Scerri, *The Periodic Table, Its Story and Its Significance*, New York, Oxford University Press, 2020.

Figure 6.1 John Nicholson

must be emphasized that Nicholson arrived at this conclusion independently of Rutherford and the other physicists just mentioned.

In fact, Nicholson's model had more in common with that of Thomson, which regarded the electrons as being embedded in the positive charge that filled the whole of the volume of the atom. Thomson's later models envisaged electrons as circulating in rings, but still within the main body of the atom. More specifically, the way in which Nicholson's model resembled those of Thomson lies in the mathematical analysis and the concern for the mechanical stability of the system.

Where Nicholson's model differed from all previous ones, regardless of whether planetary or not, was in his emphasis on astronomical data. He postulated a series of proto-atoms, as he called them, that would combine to form the familiar terrestrial elements. Nicholson believed that the proto-atoms, and the corresponding proto-elements, existed only in the stellar regions and not on the earth. This way of thinking was part of a British tradition that included William Crookes and Norman Lockyer, each of whom believed in the evolution of the terrestrial elements from matter present in the solar corona and the astronomical nebulae. Like Crookes

element	symbol	nuclear charge	atomic weight
coronium	Cn	2e	0.51282
hydrogen*	H	3e	1.008
nebulium	Nu	4e	1.6281
Protofluorine	Pf	5e	2.3615

Figure 6.2 Nicholson's proto-elements.

Note: hydrogen* does not represent terrestrial hydrogen but hydrogen the proto-element.

and Lockyer, Nicholson was an early proponent of the study of spectra for gaining a deeper understanding of the physics of stars as well as the nature of terrestrial elements.

The particular details of Nicholson's proto-atoms were entirely original and are represented in the form of a table (Figure 6.2). The first feature to notice is a conspicuous absence of any one-electron atom.[11] This is because Nicholson believed that such a system would be unstable according to his electromagnetic analysis.[12]

For Nicholson, the identity of any particular atom was governed by the number of positive charges in the nucleus, regardless of the number of orbiting electrons present in the atom. Nicholson may thus be said to have anticipated the notion of atomic number that was later elaborated by van den Broek and Moseley. Nicholson argued that a one-electron system could not be stable since he believed this would produce a resultant acceleration towards the nucleus. By contrast, Nicholson assumed that two or more electrons adopted equidistant positions along a ring so that the vector sum of the central accelerations of the orbiting electrons would be zero. The smallest atom therefore had to have at least two electrons in a single ring around a doubly positive nucleus.[13]

By using his proto-atoms, Nicholson set himself the onerous task of calculating the atomic weights of all the elements and of explaining the unidentified spectral lines in some astronomical objects such as the Orion nebula and the solar corona. It is one of the distinctive features of Nicholson's work that

[11] Nicholson rejected a one-electron atom because he believed that at least one more electron was needed to balance the central acceleration of a lone electron. See p. 163 of McCormmach, "The Atomic Theory of John William Nicholson," *Archives for History of Exact Sciences*, 3, 160–184, 1966, for a fuller account.

[12] Nicholson's list of proto-elements was extended to include two further members in 1914 when he added proto-hydrogen with a single electron and archonium with six orbiting electrons.

[13] By the year 1914, he had accepted the possibility of a one-electron atom. See H. Kragh, "Resisting the Bohr Atom," *Perspectives in Physics*, 13, 4–35, (2011).

his interests ranged across physics, chemistry, and astrophysics and that he placed great emphasis on astrophysical data above all other data forms.

6.3 Accounting for Atomic Weights of the Elements

Of the four proto-atoms that Nicholson originally considered, he believed that coronium did not occur terrestrially.[14] He therefore set out to accommodate the atomic weights of all the elements in terms of the three remaining proto-atoms, namely his special kind of hydrogen, nebulium, and proto-fluorine. It is important to consider the relative weights that Nicholson attributed to the proto-atoms and to delve a little further into Nicholson's theory.

Although Rutherford's planetary model had recently been proposed, Nicholson's work was much more indebted to the earlier Thomson model. As is well known, Thomson regarded the atom as consisting of a diffuse positive charge in which electrons were embedded as "plums in a pudding."[15] In a later development the electrons were seen as circulating in concentric rings but still within the main body of the positive charge.

According to Thomson, the orbital radius of any electron had to be less than the size of the atom as a whole. However, Nicholson rejected this notion for reasons that were quite independent of the arguments that were being published by Rutherford at about the same time. There is a sense in which Nicholson's atom can be said to have been intermediate between that of Thomson and the later one due to Rutherford. Nicholson retained much of the mathematical apparatus that Thomson had used to argue for the mechanical stability of the atom but required that the positive nucleus should shrink down to a size much smaller than the radius of the electrons. As a consequence, Nicholson could no longer use estimates of the size of the atom to fix the radius of the electron orbits. On the other hand, Nicholson could use his atom to give what seems to have been an excellent accommodation of the atomic weights of all the elements and some astronomical spectral lines as will be discussed later.

[14] Some of the proto-elements postulated by Nicolson had been invoked earlier by other authors. For example, Mendeleev had predicted the existence of an element called coronium and Emmerson had featured proto-fluorine in one of his periodic tables. B.K. Emmerson, "Helix Chemica, A Study of the Periodic Elements," *American Chemical Journal*, 45, 160–210, (1911).

[15] It turns out that the name "plum pudding" was never used by Thomson nor any of his contemporaries. A.A. Martinez, *Science Secrets*, Pittsburgh, PA, University of Pittsburgh Press, 2011. Because of the currency of the term I will continue to refer to it as such.

In order to see precisely how Nicholson envisaged his atom we consider his expression for the mass of an atom, which he published between 1910 and 1911 in a series of articles[16] on a theory of electrons in metals.[17]

$$m = \frac{2}{3}\left(\frac{e^2}{rc^2}\right)$$

In this expression m is the mass of an atom, e the charge on the nucleus, r the radius of the electron's orbit, and c the velocity of light. This expression can be simplified to read

$$m \propto e^2/r \qquad\qquad\qquad (i)$$

given the constancy of the velocity of light. Nicholson also assumed the positive charge, ne, for any particular nucleus with n electrons would be proportional to the volume of the atom.

$$ne \propto V$$

Next by substituting e = ne into (i) he obtained

$$m \propto n^2 e^2/r \qquad\qquad\qquad (ii)$$

He also assumed that the positive charge would be uniformly distributed throughout a sphere of volume V so that

$$ne \propto V$$

$$\text{or } ne \propto r^3 \quad (\text{since } V \propto r^3),$$

$$\text{and so} \quad r \propto n^{1/3}$$

Substituting into (i) the nuclear mass would take the form

$$m \propto \frac{n^2}{n^{1/3}}$$

or

$$m \propto n^{5/3}$$

---

[16] Nicholson's theory of metals appears in, J.W. Nicholson, "On the Number of Electrons Concerned in Metallic Conduction," *Philosophical Magazine*, series 6, 22, 245–266, (1911).

[17] Interestingly, Niels Bohr's academic career also began in earnest with his development of a theory of electrons in metals.

Coronium	Cn	=	0.51282
Hydrogen	H	=	1.008
Neptunium	Nu	=	1.6281
Proto-fluorine	Pf	=	2.3615

Figure 6.3 Relative weights of Nicholson's proto-atoms.

Gas	Formula	Calculated atomic weight	Observed atomic weight
helium	$Nu + Pf$	3.99	3.99
neon	$6(Pf+H)$	20.21	20.21
argon	$5He_2$	39.88	39.88
krypton	$5(Nu_4(Pf+H)_3)$	83.0	82.9
xenon	$5(He_4(Pf+H)_3)$	130.29	130.2

Figure 6.4 Nicholson's calculations and observed weights of the noble gases.
Note: Slightly modified table based on a report of Nicholson's presentation.
Source: *Nature*, 87, 2189-501-501, 1911.

At this point Nicholson assigned the mass of 1.008 to his proto-atom of hydrogen,[18] which allowed him to estimate the relative masses of the other proto-atoms as shown in Figure 6.3.

From here Nicholson combined different numbers of these three particles (omitting Cn) to try to obtain the weights of the known terrestrial elements (Figure 6.4). For example, terrestrial helium could be expressed as,

$$He = Nu + Pf = 3.9896,$$

a value that compares very well with the weight of helium that was known at the time, namely 3.99.[19]

---

[18] This step seems a little odd given Nicholson's statements to the effect that hydrogen the proto-atom is not necessarily the same as terrestrial hydrogen. In using a mass of 1.008 he surely seems to be equating the two.

[19] The error amounts to approximately 0.3 of one percent. Moreover, Nicholson takes account of the much smaller weight of electrons in his atoms. After making a correction for this effect he revises the weight of helium to 3.9881 (or, to three significant figures, 3.99) in apparent perfect agreement with the experimental value. Such was the staggering early success of Nicholson's calculations. See J.W. Nicholson, "A Structural Theory of the Chemical Elements," *Philosophical Magazine* series 6, 22, 864–889, 871–872, (1911).

H	H	1.008	1.008
He	Nu+Pf	3.99	3.99
Li	3Nu+2H	6.90	6.94
Be	3Pf+2H	9.097	9.10
B	2He+3H	11.00	11.00
C	2He+4H	12.00	12.00
N	2He+6H	14.02	14.01
O	3He+4H	15.996	16.00
F	3He+7H	19.020	19.00
Ne	6(Pf+H)	20.21	20.21
Na	4He+7H	23.008	23.01

Figure 6.5 Nicholson's composite atoms for the first 11 elements in the periodic table.

Nicholson's calculations of atomic weights were not confined to just the first few elements as shown in the Figure 6.3. He was able to extend his accommodation to all the elements up to and including the heaviest known at the time, namely uranium, and to a very high degree of accuracy. For example, Figure 6.4 shows his calculations as well as the observed atomic weights for the noble gases.[20] Meanwhile Figure 6.5 shows the calculated and observed weights for the first eleven elements in the periodic table.

Nicholson's contemporaries initially reacted to his work in a positive but cautious manner. For example, one commentator wrote,

> The coincidence between the calculated and observed values is great, but the general attitude of those present seemed to be one of judicial pause pending the fuller presentation of the paper, stress being laid on the fact that any true scheme must ultimately give a satisfactory account of the spectra.[21]

Nicholson promptly rose to this challenge and provided precisely such an account of the spectra of some astronomical bodies in his next publication.

This contribution involved the hypothetical proto-element nebulium, which Nicholson took to have four electrons orbiting on a single ring around a central positive nucleus with four positive charges. Like his other proto-elements,

---

[20] Figure 6.4 did not appear in Nicholson's own papers but in a 1911 article in *Nature* magazine as part of a report on the annual conference of the British Association for the Advancement of Science meeting at which Nicholson had presented some of his findings.

[21] Anonymous, *Nature*, 501, (October 12, 1911).

Nicholson did not believe that this element existed on the earth but only in the nebulae that had long ago been discovered by astronomers, such as the one in the constellation of Orion. Following a series of intricate mathematical arguments building on Thomson's model of the atom, Nicholson found that he could account for many of the lines in the nebular spectrum that had not been explained by others who had only invoked lines associated with terrestrial hydrogen or helium.

Nicholson's feat could well be regarded as a numerological trick, given that it is always possible to explain a set of known data points given enough doctoring of any theory. In fact when it was first publicly proposed at a meeting of the British Society for the Advancement of Science the reaction was indeed one of further caution. A report that appeared in the magazine *Nature* stated that,

> Dr. J.W. Nicholson contributed a paper on the atomic structure of elements, with theoretical determinations of their atomic weights, in which an attempt was made to build up all the elementary atoms out of four prolytes containing respectively 2, 3, 4 and 5 electrons in a volume distribution of positive electricity. Representing the prolytes by the symbols Cn (coronium), H (hydrogen), Nu (nebulium), Pf (protofluorine), the accompanying table indicates the deductions of the author with regard to the composition of several elements, allowance being made for the masses of both positive and negative electrons.[22]

Scientists usually demand that a good theory should also make successful predictions so as to avoid any suspicion that a theory may have been deliberately rigged in order to agree with the experimental data.[23] Surprisingly enough, Nicholson's theory was also able to make some genuine predictions. In addition to providing a quantitative accommodation of many spectral lines that had not previously been identified, Nicholson predicted several experimental facts that were confirmed soon afterward.

Nicholson assumed that each spectral frequency could be identified with the frequency of vibration of an electron in any particular ring of electrons. Furthermore, he believed that these vibrations took place in a direction that was perpendicular to the direction of circulation of the electrons around the nucleus (Figure 6.6). The model that was eventually developed by Bohr in 1913 differed fundamentally, in that spectral frequencies are regarded as resulting

---

[22] Anonymous, *Nature*, 501 (October 12, 1911).
[23] There is nevertheless a long-standing discussion in the philosophy of science regarding the relative worth of temporal predictions as opposed to accommodations, or retro-dictions as they are sometimes termed. See, S.G. Brush, *Making 20th Century Science*, New York, Oxford University Press, 2015.

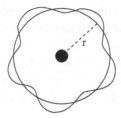

Figure 6.6 Nicholson's atomic model.

Note: Figure created by the present author; no diagram was published by Nicholson. As the electron orbits the nucleus Nicholson supposes that it oscillates in a direction at right angles to the direction or circulation.

from differences between the energies or frequencies of two different levels in the atom. Bohr's spectral frequencies do not correspond directly to any actual orbital frequency that an electron possesses. And it was this new understanding of the relationship between spectra and energy levels that provided Bohr with one of the main ingredients of his own theory. On the face of things, Nicholson was therefore simply wrong, since he based his whole theory on what we now know to be incorrect physics.

But such a view is a typical example of Whiggism and remains at the level of "right" and "wrong" that I will be aiming to move beyond. Parts of Nicholson's theory seem to have succeeded very well, given that many scientists were impressed by his explanation of the nebular spectrum and his successful prediction of new lines before they had been observed. In addition, Nicholson also proposed the notion of quantization of angular momentum, which Niels Bohr very soon embraced, and to much effect.

It is not easy to dismiss Nicholson's accommodation of so many spectral lines and especially his predictions of some unknown lines. It would not be the first time that progress had been gained on the basis of what later seemed like an insecure foundation. Perhaps just enough of Nicholson's overall view was sufficiently correct to allow him to do some useful science. After all, it would be unreasonable to expect there to be uniform progress in every single aspect of a theory. Typically some parts may be regarded as being progressive while others may be degenerating, to use the Lakatosian terminology.

And if we take an even wider perspective and consider the *longue durée* in the history of science, surely all scientific progress has been gained on the basis of what later turned out to be incorrect foundations when seen in the light of later scientific views. What really matters is that science, in the form of the scientific community, should progress as a whole. Attributing credit to a particular scientist may be important in deciding who prizes should be awarded to, but does not matter in the broader question of how the scientific community gains a better understanding of the world.

6.4 Accommodating the Spectra of Four Nebula, Including Orion Nebula

In this section the manner in which Nicholson was able to assign many unknown lines in the spectrum of the Orion nebula will be examined. First I present a figure containing the spectral lines that had been accounted for in terms of terrestrial hydrogen and helium (Figure 6.7). The dotted lines signify the spectral lines that had not yet been assigned, or identified, in any way. This situation therefore provided Nicholson with another opportunity to test his theory of proto-atoms and proto-elements.

As in many other features of Nicholson's work his approach was rather simple. He began by assuming that ratios of spectral frequencies correspond to ratios of mechanical frequencies among the postulated electron motion.[24] In mathematical terms he assumed

Nebular line	Vacuum tubes	Nebular line	Vacuum tubes
3726.4	...	4101.91	4101.89 Hδ
3729.0	...	4340.62	4340.63 Hγ
3835.8	3835.6 Hη	4363.37	...
3868.88	...	4471.71	helium
3889.14	3889.15 Hζ	4685.73	...
3965.1	3964.9 helium	4740.0	...
3967.65	...	4861.54	4861.50 Hβ
3970.23	3970.25 Hε	4959.05	...
4026.7	...	5006.89	...
4068.8	...		

Figure 6.7 Spectrum of Orion nebula showing many unassigned lines.

---

[24] The detailed calculations can be found in Nicholson's articles, J.W. Nicholson, "The Spectrum of Nebulium," *Monthly Notices of the Royal Astronomical Society, (London)*, 72, 49–64, (1911); "A Structural Theory of the Chemical Elements," *Philosophical Magazine*, 6, (22), 864–89, (1911); "The Constitution of the Solar Corona I, Protofluorine," *Monthly Notices of the Royal Astronomical Society (London)*, 72, 139–50, (1911); "The Constitution of the Ring Nebula in Lyra," *Monthly Notices of the*

Nebular line	Identification	Nebular line	Identification
3726.4	Nu+	4101.91	Hδ
3729.0	...	4340.62	Hγ
3835.8	Hη, Nu−, Nu++	4363.37	Nu
3868.88	Nu−	4471.71	helium
3889.14	Hζ	4685.73	...
3965.1	helium	4740.0	Nu−
3967.65	Nu++	4861.54	Hβ
3970.23	Hε	4959.05	Nu−−
4026.7	Helium?, Nu+	5006.89	Nu
4068.8	Nu−−	...	...

Figure 6.8 Nicholson's accommodation of 9 of the 11 unidentified lines in Figure 6.7.

$$\frac{v_{\text{spectral line 1}}}{v_{\text{spectral line 2}}} = \frac{f_{\text{rotation A}}}{f_{\text{rotation B}}}$$

where the f values emerged from his calculations, while one of the v values was obtained empirically from the spectra in question and the other one was 'predicted'.

In his 1911 article "The Spectrum of Nebulium," Nicholson also predicted the existence of a new spectral line for the nebulae in question:

> Now the case of k = −2 for the neutral atom has been seen to lead to another line which will probably be very weak. Its wavelength should be 5006.9 × .86939 = 4352.9. It does not appear in Wright's table.[25]

Remarkably enough this prediction was also soon confirmed. In a short note in the same journal in the following year, 1912, Nicholson was able to report that the spectral line had been found at a wavelength of 4353.3 Ångstroms

Royal Astronomical Society (London), 72, 176–77, (1912); "The Constitution of the Solar Corona II, Protofluorine," *Monthly Notices of the Royal Astronomical Society, (London)*, 72, 1677–692, (1912); "The Constitution of the Solar Corona III," *Monthly Notices of the Royal Astronomical Society (London)*, 72, 729–39, (1912).

[25] J.W. Nicholson, "The Spectrum of Nebulium," *Monthly Notices of the Royal Astronomical Society*, 72 (1), 49–64 (1911). https://doi.org/10.1093/mnras/72.1.49.

with an error of just 0.009% or roughly 1 in 11,110 by comparison with his prediction.

At the meeting of the Society of 1912 March the writer announced the discovery of the new nebular line at λ4353 which had been predicted in his paper on "The Spectrum of Nebulium." A plate of the spectrum of the Orion nebula, on which the line was found, and which had been taken with a long exposure at the Lick Observatory in 1908 by Dr. W.H. Wright, was also exhibited. In the meantime the line has been recorded again by Dr. Max Wolf, of Heidelberg, who has, in a letter, given an account of its discovery, and this brief note gives a record of some of the details of the observation.

The plate on which the line is shown was exposed at Heidelberg between 1912 January 20 and February 28, with an exposure of 40h 48m. The most northern star of the Trapezium is in the center of the photographed region, and the new line is visible fairly strongly, especially in the spectrum of the star and on both sides.

The wave-length in the Orion nebula, obtained by plotting from an iron curve, is 4353.9, which is of course, too large, as all the lines in this nebula are shifted to greater wave-lengths, on account of the motion of the nebula. But the correction is not so large as a tenth-metre.

The wave-length of the line on the Lick plate, as measured at the Cambridge Observatory by Mr. Stratton, is 4353.3, the value calculated in the paper being 4352.9.[26]

Nicholson experienced a similar triumph with the prediction of a new spectral line, which he believed was due to proto-fluorine and which he estimated to have a wavelength of 6374.8 Ångstroms. A new spectral line was soon discovered in the solar corona with a wavelength of 6374.6.

Considered together, these successes by Nicholson are indeed rather remarkable. Just to recap, he accounted for 9 of 11 previously unidentified lines in the spectrum of the Orion Nebulae; 14 of the unidentified spectral lines in the solar corona. In addition, and perhaps more impressively, Nicholson predicted two completely unknown lines, one in each of these spectra, both of which were promptly discovered and found to have almost exactly the wavelengths he predicted:

Nebulium prediction:	4352.9A,	observation: 4353.3A,	error: 1 in 11,111
Solar corona prediction:	6374.8A,	observation: 6374.6A,	error: 1 in 31,745

[26] J.W., Nicholson, "On the New Nebular Line at λ4353," *Monthly Notices of the Royal Astronomical Society (London)*, 72, 693, (1912).

6.5 Nicholson's Calculations on the Spectrum of the Solar Corona

Nicholson next turned his attention to the spectrum of the solar corona that had been much studied and that showed numerous lines that had not yet been accounted for (Figure 6.9). In this study Nicholson was even more successful than he had been with the spectrum of the Orion nebula,

1900	1901	1905	Mean	Intensity
...		5535.8	5535.8	2
...	5304	5303.1	5303.5	20
...	...	5117.7	5117.7	2
5073	...	...	5073	1
4779	...	...	4779	1
4725	...	...	4725	...
4722	...	...	4722	...
4586.3	...	...	4586.3	4
4566.5	4565	...	4566	6
4400	...	...	4400	1
4358.8	...	...	4359	4
4311.3	...	...	4311	2
4230.6	4230.9	4231.1	4231.0	5
4130	...	...	4130	...
...	...	4087.4	4087	...
3987.2	3987	3987.1	3987.1	3
...	3891.2	...	3891	...
3800.8	3801.1	3800.8	3800.9	3
3642.9	...	3642.0	3642.5	2
...	3505	...	3505	...
3461.3	...	...	3461	1
...	3454.4	...	3454	9
...	3387.9	...	3387.9	12
...	3361	...	3361	...

Figure 6.9 Observed lines in the solar corona spectrum.

Coronal line	Suggested origin	Intensity
5535.8	Pf, –2e	2
5073	Pf, –3e (?)	1
4586	Pf, +2e	4
4566	Pf, –e	6
4359	Pf, –2e	4
4311	Pf, +2e	2
4231	Pf, +e	5
4087	Pf, +3e (?)	...
3987.1	neutral atom, +3e (?)	3
3800.9	Pf, +e	3
3642.5	Pf, –e	2
3454	Pf, neutral atom	9
3387.9	Pf, neutral atom	12
3361(?)	Pf, –3e (?)	...

Figure 6.10 Nicholson's accommodation of 14 of the lines from Figure 6.9 using proto-fluorine and ionized forms of this proto-atom.

because he succeeded in accounting quantitatively for as many as 16 unexplained lines.

Figure 6.10 shows the observed frequencies of the lines, along with Nicholson's assignments in terms of the atom of proto-fluorine and various ionized forms of the same atom.

6.6 Nicholson and Planck's Constant

The manner in which Nicholson arrived at the all-important Planck constant was by calculating the ratio of the energy of a particle to its frequency and finding that this ratio was equal to a multiple of Planck's constant. Nicholson concluded that this constant therefore had an atomic significance and indicated that angular momentum could only change in discrete amounts when electrons leave or return from an atom. The relevance of this finding lies in the fact that up to this point the quantum had only been associated with energy and not with angular momentum. Nicholson was the first person to make this association, in what would soon become an integral aspect of Bohr's theory of the hydrogen atom. In the case of the proto-fluorine atom, Nicholson calculated the ratio of potential energy to frequency to be approximately

$$\text{Potential energy/frequency} = 154.94 \times 10^{-27} \text{erg seconds} = 25 \text{ h}$$

In arriving at his result Nicholson had used the measured values of e and m, the charge and mass of the electron. However, his method still did not provide a means of estimating the radius of the electron, and he was forced to eliminate this quantity from his equations, a problem that he would overcome a little later.

Nicholson proceeded to calculate the ratio of potential energy to frequency in proto-fluorine ions with one or two fewer electrons and found 22 h and 18 h, respectively. He noted that the three values for Pf, Pf^{+1}, and Pf^{+2} were members of a harmonic sequence,

$$25, 22, 18, 13, 7, 0.$$

By dividing each value by the number of electrons in the atom he found the Planck units of angular momentum per electron to be,

$$5, 5.5, 6, 6.5, \text{and } 7.$$

Nicholson thus arrived at the general formula for the angular momentum of a ring of n electrons as

$$\tfrac{1}{2}(15 - n) \, n$$

This formula then allowed him to fix the values of the atomic radius in each case, and since angular momentum did not change gradually, he took this to mean that atomic radius would also be quantized.

Several authors have traced the manner in which Bohr picked up this hint of quantizing angular momentum.[27] This feature was not present in Bohr's initial atomic model, and he only incorporated it over a series of steps following a close study of Nicholson's papers. Bohr also spent a good deal of time trying to establish the connection between his own and Nicholson's atomic theory.

---

[27] J. Heilbron, T.S. Kuhn, "The Genesis of Bohr's Atom," *Historical Studies in the Physical Sciences*, 1, 211–90, (1969).

6.7 Reactions to the Work of Nicholson

As the historian of physics John Heilbron stated in a recent plenary lecture to the American Physical Society, the success of Nicholson's work on nebulium had been "spectacular." Heilbron also commented on how it had served as a motivation for Bohr's work. But looking at the literature in physics and the history of physics one finds a remarkable range of views expressed concerning Nicholson's work. The following is a brief survey of these varied reactions.

Initially the commentators tended to praise Nicholson. For example, after a meeting held in Australia in 1914, W. M. Hicks remarked,

> Nicholson's calculated frequencies and the observed lines were "so close and so numerous as to leave little doubt of the general correctness of the theory . . . Nicholson's theory stands alone as a first satisfactory theory of one type of spectra."[28]

In a paper published at the end of 1913, William Wilson observed that Nicholson had

> used the quantum hypothesis with extraordinary success in his valuable investigations of the sun's corona.[29]

Physics historian Abraham Pais saw the relationship between Bohr and Nicholson sometime later in the following terms,

> Bohr was not impressed by Nicholson when he met him in Cambridge in 1911 and much later said that most of Nicholson's work was not very good. Be that as it may, Bohr had taken note of his ideas on angular momentum, at a crucial moment for him . . . He also quoted him in his own paper on hydrogen. It is quite probable that Nicholson's work influenced him at that time.[30]

Returning to Heilbron:

> The success of Nicholson's atom bothered Bohr. Both models assumed a nucleus, and both obeyed the quantum; yet Nicholson's radiated—and with unprecedented accuracy—while Bohr's was, so to speak, spectroscopically mute.

---

[28] McCormmach, "The Atomic Theory of John William Nicholson," *Archives for History of Exact Sciences*, 3, 160–184, (1966), p. 183.

[29] Ibid, p. 184.

[30] A. Pais, *Niels Bohr's Times, In Physics, Philosophy, and Polity*, Oxford, Oxford University Press, 1991, p. 145.

By Christmas 1912, Bohr had worked out a compromise: his atoms related to the ground state, when all the allowed energy had been radiated away; Nicholson's dealt with earlier stages in the binding. . . . Just how a Nicholson atom reached its ground state Bohr never bothered to specify. He aimed merely to establish the compatibility of the two models. The compromise with Nicholson was to leave an important legacy to the definitive form of the theory.[31]

Later in the same paper Bohr proposed other formulations of his quantum rule, including, with full acknowledgement of Nicholson's priority, the quantization of the angular momentum.[32]

Another historian-philosopher of physics, Max Jammer, writes

It should also be pointed out that Nicholson's anticipations of some of Bohr's conclusions were based, as Rosenfeld has pointed out, on *the most questionable and often even fallacious reasoning.*[33] (emphasis added)

Now for one last commentator, Leon Rosenfeld, who reveals some further aspects of how Nicholson has been regarded. In his introduction to a book by Niels Bohr to celebrate the 50th anniversary of the 1913 theory of the hydrogen atom, Rosenfeld writes:

The ratio of the frequencies of the two first modes happens to coincide with that of two lines of the nebular spectra: this is enough for Nicholson to see in this system a model of the neutral "nebulium" atom; and as luck would have it, the frequency of the third mode, which he could then compute, also coincided with that of another nebular line, which—to make things more dramatic—was not known when he made the prediction in his first paper, but was actually found somewhat later. . . .

From the mathematical point of view Nicholson's discussion of the stability conditions for the ring configurations and of their modes of oscillation is an able and painstaking piece of work; but the way in which he tries to apply the model . . . must strike one as unfortunate accidents.[34]

In the third paper, however, published in 1912, occurs the first mention of Planck's constant in connection with the angular momentum of the rotating

[31] J. Heilbron, "Lectures in the History of Atomic Physics, 1900–1922." In *History of Twentieth Century Physics*, C. Weiner, ed., 40–108, Academic Press, New York, 1977, p. 69.

[32] Ibid., p. 70.

[33] M. Jammer, *The Conceptual Development of Quantum Mechanics*, New York, McGraw-Hill, 1966. p., 73.

[34] L. Rosenfeld, in preface to N. Bohr, *On the Constitution of Atoms and Molecules*, New York, W.E. Benjamin, 1963, p. xii.

electrons: again here there is no question of any physical argument, but just a further display of numerology....

Bohr did not learn of Nicholson's investigations, as we shall see, before the end of 1912, when he had already given his own ideas of atomic structure their fully developed form.[35]

By contrast [with Nicholson] the thoroughness of Bohr's single-handed attack on the problem and the depth of his conception will appear still more impressive.[36]

There is clearly no "fence sitting" here to give Nicholson any benefit of the doubt. Rosenfeld does not even believe that Nicholson's apparent early success was due to a cancellation of errors. But perhaps some aspects of Rosenfeld's life serve to explain some of his reaction. Rosenfeld was without doubt one of Bohr's leading supporters and also acted as a leading spokesperson for Bohr's Copenhagen interpretation of quantum mechanics for the last 30 or so years of Bohr's life. Rosenfeld is also known to have been an especially vitriolic and harsh critic in spite of his apparently shy and retiring personality. His fellow Belgian and one-time collaborator, the physical chemist Ilya Prigogine, described him as a "paper tiger."[37] It is hardly surprising therefore that Rosenfeld championed Bohr against any claims from anyone, such as Nicholson, who he regarded as imposters, or anyone who might try to steal some of the thunder from Bohr.

The views of Rosenfeld can be contrasted this with those of Kragh, the contemporary historian and a Dane like Bohr,

No wonder Bohr, when he came across Nicholson's atomic theory found it to be interesting as well as disturbingly similar to his own ideas. Nicholson's atom was a rival to Bohr's and Nicholson was the chief critic of Bohr's ideas of the quantum atom.[38]

But let us say, for the sake of argument, that Rosenfeld is right and that Nicholson's work was completely worthless. Even if this were true, I contend that Nicholson's publications contributed to Bohr's developing his own atomic theory for the simple reason that Nicholson served as his foil. In some places Bohr is quite dismissive of Nicholson's work, such as when writing to his Swedish colleague Carl Oseen, where he describes Nicholson's work as "pretty crazy" while adding,[39]

---

[35] Ibid., p. xiii.

[36] Ibid., p. xiii–xiv.

[37] Prigogine was born in Russia but emigrated to Belgium.

[38] H. Kragh, *Niels Bohr and the Quantum Atom: The Bohr Model of Atomic Structure 1913–1925*, Oxford, Oxford University Press, 2012, p. 27.

[39] This comment by Bohr refers to Nicholson's earlier theory of electrons in metals, although one gathers the impression that Bohr continued to hold this crucial view about Nicholson. I am grateful for a reviewer for drawing my attention to this qualification.

I have also had discussion with Nicholson: He was extremely kind but with him I shall hardly be able to agree about very much.[40]

In other places Bohr shows Nicholson considerably more respect, such as when writing to Rutherford while he was on the point of submitting his famous trilogy paper that was published in 1913.

It seems therefore to me to be a reasonable hypothesis, to assume that the state of the systems considered in my calculations is to be identified with that of atoms in their permanent (natural) state . . . According to the hypothesis in question the states of the system considered by Nicholson are, on the contrary, of a less stable character; they are states passed during the formation of the atoms, and are states in which the energy corresponding to the lines in the spectrum characteristic for the element in question is radiated out. From this point of view systems of a state as that considered by Nicholson are only present in sensible amount in places in which atoms are continually broken up and formed again; i.e. in places such as excited vacuum tubes or stellar nebulae.[41]

In another passage from a letter to his brother Harald, Niels Bohr writes

Nicholson's theory is not incompatible with my own. In fact my calculations would be valid for the final chemical state of the atoms, whereas Nicholson would deal with the atoms sending out radiation, when the electrons are in the process of losing energy before they have occupied their final positions. The radiation would thus proceed by pulses (which much speaks well for) and Nicholson would be considering the atoms while their energy content is still too large that they emit light in the visible spectrum. Later light is emitted in the ultraviolet, until at last the energy which can be radiated away is lost.[42]

After Bohr had published his three-part article, Nicholson continued to press him in a number of further publications. If we must speak in terms of winners and losers, Bohr would be regarded as a winner and Nicholson as a loser. But as in all walks of life, it is not just about winning, but more about partaking. There would be no athletic races for spectators to watch if the losers were not even to

[40] Bohr to Oseen, December 1, 1911, cited in N. Bohr, L. Rosenfeld, E. Rüdinger, & F. Aaserud, "Collected Works." In *Early Work (1905–1911)*, Vol. 1, J.R. Nielsen, ed., North-Holland Publishing, Amsterdam, 1972, pp. 426–431.
[41] N. Bohr to Rutherford, 1913, cited in L. Rosenfeld, preface to N. Bohr, *On the Constitution of Atoms and Molecules*, New York, W.E. Benjamin, 1963, p. xxxvii.
[42] Bohr to Harald, in N. Bohr, L. Rosenfeld, E. Rüdinger, & F. Aaserud, *Early Work (1905–1911)*. In both instances where I have written "my," Bohr had actually written "his"—which must surely be typos.

participate in the race. The very terms winner and loser are necessarily code-pendent in the context of any scientific debate, as were the roles of Bohr and Nicholson.

Now this picture that I have painted would seem to raise at least one obvious objection. If all competing theories are allowed to bloom because there is no such thing as a right or wrong theory, how would scientists ever know which theories to utilize and which ones to ignore? Indeed this is the kind of criticism that was levelled against Feyerabend[43] when he claimed that "anything goes" and was promptly criticized by numerous authors.[44] I think the answer to this question can be found in evolutionary biology. Nature has the means of finding the best way forward. Just as any physical trait with an evolutionary advantage eventually takes precedence, so the most productive theory will eventually be adopted by more and more scientists in a gradual, or trial and error fashion. The theories that lead to the most progress will be those that garner the largest amount of experimental support and which provide the most satisfactory explanations of the facts. This entire process will not be rendered any the weaker even if one acknowledges an anti-personality and anti–"right or wrong view" of the growth of science.

More generally, I believe that the two aspects can coexist quite happily. Scientists can, and regularly do, argue issues out to establish the superiority of their own views, as well as their claims to priority. But the march of progress, to use an old-fashioned term, does not care one iota about these human squabbles. And it is the overall arc of progress that really matters, not whose egos are bruised or who obtains the greater number of prizes and accolades.

6.8 How Was Any of the Success Possible Given the Limitations of Nicholson's Theory?

Having examined the apparent successes of Nicholson's theory we must still ask how any of this was even possible given what we now know of his ideas. Here is a brief list of what seems to be problematical in Nicholson's scheme: First of all, the proto-elements like nebulium that he postulated do not exist. Second, he identified mechanical frequencies of electrons with spectral frequencies and assumed that these oscillations took place at right angles to the direction of electron circulation. Third, Nicholson's electrons were all in one single ring, unlike the subsequent Bohr model in which they are distributed across different rings or shells.

---

[43] P. Feyerabend, *Against Method*, London, Verso, 1975.

[44] J. Preston, Paul Feyerabend entry in Edward N. Zalta, ed., *The Stanford Encyclopedia of Philosophy*, Winter 2016 Edition. https://plato.stanford.edu/archives/win2016/entries/feyerabend/.

So in the light of modern knowledge, Nicholson seemed to be making several false assumptions. According to the traditional way that a realist might regard such matters, Nicholson's ideas were not even approximately correct since they would be regarded as downright false.[45] And yet he achieved remarkable success, at least according to most of the commentators whose views were quoted here. Are cases such as Nicholson exceptional or can other examples be found in the history of science? If one accepts that all, or most, theories are eventually refuted, one has to concede that the progress of science implies that "wrong theories" regularly lead to progress!

Having gone into some details about Nicholson's work in early 20th century atomic physics as well as astrophysics, I would now like to return to the general philosophical question and try to make some sense of the apparent success his view seemed to enjoy, at least initially.

I am proposing an evolutionary theory of the development of science in a literal biological sense.[46] While appealing to some of the ideas that have emerged from evolutionary epistemology as supported by Donald T. Campbell and many others, my account also involves a new departure as I will be explaining later. The general idea that is common to most approaches in evolutionary epistemology is that since evolution drives all biological development, it also drives the way in which we human agents think and how we develop theories and experimentation. It is a central tenet of evolutionary epistemology that all the knowledge of the natural world that we possess is essentially determined by evolutionary biology. Consequently, it seems plausible to further suggest that scientific knowledge is neither right nor wrong at any epoch in history but instead better or worse suited to the environment that science finds itself working in, namely the way in which scientific theories are suited to or correspond with the natural world as revealed by experimentation.

Needless to say, I do not dispute such scientific facts as the view that the earth is round rather than flat. Of course I accept that there are truths with a small "t" such as it is true that the earth is round or that grass is green. My concern is more with the notion that there may be a scientific truth with a capital "T." It is also important to distinguish theories from facts or observational statements. Whereas it is true that grass is simply green, this is a far cry from the question of whether Bohr's theory or the theory of evolution is true or correct. Theories, especially those in the physical sciences, consist of mathematical relationships. They do not just assert whether the earth circles the sun or vice versa. Similarly, Newton's theory is based on his famous three mathematical laws, rather than making a

---

[45] As I have already indicated I reject such an over-simplification, especially when it comes to assumptions and theories.

[46] E.R. Scerri, *A Tale of Seven Scientists and A New Philosophy of Science*, New York, Oxford University Press, 2016.

simple assertion on whether or not the moon is made of blue cheese or anything quite so specific.

I take it that one would never wish to claim that biological evolution is either right or wrong, but rather that any biological developments are either suited to their environment or not. Those developments that are suited to their biological niches result in their being perpetuated in future generations while those that are not simply wither away. So, I claim, it is with the development of science.

In passing, I should note some kinship with the views of the influential anti-realist philosopher Van Fraassen, who supports an evolutionary view of the progress of science when he writes,

> For any scientific theory is born into a life of fierce competition, a jungle red in tooth and claw. Only the successful theories survive—the ones which in fact latched onto the actual regularities in nature.[47]

This view that has been criticized in particular by Kitcher,[48] and Stanford,[49] although it would take me too far afield to comment on this exchange here.

My own brand of evolutionary epistemology also holds that the study of scientific development should not be approached by concentrating on individual discoverers nor through individual theories. I claim that science essentially develops as one unified organism, while fully anticipating that this aspect will meet the greatest resistance from critics. Needless to say, many people have realized the societal/collective nature of science, and I cannot claim any originality on that score. For example, there have been many programs such as the Strong Program, Science Studies, and the Sociology of Science, which all take a more holistic approach to studying how science develops. However, these programs generally hold that social factors determine scientific discoveries. My interest lies primarily in the actual science, while at the same time maintaining a radical form of sociology that considers scientific society to be a single organism which is essentially developing in a unified fashion.

I propose that scientific research is conducted by a tacit network of scholars, and researchers, who frequently appear to be at odds with themselves, and frequently are involved in bitter priority disputes, while unknowingly partaking in the same overall process. For example, in the case of John Nicholson, and several others that I have discussed in a recent book, the protagonists were seldom in direct contact with the Bohrs, Paulis and other luminaries in the world of early

[47] B. Van Fraassen, *The Scientific Image*, Oxford, Clarendon Press, 1980, p. 40.
[48] P. Kitcher, *The Advancement of Science: Science without Legend, Objectivity without Illusions*, New York, Oxford University Press, 1993.
[49] K. Stanford, An Antirealist Explanation of the Success of Science, *Philosophy of Science*, 67, 266–84, (2000).

20th century atomic physics. Nevertheless, one sees an entangled, organic development in which ideas compete and collide with each other, even if one famed individual is eventually associated with any particular scientific discovery or episode.

I regard science as very much proceeding via modification by trial and error rather than via "cold rationality." My view is aligned with biological evolution rather than with an enlightenment view of the supremacy of rationality and the powers of pure deduction. I regard scientific development as being guided by evolutionary forces, blind chance, and random mutation. When seen from a distance science appears to develop as one unified organism. Under a magnifying glass there may be little sign of unity, so much so that some philosophers and modern scholars have been driven to declare the dis-unity of science, a view that I believe to be deeply mistaken.[50]

From the perspective that I propose, one can better appreciate how what a realist regards as wrong theories, as in the case of most of John Nicholson's scientific output, can lead to progress. I do not need to explain wrong theories away in my account. Nor do I believe that simultaneous or multiple discoveries deserve to be explained away as being aberrations. Similarly, priority disputes, which are so prevalent in science, can be seen as resulting from the denial of the unity of science and an underlying body-scientific that I allude to.

Developments that might normally be regarded as wrong ideas frequently produce progress, especially when picked up and modified by other scientists. Nicholson's notion of angular momentum quantization, which he developed in the course of what has generally been regarded as an incorrect theory, was picked up by Niels Bohr who used it to transform the understanding of atomic physics. The image of science that I envisage resembles a tapestry of gradually evolving ideas that undergo mutations and collectively yield progress. In some cases the source of the mutation is random, just as mutations are random in the biological world. This view stands in sharp contrast with the traditional notion of purposive and well thought out ideas on the part of scientists.[51]

Consequently, the kinds of inconsistent theories that Peter Vickers and others have examined no longer seem to be in such urgent need of explanation and should not be regarded as being so puzzling. Indeed, such developments should

---

[50] P. Galison, D. Stump (eds.), *The Disunity of Science*, Palo Alto, Stanford University Press, 1996.

[51] I regard mutations in scientific concepts to occur on the level of individual scientists and not primarily in entire social groups. My evolutionary epistemology thus also differs from Kuhn's in this respect since his unit of evolutionary change is the social group of scientists. See some interesting discussions regarding Kuhn's evolutionary account of scientific progress in B.A. Renzi, "Kuhn's Evolutionary Epistemology and Its Being Undermined by Inadequate Biological Concepts," *Philosophy of Science*, 76, 143–59, (2009); T.A.C. Raydon, P. Hoyningen-Huene, "Discussion: Kuhn's Evolutionary Analogy in *The Structure of Scientific Revolutions* and 'The Road since Structure,'" *Philosophy of Science*, 77, 468–76, (2010).

be regarded as the rule rather than exceptions. I agree completely with Kuhn that ideas and theories that survive should be regarded as facilitating progress and not as tracking the truth.

As so many evolutionary epistemologists have urged, I regard knowledge as being presumptive, partial, hypothetical, and fallible. If scientific progress is far more organic than usually supposed, we can make better sense of why Nicholson was able to contribute to the growth of science. None of the scientists I have examined really knew what they were doing, in a sense.[52] Their crude ideas developed in an evolutionary trial and error manner instead of in a strictly rational manner, or so I propose.

Nicholson and many people like him contributed very significantly to the development of science. Nicholson was not simply wrong, since he inadvertently helped Bohr to begin to quantize angular momentum. Nicholson is as much part of the history as Bohr is, and minor contributors matter as much as the well-known ones. In anticipation of the following section, I believe that the early Kuhn's focus on revolutions may have served to diminish the importance of such marginal figures.

6.9 Is the Proposed View Compatible with Kuhnian Revolutions?

Is an evolutionary view of the kind I propose compatible with the revolutions for which Kuhn is so well known? First of all, as many authors have claimed, Kuhn's own work on historical episodes sometimes points to continuity rather than revolution, such as in the case of quantum theory and the Copernican Revolution.[53]

Of course, Kuhn also acknowledges the role of normal science that in very broad terms can be seen as upholding the role of minor historical contributors to the growth of science. Nevertheless, as I see it, there was no sharp revolution in the development of quantum theory only an evolution. Paradoxically, such an evolution is easier to appreciate from the wider perspective of science as one unified whole than from the perspective of individual contributors or individual theories. Viewing theory-change as *revolutionary*, on the other hand, may mask the essentially biological-like growth of science that I am defending.

Nevertheless, I am in complete agreement with Kuhn in supposing that science does not drive toward some external truth. In this respect I am on the side of anti-realism. I prefer to regard scientific process as driven from *within*, by

---

[52] Scerri 2016.

[53] T.S. Kuhn, *Black-Body Radiation and the Quantum Discontinuity, 1894–1912*, Chicago, Chicago University Press, 1987; J. Heilbron, T.S. Kuhn, "The Genesis of the Bohr Atom," *Historical Studies in the Physical Sciences*, 1, 211–290 (1969).

evolutionary forces which look back to past science.[54] This is not to say that the world does not constrain our theorizing. Kuhn just wants us to see that the scope of our theories is not determined by nature in advance of our inquiring about it. Indeed, I find that I agree with Kuhn on a great number of issues, the main exception being the occurrence of scientific revolutions.

As scholars generally concur, Kuhn's later work was aimed toward developing an *evolutionary* epistemology.[55] As Kuhn developed his epistemology of science, he saw increasingly more similarities between biological evolution and scientific change. Consequently as he developed his epistemology of science it became a more thoroughly evolutionary epistemology of science. Wray makes the very perceptive remark that while Kuhn was one of the key philosophers of science who initiated the *historical turn* in the philosophy of science in the early 1960s, he later came to adopt what he termed a *historical perspective*. According to Wray this historical perspective, or developmental view, is nothing less than an *evolutionary perspective* on science. In accord with Wray and others, Kuukkanen (2013, 134) points out that the later Kuhn felt that his evolutionary image of science did not get the amount of attention that it deserved. In the course of his last interview, Kuhn deplored this situation while saying, "I would now argue very strongly that the Darwinian metaphor at the end of the book [SSR] is right and should have been taken more seriously than it was."[56]

Just as evolution lacks a *telos* and is not driven towards a set goal in advance, science is not aiming at a goal set by nature in advance. Kuhn continued to regard his evolutionary view as important to the end of his life, or as Wray writes, "Whatever else he changed he did not change this aspect." Kuhn also claims that in the history of astronomy, the earth-centered models held the field back for many years. Similarly, he claimed that, truth-centered models of scientific change were holding back philosophy of science. Somewhat grandiosely, Kuhn notes a similarity between the reception of Darwin's theory, and the reception of his own theory his own view on the evolution of science, in that both meet the greatest resistance on the point of elimination of teleology.

---

[54] Kuhn prefers the phrase "pushed from behind." I believe that "driven from within" better conveys the way that scientific knowledge is being generated in an outward fashion rather than being "pulled" toward the truth.

[55] B. Wray, *Kuhn's Evolutionary Social Epistemology*, Cambridge, UK: Cambridge University Press, 2011, 84.

[56] K. Jouni-Matti, "Revolution as Evolution: The Concept of Evolution in Kuhn's Philosophy." In *Kuhn's The Structure of Scientific Revolutions Revisited*, Theodore Arabatzis, ed., 134–152, Routledge, New York, 2013, p. 134. See T.S. Kuhn, J. Conant, & J. Haugeland, *The Road Since Structure: Philosophical Essays, 1970-1993* (with an autobiographical interview), 2000, Chicago: University of Chicago Press. Kuhn's first thoughts on epistemology based on evolutionary lines appear at the end of *The Structure of Scientific Revolutions*, 1962, Chicago: University of Chicago Press, 169–172.

Whereas Kuhn did not initially have an explanation for the success of science, he later proposed that *scientific specialization* was the missing factor. Kuhn argued that just as biological evolution leads to an increasing variety of species, so the evolution of science leads to an increasing variety of scientific sub-disciplines and specializations.[57] For Kuhn science is a complex social activity, and the unit of explanation is the group, not the individual scientists. This resonates with my own view that I alluded to earlier that the growth of science is not successfully tracked by considering individual scientists or individual theories. Kuhn asks us to judge changes in theory from the perspective of the research community rather than the individual scientists involved. Both early converts to a theory and holdouts aid the community in making the rational choice between competing theories. This, I suggest, is how the work of Nicholson should be viewed, namely as a step toward the new theory in the context of Bohr and others.[58]

However, there is another respect in which I would want to qualify this view to also reaffirm the importance of change on the level of individual scientists rather than change predominantly within social groups. Kuhn's evolutionary account of the growth of science appears to be somewhat half-hearted. He does not say very much on the mechanism of evolutionary change in scientific theories except for his talk of specialization. Kuhn believes that the development of new sub-disciplines such as the emergence of biochemistry from chemistry, for example, is analogous to the evolution of new species which become incapable of mating with members of the species from which they have evolved. Similarly Kuhn holds that the members of the new scientific sub-discipline, which has branched off from an older one, are unable to communicate with members of the mother discipline. Kuhn even makes a virtue of this reconceptualized incommensurability in suggesting that such isolation allows for a greater development within a particular sub-discipline.

While authors like Renzi have criticized Kuhn's evolutionary analogy, Raydon and Hoyningen-Huene defend him, in suggesting that Kuhn did not intend his analogy to be taken literally. Be that as it may, I *do* intend the evolutionary analogy to be literal. Since the evolution of scientific theories is part of an underlying biological evolution of the human species I do not regard this suggestion to be too far-fetched, even though these evolutionary processes may be occurring at levels that are far removed from each other.

[57] Out of physics and chemistry there emerges physical chemistry. Biology and chemistry give rise to biochemistry. Biochemistry in turn gives rise to physical-biochemistry.

[58] This project has shown me that contrary to what I believed for about 25 years, there is much merit in Kuhn's work. I had been too stuck on the cartoon Kuhn (the best-seller Kuhn) who is supposed to deny progress and who is often taken to be at the root of all evil, such as science wars and the sociological turn—but only because I arrived at the idea of an evolutionary epistemology through my own work in asking how a "wrong" theory can be so successful in so many cases.

In the case of the evolution of organisms, modern biology has taught us that the underlying mechanism is one of random mutation on the level of errors that occur in the copying of DNA sequences. The more accurate biological analogy for the development of scientific theories would seem to be for the "mutation" of ideas to occur in the minds of individual scientists such as Bohr or Nicholson. The unit involved in the evolution of scientific theories would therefore be individual scientists and not the social group that scientists belong to. Here then is where I depart from Kuhn's evolutionary epistemology. In my account the evolution of scientific theories takes place in a literally biological sense and is mainly situated at the level of individuals rather than social groups.

6.10 Can Kuhn Have it Both Ways?

Can revolution coexist with evolution in science as Kuhn seems to believe? In the later Kuhn, revolutions are no longer paradigm changes. They are now taxonomic or lexical changes, thus raising the question of whether the revolutions he gave as examples no longer count as revolutions.[59]

> The early Kuhn's wholesale and psychologically drastic revolution becomes a gradual and piecemeal communitarian evolution in the later Kuhn, something that may show simultaneous continuity and discontinuity between prerevolutionary and revolutionary stages.[60]

Brad Wray has claimed that revolutions are essential to Kuhn because they are incompatible with the view that scientific knowledge is cumulative and that scientists are constantly marching ever closer to the truth. I must say that I disagree with this view. Scientists may not be moving toward a fixed truth, but the development may still be gradual and not revolutionary. After all, biological evolution is not teleological but is generally thought to be gradual (pace Stephen Jay Gould et al.). To me the main insight from Kuhn is his evolutionary epistemology not his notion of discontinuous theory-change.

For example, why consider the quantum revolution to have ended in 1912 as Kuhn does? Surely an equally important revolution due to Bohr *began* around this time of 1912. Or was the true revolution the coming of quantum mechanics à la Heisenberg and Schrödinger in 1925–26 or maybe QED sometime later in the

---

[59] A quick reply to this question is that Kuhn gives very few examples of what he considers to be scientific revolutions in the sense of lexically driven revolutions.

[60] J.M. Kuukkanen, "Revolution as Evolution." In, *Kuhn's The Structure of Scientific Revolutions Revisited*, eds., V. Kindi and T. Abratzis, Routledge, London 2012.

late 1940s or QCD in the 1970s? Or could it be that there are so many revolutions that the very concept ceases to be helpful?

6.11 Conclusions

Nicholson is somewhat neglected in the history of science, but I don't believe it is because he was simply wrong in many of his basic assumptions.[61] In spite of holding some assumptions concerning the structure of the atom that were subsequently abandoned, Nicholson was still able to make a number of highly successful accommodations of known data and predictions of completely unknown data. Furthermore, Nicholson's idea of the quantization of angular momentum was key to Bohr's subsequent progress in the development of atomic physics. Nicholson was part of the organic manner in which science evolves or, in this case, the way that atomic physics evolved. He was an important "missing link" between the old classical physics and the new quantum theory and the way that it was applied to the atom.

Needless to say, if Nicholson had not been the first to propose the quantization of angular momentum somebody else would probably have done so. I am not trying to rehabilitate Nicholson's role, but merely wishing to highlight the crucial and catalytic role that is often played by the "little people" in science. Moreover, it is quite conceivable that the history of atomic physics might have taken a different path, perhaps one not involving the quantization of angular momentum. The fact remains that it did, and that Nicholson played an undeniable role in what actually took place. My main point is to try to highlight the organic and evolutionary way in which science develops and that it is only in retrospect that priority is attributed to certain contributors. Given our limitations in attempting to reconstruct such an organic and interconnected growth process, it is hardly surprising that we tend to simplify the story by latching on to the leading players in any particular scientific episode. As the reader will have noticed, I am not unduly puzzled by what may appear to be inconsistent theories and, in the case I have examined, even some bizarre theories can lead to scientific progress. If one accepts the notion that scientific work is at root a collective exercise even if not literally so, then inconsistencies can be thought of as temporary road-blocks which delay progress but which get ironed out in the *longue durée*.

---

[61] I am speaking mainly of popular science and science textbooks. Professional historians of science know better of course.

7

Selective Scientific Realism
and Truth-Transfer in Theories
of Molecular Structure

Amanda J. Nichols[1] and Myron A. Penner[2]

[1]Division of Natural and Health Sciences
Oklahoma Christian University
amanda.nichols@oc.edu
[2]Department of Philosophy
Trinity Western University
myron.penner@twu.ca

7.1 Introduction

According to scientific realists, the predictive success of mature theories provides a strong epistemic basis for thinking that such theories are approximately true. However, we know that many theories once regarded as well confirmed and predictively successful were eventually replaced with successor theories, and some claim this undermines the epistemic confidence we should have in the approximate truth of current science. Selective scientific realists in turn argue that if one can show that the predictive success of some rejected theory T is a function of theoretical claims consistent with current science, then T's failure doesn't undermine the claim that current successful theories are approximately true. As such, Selective Scientific Realism (SSR) can be tested through historical examples. Showing that the predictive success of a failed theory is the result of theoretical features later rejected provides a counterexample to SSR. Conversely, SSR is supported if its explanation of the predictive success of failed theories—namely that the factors which lead to predictive success in the failed theory survive in the successor theory—is able to handle a wide array of historical cases.

In what follows, we look at one such historical case not currently discussed in the literature on SSR: theoretical advances in understanding molecular

Amanda J. Nichols and Myron A. Penner, *Selective Scientific Realism and Truth-Transfer in Theories of Molecular Structure* In: *Contemporary Scientific Realism.* Edited by: Timothy D. Lyons and Peter Vickers, Oxford University Press.
© Oxford University Press 2021. DOI: 10.1093/oso/9780190946814.003.0007

structures at the turn of the 20th century which resulted from the Blomstrand-Jørgensen/Werner Debate about the structure of cobalt complexes. Both the Blomstrand-Jørgensen chain theory and Werner's coordination theory, which ultimately replaced it, make predictions about the number of ions that will be dissociated when the molecule undergoes a precipitation reaction. While chain theory makes correct predictions in many cases, Werner's coordination theory correctly predicted the number of ions in cases where chain theory failed. If the predictive success of chain theory depended on features that were later rejected in coordination theory, we'd have a case which undermines SSR's explanation of truth transfer between failed theories and their successors. Conversely, the rise of coordination theory supports SSR if the features of chain theory that resulted in its predictive successes were also a part of coordination theory. It turns out the latter is the case. We conclude by applying the lessons from the debate about the structure of cobalt complexes to philosophical debate about scientific realism.

In brief, the chain theory utilized by Jørgensen involved three rules for modeling the structure of inorganic compounds. Jørgensen applied this framework with some experimental success: his models predicted certain outcomes of ion dissociation which were confirmed in a series of precipitation experiments. However, there were some experiments in which Jørgensen's model failed to accurately predict ion dissociation. During the same time period, Werner was developing a geometric model for cobalt complexes much different from Jørgensen's. It turns out that Werner's model was able to succeed experimentally where Jørgensen's had failed. This led, ultimately, to Blomstrand-Jørgensen chain theory being replaced by Werner's coordination theory.

Those who are skeptical about whether experimental success indicates the truth of scientific theories might look at this historical episode as one which justifies such skepticism. After all, in retrospect we can see that Jørgensen achieved a degree of experimental success even though he was mistaken about the structure of the cobalt complexes in question. Why should we think that the experimental success of contemporary scientific theories is in virtue of their truth, when experimental success obviously does not guarantee truth?

However, we think that for this particular historical episode, SSR provides the correct historical analysis. It turns out that the factors which led to Jørgensen's predictive success are also features of Werner's coordination theory that ultimately replaced Jørgensen's chain theory. Jørgensen's predictive success *was* in virtue of him latching on to certain truths about how cobalt complexes are structured. We further argue that the lessons from this historical case can illuminate how two contemporary objections to realism—P. Kyle Stanford's Problem of Unconceived Alternatives and Timothy D. Lyons' pessimistic modus tollens argument—fall short as arguments against realism.

7.2 Molecular Structures and 19th Century Chemistry

By the middle of the 19th century, chemists were routinely investigating both physical and chemical properties of different compounds as well as identifying ratios between elements in different compounds. Masses for different elements were well established by this time, which meant that chemists could carefully weigh samples before and after experiments involving a reaction and isolation of a compound in order to determine the loss or gain of mass. An active area of inquiry was molecular structure and trying to determine how atoms are arranged and connected to each other. Since J. J. Thomson didn't discover electrons until 1897, the modern theory of molecular bonds consisting of electrons had not yet emerged.[1] While chemists at this time did not have a theory explaining how atoms connect, there was consensus that atoms connected in some way to form molecules. Moreover, chemists were drawing connections from the physical and chemical properties to structural features of the molecule, primarily in organic chemistry because of the work being done with carbon.[2] It was discovered that the way atoms were connected to one another seemed important, and the arrangement that atoms have within space—their molecular geometry— also seemed to affect chemical and physical properties. For example, Jacobus Henricus van't Hoff's work demonstrated that there are compounds that have the same chemical formula with the same connectivity between atoms, but their three-dimensional orientations differ, making the compounds behave differently in polarized light. These optically active compounds introduced the idea of the geometry of a molecule.[3] Molecules were discovered to be not mere planar entities, but rather three-dimensional structures where each constituent atom's orientation in space can change the structure of the whole molecule.

Chemists had an early idea of valence, which can be understood as the affinity an atom has to combine with other atoms. Valence is expressed as a quantity reflecting the number of bonds an atom may have. For instance, carbon is said to have a valence of 4 (tetravalence) because it has four connecting points. Hydrogen is univalent because it has one connecting point. Therefore, one carbon atom will combine with four hydrogen atoms forming the compound methane (CH_4) (Figure 7.1). How elements bond to form compounds is a function of each constituent element's valence, because not anything can bond to just anything. Valence helps to predict molecular structure. Drawing on both these valence patterns and the more developed branch of organic chemistry, Friedrich August Kekulé predicted that the arrangement of atoms in molecules could be

---

[1] See Douglas and McDaniel (1965: 338).
[2] See Day and Selbin (1962: 183).
[3] See Moore (1939: 294–295).

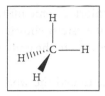

Figure 7.1 Methane (CH$_4$).

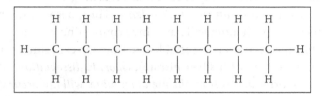

Figure 7.2 Octane (C$_8$H$_{18}$).

known. In 1858, he claimed that some carbon compounds consist of hydrocarbon chains: carbon atoms are connected together with a certain number of hydrogen atoms on each carbon because each carbon atom is tetravalent.[4] Basic organic molecules like octane follow this chain-like structure (Figure 7.2).

7.3 Cobalt Complexes: Chain-Like Structures?

Christian Wilhelm Blomstrand was an inorganic chemist who worked with metal ammines, specifically compounds that had a metal combined with ammonia molecules and halogen atoms, such as chlorides. In 1871, Blomstrand proposed that ammonia molecules are structured in chains like hydrocarbons where ammines are composed of a metal atom, a chain of ammonia molecules, and a certain number of chlorides. Drawing on the bonding principles of organic chemistry, Blomstrand assumed that the whole metal ammine would have a chain-like structure. But how is the metal ammine "chain" put together? Depending on the metal's valence, Blomstrand reasoned that a certain number of ammonia molecules could form a chain off of the metal. Chlorides would then be connected to either the ammonia chain or directly to the metal itself. Blomstrand performed precipitation experiments in order to figure out whether the chlorides were bonded either to the metal or to the ammonia

---

[4] See Farber (1952: 164–165).

chain.[5] He thought that if a chloride came off the compound easily to react with silver nitrate (and precipitate silver chloride as a product) it was because the chloride was directly connected to the end of the ammonia chain, not the metal.[6]

Sophus Mads Jørgensen continued the work with metal ammines in the 1890s, specifically with cobalt ammine complexes. A cobalt ammine complex is a molecule made up of the metal cobalt, ammonia groups, and chlorides. Jørgensen followed three bonding rules for structures when theorizing the structure of these cobalt complexes. First, according to the **valence rule,** *bonding is constrained by an element's valence.* Thus, because the metal cobalt is trivalent, a cobalt atom will have three connecting points. Second, according to the **dissociation rule** (established from lab experiments involving precipitation reactions with silver nitrate), *chloride dissociation, that is, the separation of chloride atoms from the metal ammine, will not occur if chloride is directly bonded to metal.* And third, according to the **chain rule,** *the entire structure of the compound would be chain-like,* and thus, ammine chains will form off a cobalt atom. While we are not claiming that Jørgensen explicitly formulated these rules in his model for cobalt complexes, it is clear from his dependence on Blomstrand's work and his own published research that Jørgensen followed what we are calling the valence, dissociation, and chain rules. Much of Jørgensen's work centered on the compounds that are known today as coordination complexes, and he was one of the first chemists to provide experimental evidence that the cobalt ammine complexes were not dimers, contrary to what Blomstrand hypothesized. Jørgensen's 1890 paper outlines this evidence and treats cobalt as trivalent, demonstrating that Jørgensen is following the valence rule.[7] Kauffman points out that in 1887 Jørgensen published his work using the precipitation experiment (discussed in the next section), reasoning that if the chloride is directly bonded to a metal atom, the chloride will not dissociate. This is what we have dubbed the dissociation rule.[8] Continuing the ideas originally proposed by Blomstrand, Jørgensen thought of the ammines to be in a chain structure (the chain rule), and Werner responds to these ideas directly in his paper proposing his new coordination theory.[9,10] This is relevant to what we later will argue

---

[5] See Kauffman (1965).
[6] See Meissler and Tarr (2011: 322).
[7] See Jørgensen (1890: 429) cited in Kauffman (1965).
[8] See Jørgensen (1887: 417) cited in Kauffman (1965).
[9] See Kauffmann (1965: 186).
[10] See Werner (1893: 267–330) translated by Kauffman (1968: 16–20).

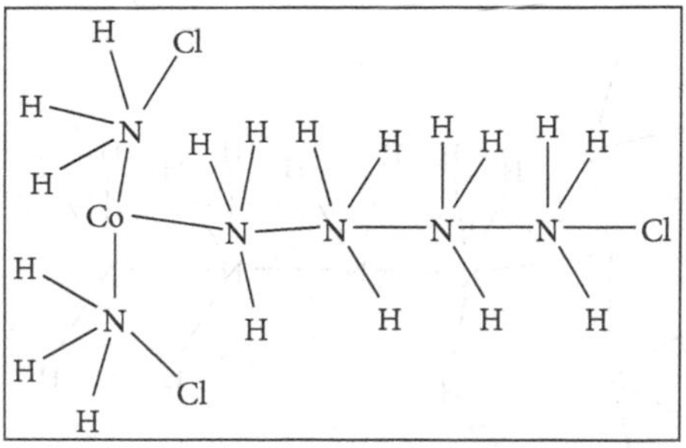

Figure 7.3 Pre-experiment of hexammine complex: Jørgensen's proposed structure.
Note: Here, Jørgensen's work involved cobalt(III) hexammine chloride $[Co(NH_3)_6]Cl_3$ complexes.

in establishing both continuity and discontinuity between Blomstrand-Jørgensen chain theory and Werner's coordination theory.

7.4 The Precipitation Experiment

Applying these three rules results in the Blomstrand-Jørgensen chain theory of cobalt ammine complexes, which includes models for structures of different compounds; the accuracy of these proposed structures can be tested in precipitation experiments. During a precipitation experiment, cobalt complex molecules split into ions (separated charged particles). The experiment begins by mixing a cobalt complex solution with a silver nitrate solution. A solid (the precipitate) forms by chloride separating from the cobalt complex and bonding with the silver; the resulting precipitate is silver chloride. Jørgensen's three bonding rules make predictions with respect to the number of ions that will result from a precipitate experiment.

An early example from Jørgensen is his postulated chain structure for cobalt (III) hexammine chloride, further referred to as the hexammine complex. Jørgensen's structure of this cobalt complex, following the valence rule, consists of a cobalt center with three groups connected to it (Figure 7.3). The three chloride atoms are connected to ammine groups (NH_3Cl), but none of the chloride atoms are attached directly to the cobalt atom. Following the chain rule, the third group is a chain of ammines with a chloride on the end. The dissociation rule predicted that there should be four ions produced in the precipitation experiment: all three chlorides will dissociate because they are each connected to an ammine group, not the cobalt atom. Three chlorides plus the leftover part of the molecule would result in four ions. That prediction was supported by measurement when four ions were confirmed after the hexammine complex underwent a precipitation experiment (Figure 7.4).

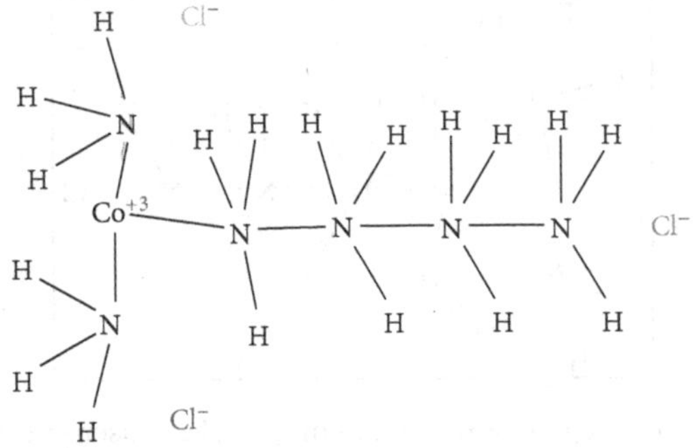

Figure 7.4 Post-experiment of hexammine complex. The precipitation experiment confirmed Jørgensen's structure: 4 ions result.

Alfred Werner, a younger chemist only 26 years old, responded to Jørgensen's work with a different possible structure for the hexammine complex: a geometrically symmetrical coordination compound.[11] Werner proposed two unique features for his model: (1) an octahedral arrangement around the cobalt center and (2) an additional type of valence called secondary valence where the metal center is directly attached to six ammine groups instead of three chains—making an inner sphere. He considered the inner sphere of the compound as "a discrete unit, a complex cation."[12] A cation is a positively charged ion, and a charged ion can be any charged molecular or atomic entity. These six attachments between the cobalt atom and an ammine group are examples of the additional valence that Werner suggests. This additional valence is referred to as secondary valence or coordination number. Werner, like Jørgensen, followed the valence rule in his model as well, according to which *bonding is constrained by an element's valence*. Because cobalt is trivalent, three chlorides are needed to fulfill the valence rule.[13] However, in Werner's model, cobalt's trivalence is structured where the three chlorides are attached to the inner sphere, but not attached in the same way as the ammine groups; he described these chlorides as being "at large." Following the dissociation rule, Werner's proposed structure predicted four ions: chlorides would only dissociate if they were not directly connected to the cobalt atom (Figure 7.5). The experimental work supported this structure as well because all three chlorides can be precipitated off, also resulting in a total of four ions (Figure 7.6).

Note that at this time in the history of chemistry, there are two mutually exclusive models for the structure of this particular hexammine complex—and thus at

[11] See Werner (1893: 267–330) translated by Kauffman (1968: 48).
[12] See (Kauffman 1965: 188).
[13] This primary type of valence is later understood to be an ionic bond where the positive part, the cobalt ion, is attracting the negative parts, the chlorides. Ammines are neutral in charge, and their attachment to the cobalt is not like the primary valence.

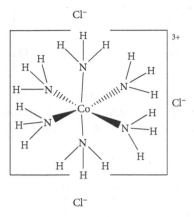

Figure 7.5 Pre-experiment of hexammine complex: Werner's proposed structure.

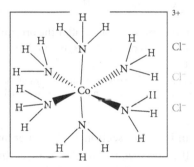

Figure 7.6 Post-experiment of hexammine complex. The precipitation experiment confirmed Werner's structure: 4 ions result.

least one model must be mistaken. But note also, that both models at this stage are making successful predictions—entailing results that are subsequently confirmed by experiment. And it's this phenomenon—the phenomenon of incorrect models making correct predictions—that seems to provide some prima facie justification for skepticism about a realist interpretation of current science. "Why," so the thinking goes, "should we be confident that the successful predictions of current science are a marker of truth, if we have all these examples from history of the predictive success of failed models?" In fact, one might extend this worry about the shortcomings of current science to include the potential superiority of "unconceived alternatives" to current theories.[14] We'll look at specific arguments

---

[14] See Stanford (2006).

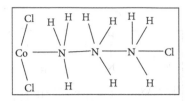

Figure 7.7 Pre-experiment of triammine complex: Jørgensen's proposed structure.
Note: Here Jørgensen's work involved cobalt(III) triammine chloride [Co(NH$_3$)$_3$Cl$_3$] complexes.

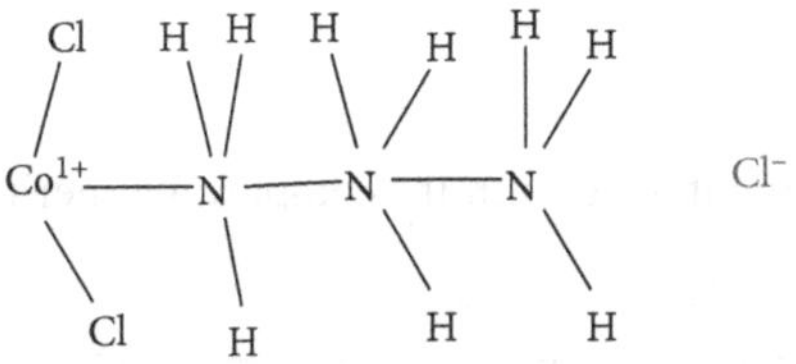

Figure 7.8 Prediction of triammine complex after the precipitation experiment.
Jørgensen's proposed structure predicted 2 ions, but no ions resulted from the
precipitation experiment.
Note: Jørgensen still thought of the cobalt as trivalent, but in terms of charge balance, after one
chloride dissociates, that leaves a positive charge on the leftover part of the complex.

against scientific realism in a later section of the paper and address these lines
of thinking more directly. However, it's worth raising at this juncture in order to
remind ourselves what's at stake, philosophically, by examining the 19th century
debate over the structure of cobalt complexes.

During this same time through 1899, Jørgensen and Werner worked
separately in synthesizing and characterizing other metal ammines. Like
experiments with hexammines, work with penta- and tetra-ammines yielded
experimental data that supported both Jørgensen's chain theory and Werner's
coordination theory. However, it was the triammines that forced the triumph
of one theory over the other.[15] The metal triammine has a cobalt atom, three
ammine groups, and three chlorides. Jørgensen predicted a chain structure
for the triammine complex with a chain of three ammines: one chloride on
the end of the ammine group, and two chlorides directly attached to the cobalt atom (Figure 7.7). Again following the dissociation rule, his structure
predicted two ions after precipitation: the chloride off the ammine chain
and then the remaining complex (Figure 7.8). Experimental work did not

---

[15] See Kauffman (1965).

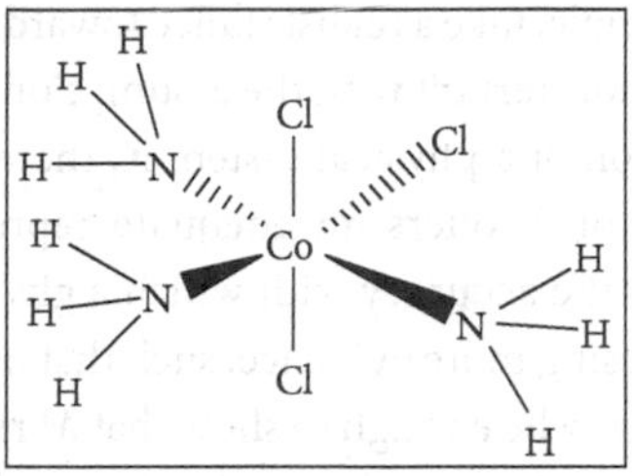

Figure 7.9 Pre-experiment and post-experiment of triammine complex

support Jørgensen's chain structure. It turns out that zero ions precipitated during the experiment.

However, Werner's proposed structure for the triammine complex successfully predicted the number of ions for the precipitation experiment. Keeping with the same two features of the hexammine model, Werner implemented an octahedral, symmetric shape and secondary valence in his triammine model. The cobalt atom is attached to three separate ammine groups and three separate chlorides. The primary valence of cobalt is satisfied with the cobalt atom being connected to three chlorides (Figure 7.9). Because the chlorides are directly attached to the cobalt, they should not separate during precipitation. Therefore, no ions are predicted, and precipitation experiments confirmed this prediction: no ions were measured.

While we don't have space for an in-depth discussion of the epistemic status of models with respect to the physical systems being represented, a few explanatory comments are in order. Margaret Morrison notes the "problem of inconsistent models" arises when multiple inconsistent models provide useful information about a physical system, and this problem undermines the degree to which one should give a realist interpretation of a model—particularly if one lacks the ability to tweak models in ways that would allow for empirical confirmation/ disconfirmation. Says Morrison, "In these contexts we usually have no way to determine which of the many contradictory models is the more faithful representation, especially if each is able to generate accurate predictions for a certain class of phenomena."[16] This is exactly the situation during the early experimentation and modeling of cobalt complexes, when both Jørgensen's chain theory and Werner's coordination complex model made predictions that were supported by precipitation experiments.

However, note the conditions that Stathis Psillos presents concerning when a model M can be interpreted as a realist depiction of a physical system X:

[16] See Morrison (2011: 343).

I think one can, in principle, take a realist stance towards particular models. For although scientists do not start off with the assumption that a particular model gives a literal description of a physical system X, there may be circumstances in which a model M of X offers an adequate representation of X. These circumstances relate to the accuracy with which a given model represents the target system . . . . Amassing more evidence, such that novel correct predictions for X derived from M, may be enough to show that M represents X correctly. In sum, taking a realist attitude towards a particular model is a matter of having evidence warranting the belief that this model gives us an accurate representation of an otherwise unknown physical system in all, or in most, causally relevant respects.[17]

The conditions specified by Psillos for supporting a realist interpretation of a model M were met by Werner's model, specifically with respect to Werner correctly predicting ion precipitation in the triammine complex experiments.

Our survey of work on cobalt ammine complexes in 19th century chemistry reviewed a series of experiments that led to the replacement of the Blomstrand-Jørgensen chain theory by Werner's coordination theory. While both theories were supported by precipitation experiments involving hexa-, penta-, and tetraammine complexes, only one theory was supported after the precipitation experiment done on the triammine complex. Historians of chemistry point out that Jørgensen resisted coordination theory even after the triammine complex experiments. According to Kauffman, Jørgensen questioned Werner's experimental methods viewing coordination theory "as an ad hoc explanation insufficiently supported by experimental evidence."[18] Kauffman further states, "Jørgensen always preferred facts to bold hypotheses, and his controversy with Werner clearly reflects his attitude."[19] While Jørgensen ultimately admitted defeat when Werner was able to prepare two isomers of a different complex that could not be explained by chain theory,[20,21] Werner's coordination theory was not accepted widely until after Thomson's discovery of the electron and a more robust electronic theory of valency was able to further explain the different types of valency proposed by Werner.[22]

---

[17] See Psillos (1999: 144).

[18] See Kauffman (1965: 186).

[19] See Kauffman (1965: 185). There are larger issues at stake here—beyond the scope of this paper to discuss—concerning the practice of scientific research and the nature of theoretical justification. For example, Day and Selbin (1962: 264) suggest "Werner had no theoretical justification for his two types of valency." Werner's first paper (1893) proposing his theory outlines his justifications.

[20] See Meissler and Tarr (2011: 323).

[21] The widely-accepted idea that Jørgensen admitted defeat to Werner is debatable. Jørgensen never formally stated defeat or acceptance of coordination theory, rather, it seems he ignored the controversy after 1899. S.P.L. Sørensen reported in his eulogy that Jørgensen admitted defeat in a conversation in 1907. See Kragh (1997).

[22] See Day and Selbin (1962: 264).

7.5 Implications for Scientific Realism

In this section, we apply lessons from the scientific debate on the structure of cobalt complexes to the philosophical debate about whether some version of scientific realism is justified. Scientific realism reflects a certain cluster of intuitions and beliefs about the nature and purpose of science. At the heart of the realist sensibility is the notion that the success of scientific research and theory development consists of a theory explaining the data and making accurate novel predictions. Realists view the fit between data and theory as justifying an inference to the truth of theoretical statements. As such, science is normed and constrained by the mind-independent reality it seeks to map.

Although versions of scientific realism abound, we adopt as a starting point Stathis Psillos's elegant and comprehensive description of scientific realism as a conjunction of three stances. According to Psillos, the realist adopts a metaphysical stance that the world has a definite, mind-independent, natural kind-like structure, a semantic stance that theories are truth-governed descriptions in which putative referring terms do, in fact, refer, even to unobservable entities, and an epistemological stance according to which mature and predictively successful theories are held to be approximately true.[23] Each stance has its philosophical detractors.

Post-structuralists of various sorts will deny that the world has any accessible mind-independent structure. Non-realist philosophers of science will deny that successful science requires accepting that theoretical terms in scientific theories are successful referring terms, and will further deny that predictive success is a mark of, or best explained by, truth.

7.5.1 The Problem of Unconceived Alternatives

One of the most prominent arguments against the claim that predictive success is a marker of truth comes from P. Kyle Stanford's argument based on the possibility of unconceived alternatives. According to Stanford,

> [T]he most troubling historical pattern for scientific realism is the repeated and demonstrable failure of scientists and scientific communities throughout the historical record to *even conceive* of scientifically serious alternatives to the theories they embraced on the strength of a given body of evidence, that were

---

[23] See Psillos (1999: xix). More specifically our version of realism is best described as a type of *divide et impera* realism; for the purpose of this paper we remain open to various options concerning what, exactly, constitutes the truth-conducive constituents of scientific theories.

nonetheless also well-confirmed by that same body of evidence. This "Problem of Unconceived Alternatives" arises because the inferential engine of much fundamental theoretical science is essentially eliminative in character, proposing and testing candidate hypotheses and then selecting from among them the one best supported by the evidence as that in which we should invest our credence . . . But the reliability of such inferences requires that particular epistemic conditions be satisfied. One such condition is that we must have all of the likely or plausible alternative possibilities in view before proceeding to embrace the winner of such an eliminative competition as the truth of the matter.[24]

Stanford's idea can be explicated as follows. Suppose that for some body of evidence E, there are n number of theories being considered by the relevant scientific community to account for E. Suppose further that through a combination of testing and theoretical work, consensus emerges within the scientific community about which of the candidate theories can be eliminated and about which theory among the remaining candidates best explains E. The problem, says Stanford, is that we have good reason based on the history of science to think that the epistemic control condition for determining when eliminative inference is reliable—namely, the condition that *we must have all of the likely or plausible alternative possibilities in view before proceeding to embrace the winner*—is never satisfied. Stanford continues:

> [T]he historical record of scientific inquiry itself gives us compelling empirical grounds for doubting whether [the epistemic control condition] is typically satisfied when we formulate and test what we might call "foundational" scientific theories concerning the constitution of entities, the underlying causal mechanisms, and the dynamical principles at work in otherwise inaccessible domains of nature . . . . What the historical record reveals instead . . . is a robust pattern of theoretical succession in which such foundational theories are accepted on the strength of a given body of evidence, only to be ultimately superseded by alternatives that were also well-confirmed by that evidence, but nonetheless simply remained unconceived at the time of the earlier theory's acceptance.[25]

Stanford's Problem of Unconceived Alternatives has a direct bearing on our analysis of the Jørgensen/Werner debate about the structure of cobalt complexes. Recall that we're defending a selective scientific realist interpretation of the triumph of Werner's coordination theory that modeled molecular structures with

---

[24] See Stanford (2018: 213).
[25] Ibid.

symmetrical geometric shapes over Jørgensen's theory that modeled molecular structures in chain-like shapes. Jørgensen's theory, though ultimately rejected, did enjoy a certain amount of predictive success—a phenomenon that skeptics about whether predictively successful science is a marker of truth can point to as justification for such skepticism. However, we've argued that a better way to understand the predictive success of Jørgensen's failed theory is to recognize that the factors that lead to his theory making successful predictions (what we earlier called the valence rule and the dissociation rule) are also a part of Werner's coordination theory. As a result, on a selective realist interpretation, the transition from Jørgensen's chain theory to Werner's coordination theory is an example of the onward and progressive march of science toward truth. Jørgensen's theory, while mistaken in some respects, wasn't completely mistaken (hence the limited experimental success he did achieve). Now recall Stanford's description of eliminative inference: "proposing and testing candidate hypotheses and then selecting from among them the one best supported by the evidence as that in which we should invest our credence."[26] This type of eliminative inference is an accurate description of the process by which chemists came, eventually, to reject chain theory in favor of coordination theory—the two main theories on the table—as experimental evidence continued to mount in favor of Werner's theory.

But notice that if we take into account Stanford's Problem of Unconceived Alternatives, then the eliminative inference undertaken by chemists in the late 19th century is incomplete and necessarily constrained by the conceptual limits of their time. Nineteenth-century chemists working in this area were debating whether cobalt complexes were chain-like structures or symmetrical geometric shapes, and the differing proposed structures reflected theoretical views about valence and the nature of chemical bonds. But according to the Problem of Unconceived Alternatives, a lesson from the history of science is that we need to acknowledge the possibility that there exists some alternative theory distinct from both chain theory and coordination theory—an alternative such that it (a) gives a better explanatory framework for understanding the structure of cobalt complexes, and (b) is conceptually inaccessible to chemists both in the 19th century and the present day. Thus, even though in this case, selective scientific realists can tell a plausible story about truth-transfer from Jørgensen's theory to Werner's, the Problem of Unconceived Alternatives shows that the realist's story is incomplete and doesn't support a realist picture for current theories concerning the structure of cobalt complexes.

Stanford's significant objection to scientific realism has generated much discussion in the past decade and raises interesting questions in epistemology in general, as well as for the epistemology and metaphysics of science. While it's

[26] Ibid.

beyond the scope of this paper to address all the nuances of Stanford's project, we feel that there are strong defenses that the realist can present in response to Stanford. We further note that whether or not Stanford's Problem of Unconceived Alternatives is a successful objection to scientific realism will depend in large measure on more fundamental epistemological assumptions, including whether the logical possibility of an unconceived alternative does in fact constitute a defeater for evidence one has in hand.

A charitable interpretation of Stanford's philosophical objection to scientific realism has an empirical consequence that seems implausible, a consequence that is part of what we're calling the No Final Theory Objection to the Problem of Unconceived Alternatives. Suppose it's always true, as Stanford maintains, that the eliminative inference scientists perform at any given time is necessarily constrained by the inability to consider unconceived alternative theories which provide a better explanation for the data in question. This can be stated with the following:

(1) For any current theory T that purports to explain some evidence E, there is another theory T^* such that:
  (i) T^* is unconceivable at present.
  (ii) T and T^* are inconsistent—they cannot both be true. And,
  (iii) T^* is a better explanation E than T.

But notice a consequence of (1) is that there can be no "final theory," a term that can be used to denote different things. Some, following Nobel Laureate Steven Weinberg, use "final theory" to denote a final physical theory that provides a unified explanation for gravity, quantum mechanics, and the strong and weak nuclear forces.[27] However, suppose, as suggested by Weinberg himself, possessing a final theory that provides a unified explanation of fundamental *forces* might still leave problems about consciousness and mind/body interaction unsolved.[28] In that case, having a final theory wouldn't constitute having a "theory of everything," if "everything" included consciousness. For ease of reference, we'll use final theory to denote the type of unifying theory that Weinberg has in view—nothing of consequence in our objection turns on whether one has a wider scope in view for final theory. Thus, considering candidates for a final theory as a substitution instance for the schema presented in (1) it follows that:

[27] See Weinberg (1994).
[28] Weinberg interviewed by John Horgan, https://blogs.scientificamerican.com/cross-check/nobel-laureate-steven-weinberg-still-dreams-of-final-theory/.

(2) For any candidate for a final theory F, there will be a theory F* such that:
   (i) F* is unconceivable at present.
   (ii) F and F* are inconsistent—they cannot both be true. And,
   (iii) F* is a better explanation than F.

Thus, if a final theory is one that provides a complete and accurate unification of whatever fundamental forces there be, from (1) and (2) it follows that:

(3) There is no final theory.

However, why should we think (3) is true? Why should epistemological considerations based on the history of science tell us whether there is an accurate and unified way of explaining how fundamental forces interact? Instead, it seems that (1)–(3) serve as premises in a reductio against the Problem of Unconceived Alternatives, for the negation of (3) seems much more plausible—asserting (3) commits one to accepting a fundamental inconsistency between physical reality and the logical possibility of, ultimately, representing physical reality through comprehensive theory. If Stanford is right, not even an omniscient being could have a final theory—which indicates that something is amiss in the initial premises.

To be sure, the truth of a theory isn't the only available explanation for a theory's predictive success. Suppose a theory T generates a testable observation statement O and we do in fact observe that O. Observing O certainly doesn't entail that T is true. But would observing that O provide an epistemologically respectable reason for thinking that T is true, or approximately true? A defeasible reason, but a reason nonetheless?

Timothy D. Lyons doesn't think so. In his 2017 "Epistemic selectivity, historical threats, and the non-epistemic tenets of scientific realism," Lyons presents two arguments against what he terms the realist meta-hypothesis:

Realist Meta-Hypothesis: Those constituents that are genuinely deployed in the derivation of successful novel predictions are approximately true.[29]

Lyons's first argument concludes that the meta-hypothesis is false; we've dubbed it the False Basis Argument, because it employs the observation that successful predictions are sometimes derived from constituents of theories that are not true. In those cases, successful predictions are derived from theoretical bases that turn out to be false, and the successful predictions made by the Blomstrand-Jørgensen

---

[29] See Lyons (2017).

chain theory are examples of this.[30] Here's a slightly modified version of Lyons's argument:

7.5.2 The False Basis Argument

1. If the realist meta-hypothesis were true, then all the constituents genuinely deployed in the derivation of successful novel predictions are approximately true.
2. However, we do find constituents genuinely deployed toward success that cannot be approximately true.
3. Therefore, the realist meta-hypothesis is false.

In the False Basis Argument, much is going to turn on what, precisely, is meant by "genuine deployment." Notice that we can distinguish between those constituents one *takes to be relevant* in generating an inference or a novel prediction and those constituents that *are in fact relevant* in generating an inference or a novel prediction.

Suppose S believes that the conjunction of p, q, and r entails some observation O. Suppose also that, unknown to S, it is in fact the conjunction p and q that entails O, and r is an explanatory free-rider; r is in fact predictively idle. It's a good thing, too, because it turns out that r is false and later comes to be rejected. Now, in S's mind, she is genuinely deploying each constituent in the conjunction when deriving O. Deploying r does some psychological work for S, and factors into the story S tells as to why one should expect O, given p, q, and r. So there is a psychological sense of "genuine deployment" according to which S is genuinely deploying r.

But there's also a more causally robust sense according to which r isn't being genuinely deployed in deriving O, for r doesn't genuinely factor into the actual causal explanation for O. And as far as the realist meta-hypothesis is concerned, from a realist's perspective, "genuine deployment" should be understood in the causally robust sense of deployment.

---

[30] One might think that the phenomenon of false theories making successful predictions parallels what's going on in Gettier cases in epistemology. In Gettier cases, we have examples of justified true beliefs, but such that the truth of the belief is a "happy accident." For example, the successful predictions made by Jørgensen might be Gettier-like in that regard because it's merely a "happy accident" that the false theory correctly predicted the number of ions that were dissociated in precipitation experiments. However, the selective scientific realist wants to say that that is not so—rather, what led to the predictive successes of Jørgensen was the fact that he followed accurate rules that survived into the successor theory. Thus, while there's a superficial similarity to Gettier cases (in the appearance of truth being a function of "mere coincidence"), the similarity disappears on closer inspection.

What we mean by "causally robust" is similar to Peter Vickers's distinction between "working posits" and "idle posits" that are used within a theory to derive a novel prediction.[31] Vickers's clear and comprehensive paper looks to flesh out the general *divide et impera* position of SSR while taking seriously anti-realist criticisms.[32] In analyzing historical cases, the selective scientific realist attempts to determine whether the predictive success of failed theories is a function of portions of the failed theory which both approximated truth and survived into successor theories. Vickers notes that when analyzing the role of theory in making predictions, we can identify Derivation External Posits (DEPs) as those posits that might "play some role in guiding the thoughts of scientists but do not deserve realist commitment because they cannot be considered part of the derivation of the prediction in question."[33] Vickers contrasts DEPs with Derivation Internal Posits (DIPs), where DIPs are the posits the scientist uses to derive the prediction. However, DIPs can be further subdivided into working posits and idle posits.[34] Idle posits in Vickers's sense are what we mean by *r* in our earlier example, where *r* is an explanatory free-rider.

With these distinctions in hand, let's review the Blomstrand-Jørgensen/ Werner debate about the structure of cobalt complexes. In order to do so, it is helpful to point out that Werner's theory is still accepted today within the larger theory of chemical bonding. Though Werner's notion of secondary valence, the inner sphere around the cobalt, was required for the derivation of his octahedral model, there was not a clear understanding of different types of bonding. While both Jørgensen and Werner observed the dissociation rule, *that is, the separation of chloride atoms from the metal ammine, will not occur if chloride is directly bonded to metal*, they both recognized that this did not fit with observations about all metal salts. Many, but not all, metal salts dissociate in water. Chemists during Jørgensen and Werner's time would probably consider the metal salt, sodium chloride (NaCl), as a molecule with chloride directly bonded to the sodium metal atom. Observations for these types of metal salts showed that the metal and chloride would readily dissociate. Metal ammines, like the cobalt ammine complexes Werner and Jørgensen studied, have cobalt, ammines, and chlorides, and follow the dissociation rule. While this dissociation rule for metal ammines is still part of the larger theory of chemical bonding today, as Werner points out, neither he nor Jørgensen had an explanation as to why cobalt ammine complexes behaved differently than the other type of metal salts.[35] Modern theory of

---

[31] See Vickers (2013).
[32] The classic presentation of *divide et impera* is Psillos (1999:108–114); for an alternative view, see Lyons (2006).
[33] See Vickers (2013).
[34] Ibid.
[35] See Werner (1893: 267–330) translated by Kauffman (1968: 16–17).

chemical bonding explains the difference.[36] Metal salts that would dissociate in water are classified as ionic, where opposite charges (e.g. the positive sodium and negative chloride) attract. Another type of bonding is classified as covalent, where the electrons in the bond are shared (either equally or unequally), usually seen in bonding between two nonmetals. Transition metal complexes, like the cobalt ammine complexes, have both types of bonding. Werner's secondary valence, the inner sphere around the cobalt, are covalent bonds whereas the primary valence (e.g. the chlorides in the outer sphere) are ionic bonds. Therefore, the chlorides that will dissociate are the same type of bonds seen in other metal salts: ionic. Ligand Field Theory further expands this theory,[37] and one can say that Werner's coordination theory is approximately true, exhibiting novel predictive success, and has been preserved in present-day chemistry.

Jørgensen was deploying the valence rule, dissociation rule, and the chain rule in order to, ultimately, make predictions about the number of ions that would separate during precipitation experiments. In the psychological sense of deployment, he was genuinely deploying all three rules. But what turns out to be the case is that the chain rule was an explanatory free-rider—it didn't factor into the actual causal story concerning the number of ions that separated during the precipitation experiments. Thus, this particular historical example doesn't serve as an example referenced in premise (2) of the False Basis Argument, for the relevant sense of genuine deployment doesn't apply. Moreover, as surveyed in the previous paragraph, the successful working posits within Blomstrand-Jørgensen chain theory survived into Werner's model and through to the present day.

We have only looked at one case where it's plausible to interpret the historical sequence of events in a way that is friendly to scientific realism. In order to seriously undermine (2) of the False Basis Argument, more work would need to be done in order to look at the historical cases cited by anti-realists in support of the pessimistic induction.[38] However, we think that the transition from Blomstrand-Jørgensen chain theory to Werner's geometrical structures and the coordination theory that support it can be seen as the ever-increasing glory of science at work. Blomstrand and Jørgensen were wrong, importantly wrong, about some things but not everything. And Jørgensen did deploy a false constituent in generating predictions which were then observed in experiments. However, when faced with cases where false theoretical constituents lead to successful predictions, it seems like one is faced with an interpretive choice: either there's a truth-conducive explanation lurking nearby that we don't have clearly in view, or there isn't, in which case it turns out that the successful prediction happens to

---

[36] See Meissler and Tarr (2011: chapters 3, 9, 10).

[37] Ligand Field Theory is based upon crystal field theory and molecular orbital theory. Griffith and Orgel (1957), as cited in Meissler and Tarr (2011), give a complete description of their theory.

[38] For a recent list of such historical cases, see Lyons (2017: 3215).

be a mere coincidence.[39] And if one isn't inclined to attribute theoretical success to mere coincidence, it seems perfectly reasonable to look for the more robust, realistic, truth-conducive explanation lurking nearby in order to discover those causally robust features that are actually being deployed in generating the novel prediction. This is what we have done in the debate over the structure of cobalt complexes, and it is a strategy that holds much promise in adjudicating other historical examples of theory change in science and the alleged problem for scientific realists of false constituents generating novel predictions.

Lyons's second argument concludes that the realist meta-hypothesis lacks epistemic justification—it can't be justifiably believed. Here follows a slightly modified version of the argument.

7.5.3 The Weak Evidence Argument

4. We are justified in believing the realist meta-hypothesis (given the evidence) only if we do not have greater evidence for those that oppose it.
5. However, we do have greater evidence for those that oppose it.
6. Therefore, given the evidence, it is not the case that we are justified in believing the realist's meta-hypothesis.

(4) is a plausible epistemological principle which ties epistemic justification to evidence; while a deeper analysis would want to flesh out notions of evidence, we'll grant (4) for the sake of argument. What about (5)?

Lyons considers the following alternatives to the realist meta-hypothesis, dubbed by Lyons as ContraSRs, and which we subdivide accordingly:

ContraSR1: Those constituents that are genuinely deployed in the derivation of successful novel predictions are statistically unlikely to be approximately true.

ContraSR2: Those constituents that are genuinely deployed in the derivation of successful novel predictions are not-even-approximately-true.

Compare again the skeptical meta-hypotheses with the realist meta-hypothesis:

Realist Meta-Hypothesis: Those constituents that are genuinely deployed in the derivation of successful novel predictions are approximately true.

---

[39] We prefer "mere coincidence" as opposed to "miracle," for presumably miracles, if any there be, wouldn't be random, accidental occurrences.

Why does Lyons think that we have more evidence for the ContraSRs than for the realist meta-hypothesis? Avoiding language of confirmation/disconfirmation, and opting instead for a hypothesis having "correlatively precise positive instances," or "correlatively precise negative instances," it seems like Lyons supports (5) with something like the following No Instances Argument.

7.5.4 The No Instances Argument

7. A predictively successful theory that is not approximately true would be a correlatively precise positive instance of the ContraSRs.
8. We have a list of predictively successful theories that are not approximately true.
9. Thus, we have correlatively precise positive instances of the ContraSRs.
10. We have no correlatively precise negative instances of the ContraSRs.
11. Moreover, we have no correlatively precise positive instances of the realist meta-hypothesis.
5. Thus, we have greater evidence for the ContraSRs.[40]

The controversial premises in the No Instances Argument are (8) and (11). According to (8), we have a list of predictively successful theories that are not approximately true. That is, we have a list of theories where false constituents were genuinely deployed in deriving novel predictions. However, what our preceding discussion about genuine deployment in the Blomstrand-Jørgensen/ Werner case has shown, is that there's a difference between a psychological sense of genuine deployment and a causally robust sense of genuine deployment that involves the working posits within a theory. Recall that in cases where it appears that false constituents were genuinely deployed to generate predictive success, one is faced with an interpretive choice between explaining the predictive success in terms of approximate truth or mere coincidence. Either the predictive success is accounted for because there's a truth-conducive explanation lurking nearby, or there isn't (in which case the predictive success is a mere coincidence). Certainly more investigative work would need to be done, along the lines of what we've done with the story of cobalt complexes. However, if a long string of mere coincidences is itself unlikely, than we have reason to think that the evidence for (8) is not as strong as it's taken by Lyons to be.[41]

---

[40] This is just a slightly differently worded version of (5) listed earlier.

[41] Thanks to Peter Vickers for reminding us that realists can still acknowledge the existence of minor coincidences that occur on occasion. Moreover, acknowledging minor coincidences does not undermine the overall force of the claim that empirical success is a marker of theoretical truth.

What about (11)? In setting up the Weak Evidence Argument, Lyons states the following:

> Moreover, without granting victory in advance to the realist's meta-hypothesis, we have no data that stand as correlatively precise negative instances of a ContraSR. By contrast, the realist's meta-hypothesis has no correlatively precise positive instances without already establishing it . . . . I emphasize that these restricted tasks of, first, tallying such positive/negative instances for the ContraSRs and the realist meta-hypothesis and, second, recognizing the evidential superiority of the former over the latter involve no inference to, acceptance of, let alone belief in, any such empirically ampliative meta-hypothesis about successful theories.[42]

He further states:

> First, it is admittedly logical considerations that reveal that, on the one hand, instances of the correlates of the ContraSRs can be secured without presupposing the truth of the ContraSRs, and, on the other, that such a feature is not shared by the realist meta-hypothesis.[43]

Lyons's worry here seems to be that there is no non-circular way of assessing putative positive correlates of the realist hypothesis which wouldn't already presuppose the truth of the realist hypothesis. As such, those cases can't really count as evidence in favor of the hypothesis.

But can the realist hypothesis really be defeated so easily by a priori considerations in epistemology? Here the realist meta-hypothesis is in a position similar to analogous positions about the reliability of one's cognitive faculties—sense perception, say. Consider the following realist hypothesis about sense perception:

> Sense Perception Meta-Hypothesis: In the typical case, sense perception is generally reliable.

What possible evidence could one provide in support of the Sense Perception Meta-Hypothesis that doesn't assume the reliability of sense perception? Not much. But here we'll just point out that a number of philosophers, recognizing the challenge of finding non-circular justification for our belief-forming practices, have concluded that if there are any justified beliefs at all, then certain types of circularity are unavoidable and do not undermine justification. And

---

[42] See Lyons (2017).
[43] Ibid.

because it is extremely difficult to deny that there are *any* justified beliefs, scientific or otherwise, the inevitable conclusion is that some types of circularity are unavoidable but not such that they undermine justification.

For example, William Alston argues that our inability to give a non-circular justification for the reliability of sense perception does not undermine the justified deliverances of sense perception.[44] Michael Bergmann argues if one accepts the foundationalist principle that some beliefs are justified non-inferentially, one is thereby committed to approving of epistemically circular arguments; Bergmann goes on to argue that because foundationalism is vastly superior to competing accounts of the structure of knowledge, epistemic circularity can't be all bad.[45] Psillos argues that "rule-circularity"—that is, arguing for some conclusion p via an inference rule R when p entails that R is reliable—is not an epistemologically vicious type of circularity.[46] Thus, if there are benign instances of circularity, like Alston, Bergman, and Psillos agree is the case, then it's not the case as Lyons asserts in (11), that there can be no correlatively precise positive instances of the realist meta-hypothesis. True, realism will require interpreting evidentially significant cases in ways that adopt some version of circularity, but perhaps in a way that is not vicious circularity, particularly if the alternative to realist interpretations of novel predictions requires appealing to a string of mere coincidences.

In essence, what we are claiming is that there is lots of evidence *for* the realist meta-hypothesis and lots of evidence *against* the ContraSRs if one understands genuine deployment in a causally robust sense of working posits that lends itself naturally to a realist interpretation of scientific practice, including practice that involves theory change over time. Granted, there is a kind of circularity here. But suppose, as Psillos, Alston, Bergmann, and others have argued, that some instances of circularity are benign and are simply a feature of the epistemological landscape, a landscape that includes the epistemological territories of scientific practice and philosophy. In that case, more work would need to be done to establish that the type of circularity we are acknowledging is vicious, not benign.[47]

Moreover, interpreting cases of genuine deployment in ways that favor antirealism may also involve an appeal to circularity of the sort that requires some question-begging assumptions about base-rates and the probability space of

---

[44] See Alston (1986).

[45] See Bergmann (2004).

[46] See Psillos (1999: 83–85).

[47] Some of the issues relevant to this section but beyond the scope of our paper to explore in any detail involve internalism vs. externalism in epistemology. As Psillos (1999: 85) notes, those with externalist intuitions will not be bothered by a type of rule-circularity that uses an inference rule R to justify R. Moreover, externalists will be open to criteria for genuine deployment that reflect working posits which actually carve nature's joints, regardless of whether one is able to demonstrate successful deployment according to internalist criteria.

theories from which the relevant base-rates are drawn. Peter Dicken (2013) looks at ways in which the realism/anti-realism debate is subject to exemplifying instances of the base-rate fallacy.[48] Dicken uses the common example of circumstances involving medical diagnosis in order to illustrate the base-rate fallacy and then proceeds to apply the statistical lessons learned to the realism/anti-realism debate. In order to make the analogy explicit, let us define the ALPHA condition as one where the facts on the ground make it very likely that the relevant indicator conditions are present. Moreover, let the BETA condition be one where the relevant indicator conditions are taken to indicate certain facts on the ground.[49] Here then, is the analogy:

Case 1: Testing for Disease (95% Accuracy)
ALPHA: If you have the disease, then testing positive for the disease is a strong
indicator that you have the disease.
BETA: If you (a random person drawn from a population) test positive for the
disease, then whether this is a good indicator of whether you have the disease depends on the relevant base-rate—that is, the percentage of people within a population that have the disease.

Dicken illustrates the base-rate fallacy by noting that if, say, we're looking at a population of 100,000 and a base-rate of 0.1%, then the BETA condition is not a strong indicator of the presence of disease. In that scenario, a positive test result on a random subject yields only a 2% chance that the subject has the disease. This is because even though the test has a 95% success rate, given the very low number of actual disease carriers relative to the large population, the 5% rate of misdiagnosis means that the number of uninfected people misdiagnosed as having the disease (i.e. 4995) is much, much higher than the number of infected people accurately diagnosed as having the disease (i.e. 95). So, a positive test result on a randomly drawn subject from this population has a 95/4995 (i.e. 2%) chance of being accurate.

Applying the lesson to the realism/anti-realism debate, we get the following two analogous conditions:

Case 2: Testing for the Truth of Theories
ALPHA: If a theory T is true, then the chance that T will generate accurate
novel predictions is high.
BETA: If T is successful in generating accurate novel predictions, then
whether this is a good indicator that T is true will depend on the relevant

<hr />

[48] See Dicken (2013).
[49] Using ALPHA and BETA to specify these conditions is our convention, not Dicken's.

base rate—that is, the percentage of theories in the relevant "population" being true. But how should one determine the base-rate in a non-circular way? According to Dicken, the prospects are not good:

> So the initial threat of circularity remains . . . the scientific realist can no longer justify his initial philosophical argument on the basis of the reliability of our first-order inferences to the best explanation, since this is in fact precisely what his intended argument attempts to establish . . . . To put the point even more succinctly, any evidence the scientific realist can offer for the truth of our current scientific theories will be swamped by the background probability of an arbitrary scientific theory being true; therefore in order for the no-miracles argument to make any positive justificatory contribution to scientific realism, one must first assume that the base-rate likelihood of a predictively successful theory being true is actually quite high; but to assume *that* is just to assume what the scientific realist was attempting to establish all along.[50]

However, Dicken goes on to observe that a similar type of base-rate challenge effects certain forms of anti-realism, the cogency of which requires that the base-rate be determinately low. In his treatment of Stanford's problem of unconceived alternatives, Dicken states:

> Yet while the threat of unconceived alternatives may well be more robust than the threat of the standard pessimistic meta-induction, and while it may indeed avoid the need for knowing the underlying base-rate of any arbitrary scientific theory being true it seems that for the inference to be compelling we must simply presuppose a different base-rate concerning the reliability of scientists: maybe our historically unconceived alternatives are all heavily biased towards researchers working in a very specific domain of inquiry; or maybe while the history of science furnishes us with a great number of instances of scientists failing to exhaust the relevant possibilities, this is due to the relatively large number of scientific practitioners, rather than the likelihood of any arbitrarily successful researcher failing to consider another alternative theoretical formulations.[51]

Are we at a stalemate with respect to worries about circularity more broadly, and base-rate fallacies in particular? We think not, at least so far as defending SSR through our particular historical example is concerned.

---

[50] See Dicken (2013, 565).
[51] Ibid., 569.

First, notice that one key determination in establishing the relevant base-rates is determining the relevant "population" from which a candidate theory is being assessed. There are many different ways to determine the relevant population base for which one is attempting to determine base-rates. Given that there do not seem to be clear and neutral criteria for determining base-rate thresholds in ways that clearly privilege realism or anti-realism, it is unlikely for there to be such clear and neutral criteria for determining the relevant population base.

Second, note that while the analogy between the medical diagnosis case and the scientific realism case is helpful for illustrating the relevance of base-rates to the realism/anti-realism debate, notice also that the two sorts of cases are not completely analogous. With respect to medical diagnosis, the presence or absence of disease is a binary, "on/off" affair: one either has the disease or one doesn't. But with respect to the truth or approximate truth of any given theory T, things are not cast in such binary terms—at least not in binary terms that the scientific realist is obligated to accept. This is because, on the selective scientific realist framework we are adopting, theories that are, strictly speaking, false when taken in their entirety, can have true working posits. This further complicates the task of determining the relevant population for determining base-rate threshold, as determining base-rate isn't as simple as determining which members of the population "have or do not have the disease."

Third, recall where we are in the dialectic with Lyons's arguments against the realist meta-hypothesis. The main focus of this section of our paper has been to determine whether Lyons's criticisms of realism undermine our interpretation of the transition from the Blomstrand-Jørgensen model of cobalt complexes to Werner's coordination theory. We have argued that this episode of theory change in 19th century chemistry can be plausibly interpreted using the tools of SSR, noting that the features that lead to predictive success of the failed theory—its working posits that were genuinely deployed in a causally robust sense—survived and were part of the successful theory that replaced it. Lyons has identified a re- alist meta-hypothesis that is required by realists—including the *divide et impera* realism we are advocating in this paper:

> Realist Meta-Hypothesis: Those constituents that are genuinely deployed in the derivation of successful novel predictions are approximately true.

We then examined two arguments by Lyons to the conclusion that the realist meta- hypothesis is false.

According to the False Basis Argument, some theoretical constituents genu- inely deployed in deriving successful novel predictions turn out to not be true— a consequence that is inconsistent with the realist meta-hypothesis. However, we argue that if we distinguish between psychological deployment and a more causally robust sense of genuine deployment, something akin to the actual working posits in a theory, then, minimally, it's plausible to read Jørgensen's success in generating novel predictions as consistent with, and not a counterexample to, the realist meta-hypothesis. While not conclusive refutation of the False Basis Argument, we suggest that this strategy is likely to yield similar fruit in applying it to other alleged counterexamples to the realist meta-hypothesis.

According to the Weak Evidence Argument, there is greater evidence for the meta-hypotheses that contradict the realist meta-hypothesis (the ContraSRs) than there is for the realist meta-hypothesis itself. However, if one restricts genuine deployment to map onto the causally robust working posits of a theory, then *pace* Lyons, both the evidence for the ContraSRs is undermined and evidence for the realist meta-hypothesis accumulates. Moreover, it is hard to avoid some type of circularity for both realists and anti-realists in determining whether particular instances of deployment count as evidence for the ContraSRs or for the realist meta-hypothesis.

There are interesting and important epistemological issues that require further attention here related to the epistemic authority of science as a guide for what one should think is true about the world. Lyons (2016) suggests that we'd be better off and more reflective of scientific practices if we looked at science as a "mode of inquiry" and not as being "fixated on belief." He suggests that realists give up the epistemic facet of scientific realism and be "axiological realists" where science, as a mode of inquiry is aimed at truth, but stop short of being epistemic realists who claim that we should look to science for epistemic justification.

We think that conceiving of science as a mode of inquiry is exactly right, however, even conceiving of science along these lines doesn't preclude the relevance of science as an epistemic guide for belief. If, say, science as a mode of inquiry is focused on explanation, and explanations can have epistemic good-making properties, it's not a stretch to also say that the reach of scientific explanations can serve as an epistemic guide to what we should think is true about the world— that is, as an epistemic guide to belief. Retaining the epistemic feature of realism explains why this is so. Well-confirmed and mature scientific theories give us a good reason to think that the world is being accurately, approximately described by those theories. As such, we can look to well-confirmed and mature science as a fallible, but reliable, source of epistemic justification.

7.6 Conclusion

Evolving views of the structure of cobalt complexes in 19th century chemistry provide an interesting case study through which we can test SSR. It turns out that in this case the factors which lead to predictive success of the failed model survived into the successor theory, providing a positive instance in which the predictive success of a failed theory can be accounted for along plausible, realist lines.

Acknowledgments

This work was substantially improved by comments received when this paper was presented at the Canadian Society for History and Philosophy of Science 2017 annual meeting, the 2017 *Quo Vadis* Selective Scientific Realism conference, the University of Notre Dame Centre for Philosophy of Religion, and the Purdue University philosophy department colloquium. Thanks in particular for detailed written comments by Tim Lyons and Peter Vickers, as well as by anonymous reviewers. Also, work on this project was supported by a grant from the Templeton Religion Trust.

References

Alston, William P., "Epistemic Circularity." *Philosophy and Phenomenological Research* 47 [1] (Sep., 1986), 1–30.

Bergmann, Michael, "Epistemic Circularity: Malignant and Benign." *Philosophy and Phenomenological Research* 69 [3] (Nov., 2004), 709–727.

Day, M. Clyde, Jr., and Joel Selbin, *Theoretical Inorganic Chemistry*. Reinhold Publishing Corporation (1962), 183.

Dicken, Peter, "Normativity, the Base-Rate Fallacy, and Some Problems for Retail Realism." *Studies in History and Philosophy of Science* 44 (2013), 563–570.

Douglas, Bodie E., and Darl H. McDaniel, *Concepts and Models of Inorganic Chemistry.* Blaisdell Publishing Company (1965), 338.

Farber, Eduard, *The Evolution of Chemistry*. New York: The Ronald Press Company (1952), 164–165.

Griffith, J.S., and L.E. Orgel. "Ligand-Field Theory." *Quarterly Reviews, Chemical Society* 11 [4] (1957): 381–393.

Jørgensen, Sophus Mads, "Beiträge zur Chemie der Kobaltammoniakverbindungen: VII I. Ueber die Luteokobaltsalze." *Journal für praktische Chemie* 35 [2] (1887): 417, cited in Kauffman.

Jørgensen, Sophus Mads, "Zur Constitution der Kobaltbasen." *Journal für praktische Chemie* 41 [2] (1890): 429, cited in Kauffman.

Kauffman, George B., *Sophus Mads Jorgensen (1837–1914) A Chapter in Coordination Chemistry History* in *Selected Readings in the History of Chemistry*. Compiled by Aaron J. Ihde and William F. Kieffer, editor, reprinted from the 1933–1963 volumes of *Journal of Chemical Education*, Easton, Pennsylvania: Division of Chemical Education of the American Chemical Society (1965).

Kragh, Helge, "S.M. Jørgensen and His Controversy with A. Werner: A Reconsideration." *The British Journal for the History of Science* 30 [2] (1997), 203–219.

Lyons, Timothy D. "Scientific Realism and the Strategema de Divide et Impera." *British Journal for the Philosophy of Science* 57 (2006), 537–560.

Lyons, Timothy D., "Structural Realism versus Deployment Realism: A Comparative Evaluation." *Studies in History and Philosophy of Science* 59 (2016), 95–105.

Lyons, Timothy D., "Epistemic Selectivity, Historical Threats, and the Non-Epistemic Tenets of Scientific Realism." *Synthese* 194 [9] (2017): 3203–3219.

Meissler, Gary L., and Donald A. Tarr, *Inorganic Chemistry*, 4th ed. Prentice Hall (2011), 322.

Moore, F. J., *A History of Chemistry*, 3rd ed. McGraw-Hill Book Company, Inc. (1939), 294–295.

Morrison, Margaret. "One Phenomenon, Many Models: Inconsistency and Complementarity," *Studies in History and Philosophy of Science* 42 (2011): 342–351.

Psillos, Stathis. *Scientific Realism: How Science Tracks Truth*. Oxford: Oxford University Press (1999).

Stanford, P. Kyle, *Exceeding Our Grasp: Science, History, and the Problem of Unconceived Alternatives*. Oxford (2006).

Stanford, P. Kyle, "Unconceived Alternatives and the Strategy of Historical Ostension," in Juha Saatsi (ed), *The Routledge Handbook of Scientific Realism*. Routledge (2018): 212–224.

Vickers Peter, "A Confrontation of Convergent Realism." *Philosophy of Science* 80 [2] (April 2013), 189–211.

Weinberg, Steven, *Dreams of a Final Theory*. Vintage Books (1994).

Weinberg, Steven, interviewed by John Horgan, https://blogs.scientificamerican.com/cross-check/nobel-laureate-steven-weinberg-still-dreams-of-final-theory/.

Werner, Alfred, Zeitschrift *für anorganische Chemie* 3 (1893): 267–330, translated in George B. Kauffman, *Classics in Coordination Chemistry Part I: The Selected Papers of Alfred Werner*. New York: Dover Publications, Inc. (1968), 17.

8

Realism, Physical Meaningfulness, and Molecular Spectroscopy

Teru Miyake[1] and George E. Smith[2]

[1]Department of Philosophy
Nanyang Technological University
TMiyake@ntu.edu.sg
[2]Department of Philosophy
Tufts University
George.Smith@tufts.edu

8.1 Introduction

The question of the reality of microphysical entities, such as atoms, molecules, and their constituents, has always been central to the scientific realism debate. In particular, the experiments of Jean Perrin on Brownian motion (Perrin 1910) have been the focus of work by realists,[1] who present them as having successfully shown that molecules exist. On the other hand, Bas van Fraassen (2009) has recently offered an anti-realist reading of Perrin's work, and indeed the entire tradition in which it is situated. According to van Fraassen, the aim of the nineteenth- and early twentieth-century tradition from Dalton through to Perrin was not, as the realists would have it, to establish the reality of atoms and molecules at all, but the development and enrichment of theory so as to make what he calls "empirical grounding" possible for the theoretical quantities of kinetic theory.

Empirical grounding of a theory, according to van Fraassen, consists of three main features (van Fraassen 2009, p. 11): first, any significant parameter of the theory must be measurable under certain conditions; second, measurements may be theory-relative, that is, the connections between the theoretical parameters and observable parameters may be theoretically posited; and third, there must

---

[1] See, for example, Wesley Salmon (1984), Deborah Mayo (1996), Peter Achinstein (2001), Stathis Psillos (2011a, 2011b, 2014), and Alan Chalmers (2009, 2011).

Teru Miyake and George E. Smith, *Realism, Physical Meaningfulness, and Molecular Spectroscopy* In: *Contemporary Scientific Realism*. Edited by: Timothy D. Lyons and Peter Vickers, Oxford University Press. © Oxford University Press 2021. DOI: 10.1093/oso/9780190946814.003.0008 be concordance in the values of such parameters determined by different means. There are some further constraints that van Fraassen takes from Glymour (1980) to avoid ways of trivially satisfying these requirements. The empirical grounding of kinetic theory, as van Fraassen envisions it, was a great achievement, but it did not require any commitment to the claim that atoms or molecules exist. Stable concordant measurement of theoretical parameters does not, by itself, show that the parameters being measured correspond to definite properties of entities that actually exist.

Whether, in fact, van Fraassen is correct about the aims of this tradition is a rather difficult interpretational issue, which we shall not attempt to answer here. Instead, in setting the stage for the rest of this paper, we shall simply point out that even if successful, Perrin's experiments did not, in fact, give us much access at all to the structure and properties of atoms and molecules. If we take Perrin's claims at face value, he established a new method for the determination of Avogadro's number, which he claimed was the most accurate method yet. In retrospect, however, this was just one of several different methods for the measurement of this quantity that were developed in the early twentieth century. The values measured by Perrin, moreover, turned out to be 10% or more higher than other values measured, for example, by Planck (1900), Rutherford and Geiger (1908), Millikan (1911, 1913), Boltwood and Rutherford (1911), and others, which are closer to our modern values. By 1913, Bohr and the Braggs were already adopting values for molecular magnitudes that were in agreement with significantly lower values for Avogadro's number than Perrin's. In particular, in Bohr's landmark 1913 paper, the sole direct empirical test of his model compared his theoretically calculated value of Rydberg's constant, 3.1×10^{15} Hz, with the measured value, 3.29×10^{15} (Bohr 1913, p. 9). Had Bohr adopted the announced value for the fundamental charge that Perrin had inferred from his value of Avogadro's number, 4.3×10^{-10} esu (Perrin 1912, 248), instead of the value he chose, 4.7×10^{-10}, his theoretically calculated value would have been 2.0×10^{15}, too far removed from the measured value to provide much in the way of a positive test of his model. So, exactly what Perrin achieved in his determination of molecular magnitudes is more open to question than is generally realized.

Realists would, of course, be quick to point out that Perrin's experiments did not just give us another measurement of Avogadro's number. For example, a further accomplishment of Perrin's experiments, as Deborah Mayo (1996) has pointed out, is that one of them (Chaudesaigues 1908) established the stochastic nature of Brownian motion, thus indicating that Brownian motion, whatever its cause, must arise from innumerable uncoordinated momentum exchanges with the medium surrounding Brownian particles, which certainly seems to indicate that the medium is composed of uncoordinated microscopic parts. Even so, these results are consistent with such momentum-exchanging microscopic parts having only a fleeting existence. Absent any evidence that such microscopic parts retain their integrity over a significant length of time, this result falls short of establishing the existence of atoms or molecules.

An undeniably large epistemic gap exists, moreover, between having good reason to think that the medium in which Brownian particles are suspended is composed of microscopic parts in vigorous stochastic motion, and having knowledge about specific details of those parts. How much access to molecules was, in fact, gained through Perrin's experiments? According to physicists and chemists, we now know that molecules have a great deal of structure, as we shall discuss later. How much of this structure could have been inferred from measurements of Avogadro's number? The sizes of molecules could be calculated from a determination of Avogadro's number, but such calculations were based on the assumption that molecules are spherical or at least have a distinct sphere of influence. As van Fraassen points out (2009, p. 8), dependence on such unrealistic assumptions throws considerable doubt onto whether such calculations are truly yielding information about molecules. Moreover, this shortcoming was expressly recognized at the time. James Jeans adopts values of the molecular magnitudes corresponding to those of Rutherford and Geiger in the 1916 edition of his *The Dynamical Theory of Gases* and then concludes, "If, however, the molecules are *assumed* as a first approximation to be elastic spheres, experiment leads to discordant results for the diameters of these spheres, shewing that the original assumption is unjustifiable" (Jeans 1916, p. 9). Simply put, accordingly, whatever grounds the various determinations of Avogadro's number and the evidence for fine-grained stochastic motion in the substrate supporting Brownian motion had provided for molecular-kinetic theory, it had provided no information on the sizes and shapes of molecules. In the 1921 edition of *Theoretical Chemistry from the Standpoint of Avogadro's Rule and Thermodynamics*, Walther Nernst was still making a claim he made in the first edition of 1893: "At present scarcely anything definite is known regarding either the nature of the forces which bind the atoms together in the molecule and which hinder them from flying apart in consequence of the heat of motion, or regarding their laws of action" (Nernst 1895, p. 237; 1923, p. 327).

Some information can nevertheless be obtained about the structure of molecules through kinetic theory. If equipartition of energy is assumed, then kinetic energy should be distributed evenly through all the degrees of freedom of a molecule. A spherical molecule has three degrees of freedom, while a molecule in the shape of a dumbbell can rotate along two axes, so it has two additional degrees of freedom. If the parts of the molecule are free to move relative to each other, there will be additional degrees of freedom. The ratio of the specific heat of a gas at constant pressure and the specific heat at constant volume, according to kinetic theory, is a function of the number of degrees of freedom of the molecular constituents of the gas. Measurements of the ratio of specific heats might, then, provide some insight into the shapes of molecules. Unfortunately, at the beginning of the twentieth century, measurements of the ratio of specific heats of various gases could not be reconciled with what was then known about molecules. In a lecture delivered to the Royal Institution in 1900, Lord Kelvin named the specific heat anomaly one of two "Nineteenth Century Clouds over the Dynamical Theory of Heat and Light." (Kelvin 1904, pp. 486–527).

How, then, did scientists first come to gain access to the structure and properties of atoms and molecules? Our view is that this was first made possible by the development, in the late 1920s and after, of theories and techniques that enabled theory-mediated access to the structure of atoms and molecules through spectroscopy. The analysis of spectra has a history that predates quantum mechanics by more than a century, but it was the development of quantum mechanics that first allowed physicists to make stable theory-mediated measurements of parameters that they took to represent the real properties and structures of molecules. In this paper, we thus aim to answer the following question:

(1) How did spectroscopists first come to have the ability to make stable theory-mediated measurements of parameters that are taken to represent the real properties and structure of molecules?

Anti-realists such as van Fraassen will grant that such concordant measurement has been achieved but will deny that spectroscopists have thereby gained access to the real structure and properties of molecules. For that reason, we shall ask the further question:

(2) What evidence is there that when spectroscopists measure such parameters they are gaining access to the real properties and structures of molecules— in other words, that the parameters are physically meaningful and not just artifacts of the particular way spectroscopists have chosen to represent whatever lies beyond our senses?

Let us illustrate what we mean by an *artifact* by way of an example. A familiar molecular parameter from kinetic theory that is an artifact of a particular way of representing the molecular realm is *mean-free-path*. Values for this parameter were derived from Maxwell's theory of viscosity in the latter part of the nineteenth century under two assumptions that together entail that molecules have a definite spherical "collision cross-section": (1) molecules are spherical, and (2) each molecule moves entirely without constraint until it collides with the "sphere of influence" of another molecule, a "spherical surface" which it cannot penetrate at all. Both of these are now regarded as fictions (Hecht 1990, pp. 129–174). Moreover, the values for the *mean-free-paths* obtained from viscosity when extended to other phenomena—especially, to deviations from the ideal gas law and refraction of light through gases—yielded values for the sizes of the spheres that were not only discordant with one another, but also in violation of Avogadro's rule (Nernst 1895, pp. 347–352; Meyer 1899, pp. 149–221, 299–348). Van Fraassen would call this a failure to ground the theory; we see it as a representation of the molecular realm that never had clear claim to being physically meaningful. As this example suggests, the answer to the second question is likely to be rather complicated, and so in this paper we do our best to give an initial attempt to answer it. In particular, we believe that one of the keys to answering this question consists in understanding a distinction that we will make between representations that are *physically meaningful* and those that are not.

Molecular spectroscopy is a little-known topic among philosophers of science, so we will take some care in explaining how it works. Section 8.2 of this paper will be devoted to providing enough background on molecular spectroscopy to be able to answer question (1)—that is, to explain how stable theory-mediated measurement of parameters that are taken to represent properties of molecules has been achieved. In order to keep things manageable, we will limit our discussion to the molecular spectroscopy of diatomic molecules and will focus on the early period of molecular spectroscopy, from the 1920s through around 1950. In Section 8.3, we will consider ways in which one might attempt to answer question (2). We will conclude that a proper answer to question (2) will require an understanding of physical meaningfulness, which we will explain through an examination of an example from celestial mechanics. A full answer to question (2) will go beyond the scope of this paper, for it would require an examination of the history of molecular spectroscopy in the period after 1950, but we will provide, in Section 8.4, a brief look at further work in molecular spectroscopy that we believe supports our contention that spectroscopists indeed are gaining access to physically meaningful parameters of molecules.

8.2 Molecular Spectroscopy and Stable Measurement of Molecular Constants

Molecular spectroscopy is a massive field. It grew rapidly from the late 1920s, when quantum mechanics was first successfully applied to the analysis of spectroscopic data. This early period culminated with the publication of Gerhard Herzberg's landmark *Molecular Spectra and Molecular Structure*, published in four volumes between 1939 and 1979. In this paper, we will mostly focus on research in the field up through 1950, the year of publication of the second edition of Volume 1, on the spectra of diatomic molecules (Herzberg 1950). As an indication of the amount of research that was being done at the time on diatomic molecules, we note that the second edition of Volume 1 contains a table of constants for diatomic molecules that goes on for 80 pages (pp. 501–581) and lists 1574 references in its bibliography. Molecular spectroscopy has been growing steadily ever since. Volume 4 of Herzberg, published in 1979, consists solely of tables of constants for diatomic molecules. It is 716 pages long and lists its references by molecule. The ordinary hydrogen molecule, $^1H_2$, has 169 references, while $^1H^2H$, a molecule consisting of one ordinary hydrogen atom and one deuterium atom, has 51 references; $^2H_2$, or molecular deuterium, has 60 references. There is now an online database of papers on diatomic molecules called DiRef (Bernath and McLeod 2001), which covers the period after Volume 4; it contains 30,000 references for the period 1974–2000.

The molecular constants are parameters that can be determined from spectroscopic data, and they are taken to represent certain properties of molecules. When a molecule undergoes a transition between two energy states, it absorbs or releases a photon, and the frequency of the photon is taken to correspond to the difference in energy levels between these two states. Thus, an examination of the spectrum of light absorbed or released by a gas can provide information about the energy states of the molecules in the gas. The interpretation of this spectrum requires some knowledge about the possible energy states of a diatomic molecule. A diatomic molecule at room temperature is taken to be a body that is not merely in translational motion, but also undergoing vibrational and rotational motion. The two nuclei in the molecule vibrate anharmonically in accord with an asymmetric potential function that depends in large part upon the distance between the nuclei. The shape of the potential function depends upon the arrangement of the positive charges of the nuclei and the negative charges of the electrons in the molecule. The rotation has the effect of centrifugally stretching out the distance between the nuclei, but the effect is only significant at high levels of rotation, so we shall omit discussion of it here. Energy can be emitted and absorbed by the molecule through transitions between rotational, vibrational, and electronic states. Energy is quantized, with the consequence that the spectral lines correspond to a transition from one energy state to another, whether it be a change in the rotational, vibrational, or electronic state of the molecule.

The energies of these states can be calculated for idealized systems by solving the Schrödinger equation for such systems. For example, the simplest idealization of a diatomic molecule is the rigid rotor, a system consisting of two point-masses separated by a fixed distance, rotating about an axis perpendicular to the line connecting the masses (Herzberg 1950, pp. 66–73). The Schrödinger equation for such a system has solutions for only certain energy values, customarily given in units of wavenumber (cm^{-1}) by

$$F(J) = BJ(J+1), \quad where\ B = h/(8\pi^2 cI).$$

Here, J is the rotational quantum number, and can take the integral values 0, 1, 2, ..., while h is Planck's constant, c is the speed of light, and I is the moment of inertia of the rigid rotor. Thus B, called the *rotation constant*, essentially gives the inverse of the moment of inertia of the system, normalized and converted to units of wavenumber. A further analysis of the selection rules for the quantum number J shows that the spectrum of a rigid rotor should consist of a series of evenly spaced lines, the first of which lies at wavenumber $2B$, with the distance between successive lines also being $2B$. Thus, the value of B, and hence that of the moment of inertia, of a rigid rotor, can be read directly off of its spectrum, provided that the spectrum has been interpreted correctly. Given the masses of the rigid rotor, the distance between the two point-masses can be determined from the moment of inertia.

A diatomic molecule is not, however, a rigid rotor—the distance between its nuclei is constantly changing. If a diatomic molecule were, instead, a rotating harmonic oscillator, it would produce a spectrum with a single peak that, at higher resolution, contains peaks spaced uniformly corresponding to different rotational transitions. In fact, actual diatomic molecules are anharmonic oscillators, and this gives rise to non-uniformities in the spectral lines. These non-uniformities provide ways of determining constants characterizing both the anharmonicity of the vibration and the effect of vibration on rotation, as well as other effects. The values of these constants depend upon the potential functions between the two nuclei, which in turn depend upon whether the electrons in the molecule are in the ground state or excited states. In order to keep things simple, we will only be considering the ground state—that is, a state from which spontaneous transitions never occur—but no small part of the evidence that the constants are characterizing a structure of the sort specified comes from comparing the values of the constants for different electronic states.

With all of this in mind, the vibrational (G) and rotational (F_v) energy levels for a diatomic molecule may be calculated theoretically, and are customarily given by the following expressions (Bernath 2016, p. 225), again in units of wavenumber (cm^{-1}):

$$G(v) = \omega_e(v+1/2) - \omega_e x_e(v+1/2)^2 + \omega_e y_e(v+1/2)^3 + \omega_e z_e(v+1/2)^4 + \ldots$$

$$F_v(J) = B_v J(J+1) - D_v(J(J+1))^2 + H(J(J+1))^3 + \ldots$$

The vibrational energy levels $G(v)$ are thus expressed as a series expansion in $v + 1/2$, where v is the vibrational quantum number, while the rotational energy levels $F_v(J)$ are expressed as a series expansion in $J(J + 1)$, where J is the rotational quantum number. The rotation constant B_v is the first coefficient in the expansion for the rotational energy levels, and it depends on the vibrational energy state as follows:

$$B_v = B_e - \alpha_e(v+1/2) + \gamma_e(v+1/2)^2 + \dots$$

The constant B_e, the first term in the expansion of B_v, is the value that the rotation constant *would have* if the nuclei were not vibrating at all, in the manner of a rigid rotor, with the distance between the nuclei being equal to the equilibrium distance r_e, that is, the distance between the nuclei at the bottom of the anharmonic potential. We will omit discussion of coefficients of higher-order terms here. Spectroscopists have developed techniques for calculating the values of these constants from spectra. In published tables of constants, the values given are typically for the equilibrium values B_e, α_e, r_e, ω_e, $\omega_e x_e$, $\omega_e y_e$. Note that these constants are usually given in units of wavenumber, except for r_e.

Table 8.1 shows the values for some of these constants for a few diatomic molecules. The three columns on the left show constants for different isotopic forms of the hydrogen molecule. The first column shows a molecule consisting of two ordinary hydrogen nuclei, the second column shows a diatomic molecule consisting of an ordinary hydrogen and a deuterium nucleus, and the third column shows a diatomic molecule consisting of two deuterium nuclei. The two columns on the right show constants for two isotopic variants of the hydrogen

Table 8.1 Values of molecular constants for several diatomic molecules.

Molecular constant	$^1H_2$	$^1H^2H$	$^2H_2$	$^1H^{35}Cl$	$^2H^{35}Cl$
B_e (cm^{-1})	60.809	45.655	30.429	10.5909	5.445
α_e (cm^{-1})	2.993	1.9928	1.0492	0.3019	0.1118
r_e (cm $\times 10^{-8}$)	0.7416	0.7413	0.7416	1.2764	1.274
ω_e (cm^{-1})	4395.2	3817.09	3118.4	2989.74	2145.163*
$\omega_e x_e$ (cm^{-1})	117.99	94.958	64.09	52.05	27.1825*
$\omega_e y_e$ (cm^{-1})	[0.29]	1.4569	1.254	0.2243*	0.08649*

Source: From Herzberg (1950), except * = from Huber and Herzberg (1979), based on results from 1953 and after.

chloride molecule. There are complications that we are omitting for the moment, but by 1950, spectroscopists had achieved stable theory-mediated measurement of parameters that are taken to represent the structure and properties of a large number of diatomic molecules. As already noted, Herzberg (1950) contains a table that stretches over 80 pages and lists constants for a huge number of diatomic molecules. These constants vary systematically from one molecule to another, in strict correspondence to differences in their spectra.

8.3 Gaining Access to Molecular Structure

As we have mentioned, however, even if we grant that stable theory-mediated measurement of the molecular constants has been achieved, we might yet ask: What reason is there to think that those constants are giving us access to definite physical details of actual molecules, and that they are not just artifacts of the way spectroscopists have chosen to represent molecules? The answer to this question is complicated, but we will start by describing several ways in which one might try to answer it.

First, note that spectroscopists have measured the molecular constants for a large number of different molecules. In particular, as indicated in Table 8.1, the constants for isotopic variants of various molecules such as the hydrogen molecule and the HCl molecule have been measured. This gives spectroscopists one way of checking whether the constants have been interpreted correctly. The isotopic variants effectively give them a way of varying the masses in the diatomic molecule while holding other things, such as the electromagnetic field of the molecule, constant. Different isotopic variants for the hydrogen molecule should give rise to virtually the same potential function, because only the masses of the nuclei change, while no changes are made to the electromagnetic properties of the nuclei. Sure enough, if we look at the values in Table 8.1, we see that r_e, the constant that is taken to represent the equilibrium distance between the nuclei, is very nearly the same for all the isotopes of the hydrogen molecule, given in the first three columns. What *does* change are the constants that are taken to represent the frequency of vibration—ω_e, $\omega_e x_e$, $\omega_e y_e$—which is to be expected, because the frequency of vibration of the anharmonic oscillator ought to vary if the masses change. Also, B_e, which can be interpreted as the inverse of the moment of inertia of the molecule at equilibrium, changes as well, as it should, since the masses of the nuclei differ for each of the isotopic variants of the hydrogen molecule. We note further that B_e for $^2H_2$ is very nearly half that for $^1H_2$, while the value for $^1H^2H$ is very nearly 3/4 the value for $^1H_2$, which is to be expected given the interpretation of B_e as the inverse of the moment of inertia. We see a similar pattern for the isotopic variants of HCl. Of course, these are just two molecules.

We claim that if we examined all the molecules in Herzberg's table of molecular constants, we would see a similar pattern for other molecules as well. It's worth pointing out that there is a lot of work on polyatomic molecules as well. In this case, spectroscopists must deal with further molecular constants, because now there are more degrees of freedom in the molecule. Success in the interpretation of spectra for these molecules in terms of rotation and vibration ought to lead to further evidence that the given interpretation of spectra for diatomic molecules is correct. In short, one way of confirming whether the molecular constants are indeed measuring what spectroscopists say they are measuring is by comparing naturally occurring variants in which virtually the only difference is the mass of one or more of the nuclei, and confirming that the spectra change in the way they are expected to.

Another reason to think that the molecular constants are giving us access to real physical details of molecules is that the values of the molecular constants agree with facts about molecules that are known chemically or thermodynamically. For example, an approximation to the anharmonic potential function can be calculated from the equilibrium distance r_e and the vibration constants ω_e, $\omega_e x_e$, $\omega_e y_e$. From this anharmonic potential, the dissociation energy, or the amount of energy required to take a diatomic molecule in its ground state and pull the two nuclei apart, can be calculated. The energy of dissociation can also be determined empirically through chemical means, and the agreement between the energy of dissociation measured spectroscopically and chemically is evidence that the molecular constants are indeed yielding information about the actual potential of the molecule. So, in short, another check on whether spectroscopists are indeed measuring what they claim to be measuring is to compare spectroscopic measurements with chemical or thermodynamic measurements.

Finally, spectroscopic measurements have a high degree of redundancy. As mentioned earlier, the internal energy states of diatomic molecules can be separated into three types: rotational, vibrational, and electronic (Figure 8.1). The pure rotational spectra can be interpreted in terms of a rigid rotor, from which the rotational constant B and internuclear distance r may be determined from the spectra. Vibrational-rotational spectra, which are typically in the near infrared, arise due to transitions between vibrational energy levels, without any change in the electronic energy levels, and exhibit fine structure that can be interpreted in terms of a vibrating anharmonic oscillator, from which rotational and vibrational constants can be determined. There is thus redundancy in the determination of the rotation constants—they can be determined from both the pure rotation spectra and from the vibration-rotation spectra. Electronic spectra, which are typically in the visible range, arise due to transitions between electronic energy levels and have much more complicated structure. These spectra represent a great number of transitions and may appear more like bands

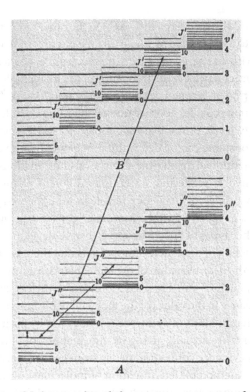

Figure 8.1 Rotational, vibrational, and electronic energy states of diatomic molecules. The three double arrows indicate a pure rotational transition (shortest double arrow, seen toward the lower left corner), a vibrational-rotational transition (medium-length double arrow), and a transition between electronic energy levels (longest double arrow).

Source: Herzberg (1971, p. 17).

than lines. High resolution is needed to resolve the structure. Because the structure in these bands corresponds to transitions between rotational and vibrational energy states within different electronic states, they contain information about the rotational and vibrational constants. The electronic spectra thus give spectroscopists another independent way of determining both the rotational and vibrational constants.

Further, Raman spectroscopy provides yet another independent way of calculating the rotational and vibrational constants. Raman spectroscopy is done by shining light of a particular wavelength, usually from a laser, at a sample and examining the scattered light. When a photon with energy $h\nu'$ is incident on a diatomic molecule, the molecule can go from a lower energy state E'' to a higher one E' by taking from the photon an amount of energy $\Delta E = E' - E''$, and then scattering a photon with an energy $hv' - \Delta E$. On the other hand, the molecule can go from a higher to a lower state by imparting the energy ΔE to the photon, scattering a photon with an energy $hv' + \Delta E$. The spectrum of the scattered light will accordingly consist of a peak at the frequency of the incident light, with lines on either side of the peak at frequencies given by $v' - (\Delta E/h)$ and $v' + (\Delta E/h)$. Thus, Raman spectroscopy provides another way of obtaining the differences between the various energy levels of the diatomic molecule, from which the rotational and vibrational constants can be determined. We note here that the spectra we have previously mentioned involve the absorption and emission of light, whereas Raman spectra involve scattering, an entirely different phenomenon. Raman spectra appear in a different frequency range, and the theoretical link between the observed lines in spectra and the molecular constants is different. In addition, there are other types of spectroscopy that were developed after the 1950s, but we omit discussion of them here. Our point is that there is a massive amount of redundancy in the determination of the molecular constants.

An anti-realist such as van Fraassen would nevertheless claim that redundancy, or concordance in the values of the molecular constants that are measured using various different kinds of spectroscopy, does not by itself show that these molecular constants are actually giving us information about the real structure of molecules. But evidence comes not merely from concordance in these values. For each different method of determination of molecular constants, the molecular structure is being accessed through a different causal pathway.

The different ways in which the molecular structure can be accessed appear as details in the molecular spectra that allow spectroscopists to decide whether they are, indeed, accessing the molecular structure that they claim to be accessing. Here is one example. The selection rules for rotational spectra imply that there should be no pure rotation spectrum for homonuclear molecules, but heteronuclear molecules should produce such a spectrum. This is because in order for there to be absorption or emission of a photon, there needs to be a changing dipole moment. Homonuclear molecules always have a dipole moment of zero, so they cannot absorb or emit photons through rotation. Spectroscopists have found that this is indeed the case—no pure rotational spectra can be found for molecules such as H_2, but such spectra can be found for molecules such as HCl. In Raman spectroscopy, however, there is a different causal pathway by which spectroscopists claim to be gaining access to the structure of molecules. The theory of Raman spectroscopy says that there ought to be rotation lines even for molecules without a dipole moment, and this is indeed what is found by spectroscopists—the Raman spectra of homonuclear molecules yields rotational lines.

Spectra are extremely complicated, and they have many more features that we have not mentioned. Spectra exhibit line splitting, shifts in the positions of the lines, broadening of the lines, variations in intensity of the lines, and so on. Without getting into specifics here, we claim that these details can often be used to check whether the molecular constants really are measuring what spectroscopists claim they are measuring, and, in fact, examination of those details can yield further information about either the molecules themselves or the causal pathways used to probe them.

With regard to the question of what reason there is to think that molecular constants are giving us access to definite physical details of molecules, and they are not just artifacts, we believe we can offer an impressive number of different ways of checking whether the interpretation assigned to spectra by spectroscopists is indeed correct. After having given all of these reasons, however, we nevertheless hesitate to answer this question affirmatively, because we think there is still a problem along a different dimension. This is in trying to answer the question of exactly what those molecular constants represent. We have so far glossed over some details in our description of the molecular constants. Take the rotation constant B, for example. If a molecule were a classical rigid rotor, the distance between the nuclei would not vary, and the rotation constant B could be interpreted straightforwardly as corresponding to the inverse of the moment of inertia. But when we go beyond the rigid rotor to a vibrating rotator with an anharmonic potential, the rotation constant is now dependent on the vibrational quantum number, and its interpretation becomes more complicated. Usually what is given in tables of constants is B_e, which is what the value of the rotation constant *would be* if the distance between the two nuclei at all times was r_e, the distance between the two nuclei at the bottom of the potential. But the molecule is never actually in a state with no vibration and hence never in a state in which the distance r_e persists for any significant length of time.[2] A way of thinking about what is happening here is that the distance between the nuclei of the molecule always fluctuates around the r_e value, and so the rotation constant, which is dependent on this distance, must be fluctuating as well. So there is some ground for saying that there is no feature of the actual molecule to which B_e corresponds, even according to the theory that enables the measurement to be made in the first place.

To take the point further, we said that the rotational, vibrational, and electronic energy states of a molecule can be separated. The separability is justified by the Born-Oppenheimer approximation and the Born adiabatic approximation, in which the assumption is made that the timescales involved in the motions of the nuclei are much slower than the timescales involved in the motions of electrons. The timescale at which electrons interact with one another and the

---

[2] This is the case even in the lowest vibrational energy state, for which the energy is not zero but a positive value, called the zero-point energy.

nuclei is of the order of 10^{-16} seconds. The period for vibrational motion is of the order of 10^{-14} seconds, while the period for rotational motion is of the order of 10^{-11} seconds. Now, consider what this means. Each time the molecule rotates, it has undergone of the order of a thousand vibrations, and since the rotation constant depends on the distance between the nuclei, this must mean that the rotation constant that is being measured can only be a time-averaged value— it does not correspond to any static or instantaneous state of the molecule. In other words, the molecular constants are a way of capturing a situation that is enormously more complicated, with time-averaging smoothing or glossing over fluctuations of much shorter periods.

Here, we claim that, nevertheless, it may be possible to argue that the molecular constants such as B_e are *physically meaningful*, even though they do not represent some actual state of a molecule. We believe that an analogy with celestial mechanics is helpful here. In celestial mechanics the orbits are characterized in terms of Keplerian parameters defining an ellipse and area swept out per unit of time. These parameters define a counterfactual situation—they describe the elliptical trajectory that *would* occur if there were only two bodies. To quote Newton, "if the sun were at rest and the remaining planets did not act on one another, their orbits *would* be elliptical, having the sun at a common focus, and they *would* describe areas proportional to the times" (Newton 1999, p. 817f, emphasis added). We contend that the Keplerian parameters, notwithstanding their counterfactual status, are, even from a modern perspective, physically meaningful.

First, an examination of the 250-year history of celestial mechanics between the times of Newton and Einstein shows that deviations between predictions made based on calculations from these orbital parameters, and actual observations of planetary motions, have been used to make inferences about the existence and location of other bodies that might be influencing the motions (Smith 2014). In other words, the Keplerian representation served as a research instrument to expose physical sources of deviations from this representation. The most famous example of this was the discovery of Neptune, but this one example grossly understates how much was discovered about the motions in our planetary system from this approach.

Second, at the time Newton began work on what became his *Principia*, he was aware of four alternatives to the Keplerian representation that fit the observations just as well as it did. Indeed, Newton was more familiar with both the planetary theories, and the tables of their motions produced by Thomas Streete (1661), Vincent Wing (1669), and Nicholaus Mercator (1676), than he was with any of Kepler's writings. These three employed three different alternatives to Kepler's rule that orbiting bodies sweep out equal areas in equal time, a rule which Newton likely had learned of from reading Mercator's systematic defense of his approach over Kepler's and Streete's (Mercator 1676, pp. 144–164). A few pages in the *Principia* after the counterfactual claim we just quoted, Newton went on to show that the then most prominent violation of Kepler's rule, namely the motion of our moon, results from the action of solar gravity alternately accelerating and decelerating the moon twice over the course of its monthly orbit, in the process markedly distorting that orbit in a way theretofore unrecognized (Newton 1999, pp. 840–848). In thus identifying a robust physical source for the deviation of the moon from the area rule, Newton gave evidence supporting the counterfactual claim that the moon *would* conform to the rule at least more closely than it does, and the orbit would be closer to a Keplerian ellipse, *were* it not for the perturbing effect of the sun.

This is but one of hundreds of examples from celestial mechanics in which the Keplerian representation of orbital motion, counterfactual though it is, revealed deviations that were then shown to have physically identifiable sources. The pre-Newton alternatives to Kepler's representation remind us that there can always be many other comparably accurate representations, but not all of them would yield deviations meeting the demand of having identifiable robust physical sources. We thus distinguish between *physically meaningful representations and parameterizations* and those that are not through the following criterion: *physically meaningful representations and parameterizations yield deviations that have physically identifiable sources.*

8.4 Realism, Physical Meaningfulness, and Molecular Spectroscopy

Are the molecular constants such as B_e physically meaningful in the same way that the Keplerian orbital elements are? That is, have deviations between predictions about spectra based on calculations from the molecular constants, and actual spectra, been informative? Have they led to the identification of further physical sources that had previously been unaccounted for? Fully answering these questions will require a detailed examination of developments in molecular spectroscopy in the period after 1950, especially microwave spectroscopy, which would exceed the scope of this paper, but we offer the following considerations.

Consider first what is giving rise to these questions. We have emphasized that the standard constants characterizing the rotation and vibration of diatomic molecules presuppose the Born-Oppenheimer and the Born adiabatic assumptions. As a consequence, these constants *represent*—in van Fraassen's sense (van Fraassen 2008)—diatomic molecules as rotating dumbbell-shaped bodies in which the two nuclei vibrate in an anharmonic fashion along a line between their point-mass centers, as if there were no interaction at all between the electrons and the two nuclei. The only respect in which the electrons are relevant is in the distinction between different electronic states, with different values of the molecular constants and different potential functions for each electronic state. (We remind readers that, for simplicity, we have restricted the discussion to the ground state, but the Herzberg tables of constants for diatomic molecules are not so restricted.) Ignoring electron-nuclei interaction allows a non-fluctuating potential function between the two nuclei to be defined for each electronic state. But, of course, there is electron-nuclei interaction, and hence this non-fluctuating potential function represents, at best, a time-averaged representation of a potential function that in fact is fluctuating at frequencies several orders of magnitude greater than those in which the molecule can be rotating and its nuclei vibrating with respect to one another. Put this way, our questions can be viewed as prompted by the question: How physically meaningful are the time-averaged potential functions?

Further complicating the situation are the spectral data themselves, from which the values of the time-averaged molecular constants are derived. The spectral lines, of course, are themselves aggregates of multiple photon-molecule interactions over time. They are, moreover, aggregates as well over the emission and absorption of photons by large numbers of molecules. The spectral data are thus, in fact, *data models*, as Patrick Suppes (1962) would put it, and so they, too, as employed in the determination of the values of the molecular constants, are *representations* in van Fraassen's sense. In saying this, we are not denying that the spectral data provide experimental access to the structure of diatomic molecules, their rotational and vibrational motions, and the relation between these motions and radiation. We are only pointing out that the experimental access they provide is not to the dynamical behavior of an individual molecule at a time scale that involves no time-averaging over still smaller time scales. To recognize this is to recognize what is giving force to our questions about the physical meaningfulness of the constants for diatomic molecules.

Some may think that the time-averaging of any physical process cannot help but be physically meaningful. One need merely turn to the history of less than successful efforts to represent turbulent fluid flow in terms of time-averaged parameters (like Reynolds stresses) to appreciate that this is not true (see Tritton 1988, p. 299ff, for example). That there is a time-averaged set of parameters that can serve to represent, for a higher-order time scale, a temporally evolving physical process involving intense unsteadiness at smaller time scales is something that has to be established. The seminal Born-Oppenheimer paper of 1927 was thus important in justifying this under certain assumptions in the case of molecules; what is missing in the case of turbulent flow is a counterpart to this paper. Born-Oppenheimer by itself, however, does not establish that the constants for diatomic molecules are physically meaningful, for this is not a theoretical question—that is, a question about the consistency of such averaging with quantum theory—but rather an empirical question about deviations from the averaging.

For discrepancies between calculation and observation to give support to the physical meaningfulness of those constants, accordingly, some sort of empirical access is needed to time scales eliminated from consideration by the Born-Oppenheimer and Born adiabatic approximations. In the decades following Herzberg, the primary approach to gaining such access has been to pursue subtle anomalies—in particular, minor violations of integral spacing of lines—in some of the rotational and rotational-vibrational spectral bands. Such anomalies are termed "perturbations" in the literature. The goal has been to link them to specific terms defining the potential function that are ignored in the Born-Oppenheimer approximation because they involve electron-nuclei interactions. This research has been summarized in two books published during the first decade of the present century, the thousand-page *Rotational Spectroscopy of Diatomic Molecules* (Brown and Carrington 2003) and the seven-hundred-page *The Spectra and Dynamics of Diatomic Molecules* (Lefebvre-Brion and Field 2004). We are not yet in a position to assess how much support the results presented in these books provide for the physical meaningfulness of the standard diatomic molecular constants. Clearly, nevertheless, these provide a source for addressing our questions about any parallel between those constants and those of Keplerian orbits.

Another approach to gaining experimental access to the time scales of electron-nuclei interactions that emerged late in the twentieth century is so-called "femtochemistry." The perturbation of molecules in this approach results from external excitation by tunable laser pulses of duration within a femtosecond (10^{-15} second) time scale. The primary focus of this research is, in the words of the leading figure of the field Ahmed Zewail, to probe "the ultrafast dynamics of the chemical bond." This process nevertheless can yield results on fluctuations in the potential function between the nuclei of diatomic molecules. For example, in one of the early results in the field, the response of NaI molecules to short-term laser pulses involved an interaction between covalent-bond and ionic-bond potential functions, yielding an oscillatory response in the 300-femtosecond range to the excitation (Zewail 1994, pp. 219–245).[3] Again, we are not yet in a position to assess how much support femtochemical results might provide for the physical meaningfulness of the standard diatomic molecular constants. That they yield short-time information about the actual potential

---

[3] The two specific papers to which we are referring are Todd S. Rose, Mark J. Rosker, and Ahmed H. Zewail, "Femtosecond Real-Time Probing of Reactions. IV. The Reactions of Alkali Halides," *Journal of Chemical Physics*, vol. 91, 1989, pp. 7415–7436; and P. Cong, A. Mokhtari, and A. H. Zewail, "Femtosecond Probing of Persistent Wave Packet Motion in Dissociative Reactions: Up to 40 ps," *Chemical Physics Letters* 172, 1990, pp. 109–113.

function does make them a possible source for such support that needs to be examined.

We have been emphasizing research after 1950 as the most instructive way for comparing the physical meaningfulness of the diatomic molecular constants with that of the Keplerian orbital parameters. Evidence supporting such a parallel between the two had emerged even before 1930. Table 8.1 shows the consistency of the values of the molecular constants obtained from spectra with the change of the isotopic mass of the hydrogen nucleus. The spectroscopic search that resulted in the 1931 discovery of deuterium was prompted by the discovery two years earlier of first one and then a second isotope of oxygen besides the ^{16}O that had previously been used, under the assumption of its being unique, as the basis for the system of atomic weights. Specifically, long-exposure absorption spectra of air had yielded an anomalous band spectrum of faint lines shifted slightly in frequency from a long-established band spectrum. Analysis showed that the shift corresponds to a $^{16}O^{18}O$ molecule under the assumption of a single time-averaged potential function (Giaque and Johnston 1929a, 1929b). Still more extended time exposure then revealed as well an even more rare ^{17}O isotope (Giaque and Johnston 1929c and 1929d). Band spectra of other molecules involving oxygen, such as NO, subsequently independently confirmed both (Naudé 1930), as did mass spectrometry (Aston 1932).

The parallel between the discovery of Neptune from anomalies in the motion of Uranus and the discovery of ^{18}O from an anomalous band in the spectrum of oxygen extends beyond this. Neptune was widely heralded at the time as evidence in support of Newtonian gravity. So too was the discovery of ^{18}O for the selection rules of the new quantum theory:

> this observation must be regarded as a particularly impressive confirmation of quantum mechanical predictions; for example, in the bands of the $O^{16}O^{18}$, all the lines appear, whereas every second line is missing for $O^{16}O^{16}$. (Herzberg 1939, p. 150)

The evidence here involved isotopic shifts only in the time-averaged values of the respective molecular constants, but some subtleties in these shifts were at least attributed back then to short-time interactions between electrons and nuclei ignored in the time-averaging (Van Vleck 1936).

Let us return to the beginning of this paper. Realists have tried to respond to van Fraassen by showing how, in the case of Perrin, concordant measurements of parameters such as Avogadro's number can be achieved and examining what this concordance of measurements is telling us about the reality of molecules. The measurements in question dated from 1908 to 1913. Whatever support for the reality of molecules resulted from them at the time, if research over the subsequent decades had failed to determine anything more about the structure and shape of molecules beyond the spherical hypothesis, this would have rendered the conclusion that molecules are real to be of, at best, limited empirical significance. In fact, however, the rotational and vibrational spectra of diatomic molecules, and in some cases of polyatomic molecules as well, have yielded seemingly impressive conclusions about detailed specifics of their structure and shape. We say "seemingly" because the values of the constants characterizing their structure and shape involve a time-averaging that glosses over extremely complicated short-term fluctuations in their structure. From the point of view of the realist-instrumentalist debate, the question therefore arises whether these constants are physically meaningful (in the sense that their values have causal implications) or whether to the contrary they amount to nothing more than a heuristically useful representation of the spectral data with no physical significance beyond these data. We claim that the question of the physical meaningfulness of these constants is now a far more appropriate focus for the realist-instrumentalist debate than are the concordant measurements of the 1908–1913 period.

We can put this point in a slightly different way that might be more helpful. We believe van Fraassen (2009) is right to emphasize that the primary aim of researchers in microphysics in the early twentieth century was making stable concordant measurement, or empirical grounding, to use van Fraassen's term, possible for microphysical parameters. Where we differ from van Fraassen is in how we view the development of evidence after stable concordant measurement has been achieved. In all the cases we know of from the history of the physical sciences, scientists have continued to pursue discrepancies from concordant measurements, no matter how small they may be, for these discrepancies have always led to discoveries of new physical sources that had, in some cases, not even been considered before. This methodology requires a distinction to be made between representations that yield deviations that have physically identifiable sources, and those that do not yield such deviations, which we term *physical meaningfulness*. We believe that a potentially important step forward in the realist-instrumentalist debate can be achieved by focusing on the questions of how exactly to characterize physical meaningfulness and how it might be possible to distinguish parameters that are physically meaningful from those that are not, even as one must acknowledge that they do not correspond to any situation that actually occurs.

Focusing the realism-instrumentalism debate on the question of the physical meaningfulness of the parameters in the standard representation of diatomic molecules has some virtues beyond those we have stressed earlier. As we have provisionally drawn the distinction, whether a parameterization is physically meaningful turns on how strong the evidence is in support of a specific physical source for each systematic discrepancy between theoretical calculation and observation. Correspondingly, one virtue of focusing on diatomic molecules is the amount of evidence that has been developed over the course of the last 90 years of research on them—at least comparable to, if not substantially exceeding, the scope of the evidence developed over the last three centuries in celestial mechanics. Another virtue is that the assessment of whether the evidence has established specific physical sources is something on which instrumentalists and realists should agree, even though their interpretations of what this physical source amounts to will likely differ. We do not see how van Fraassen can deny that the evidence has established the isotope effects we noted earlier, even though he will continue to reject the reality of atomic nuclei. Equally, we do not see how realists can deny that the diatomic molecular constants provide a *representation*, in van Fraassen's sense, that glosses over details and is in some respects counterfactual. If both sides agree that the evidence confirms that the representation is nevertheless physically meaningful, the debate comes down to whether an agreed on physical source for a discrepancy shows, for example, that the source is perturbing an actual rotation or vibration.

Thirty years ago, in an article more widely cited than heeded, Howard Stein proposed "that between a cogent and enlightened 'realism' and a sophisticated 'instrumentalism' there is no significant difference—no difference that *makes* a difference" (Stein 1989, p. 61). Focusing on the last 90 years of research on the structure of diatomic molecules might well prove to be the ideal testing ground for Stein's proposal.[4]

References

Achinstein, Peter. 2001. *The Book of Evidence*. Oxford: Oxford University Press.

Aston, F. W. 1932. "Mass-Spectra of Helium and Oxygen," *Nature*, vol. 130, p. 21f.

Bernath, Peter F. 2016. *Spectra of Atoms and Molecules, Third Edition*. Oxford: Oxford University Press.

Bernath, Peter F., and Sean McLeod. 2001. "DiRef, A Database of References Associated with the Spectra of Diatomic Molecules," *Journal of Molecular Spectroscopy*, vol. 207, p. 287.

Bohr, Niels. 1913. "On the Constitution of Atoms and Molecules," *Philosophical Magazine Series 6*, vol. 26, pp. 1–25.

Boltwood, Bertram B., and Ernest Rutherford. 1911. "Production of Helium by Radium," *Philosophical Magazine Series 6*, vol. 22, pp. 586–604.

Born, Max, and Robert Oppenheimer. 1927. "On the Quantum Theory of Molecules," tr. H. Hettema, in H. Hettema, *Quantum Chemistry: Classic Scientific Papers*. Singapore: World Scientific, 2000, pp. 1–24.

[4] Part of the research for this paper was supported by the Ministry of Education, Singapore, under its Academic Research Fund Tier 1, No. RG156/18, as well as the Radcliffe Institute for Advanced Study.

Brown, John, and Alan Carrington. 2003. *Rotational Spectroscopy of Diatomic Molecules*. Cambridge: Cambridge University Press.

Chalmers, Alan. 2009. *The Scientist's Atom and the Philosopher's Stone: How Science Succeeded and Philosophy Failed to Gain Knowledge of Atoms*. Dordrecht: Springer.

Chalmers, Alan. 2011. "Drawing Philosophical Lessons from Perrin's Experiments on Brownian Motion: A Response to van Fraassen," *British Journal for the Philosophy of Science*, vol. 62, pp. 711–732.

Chaudesaigues, M. 1908. "Le mouvement brownien et le formule d'Einstein," *Comptes Rendus*, vol. 147, pp. 1044–1046.

Giauque, W. F., and H. L Johnston.1929a. "An Isotope of Oxygen, Mass 18," *Nature*, vol. 123, p. 318.

Giauque, W. F., and H. L Johnston.1929b. "An Isotope of Oxygen, Mass 18. Interpretation of the Atmospheric Absorption Bands," *Journal of the American Chemical Society*, vol. 51, pp. 1436–1441.

Giauque, W. F., and H. L Johnston.1929c. "An Isotope of Oxygen of Mass 17 in the Earth's Atmosphere," *Nature*, vol. 123, p. 831.

Giauque, W. F., and H. L Johnston.1929d. "An Isotope of Oxygen, Mass 17, in the Earth's Atmosphere," *Journal of the American Chemical Society*, vol. 51, pp. 3528–3534.

Glymour, Clark. 1980. *Theory and Evidence*. Princeton: Princeton University Press.

Hecht, Charles E. 1990. *Statistical Thermodynamics and Kinetic Theory*. Mineola, NY: Dover.

Herzberg, Gerhard. 1939. *Molecular Spectra and Molecular Structure, I. Diatomic Molecules*. New York: Prentice-Hall.

Herzberg, Gerhard. 1950. *Molecular Spectra and Molecular Structure: I. Spectra of Diatomic Molecules, Second Edition*. Princeton: D. Van Nostrand.

Herzberg, Gerhard. 1971. *The Spectra and Structures of Free Simple Radicals: An Introduction to Molecular Spectroscopy*. Mineola, NY: Dover.

Huber, Klaus Peter and Gerhard Herzberg. 1979. *Molecular Spectra and Molecular Structure: IV Constants of Diatomic Molecules*. New York: Van Nostrand Reinhold.

Jeans, James. 1916. *The Dynamical Theory of Gases 2nd edition*. Cambridge, UK: Cambridge University Press.

Kelvin, Lord. 1904. *Baltimore Lectures on Molecular Dynamics and the Wave Theory of Light*. Cambridge: Cambridge University Press.

Lefebvre-Brion, Hélène, and Robert W. Field. 2004. *The Spectra and Dynamics of Diatomic Molecules*. Amsterdam: Elsevier Academic Press.

Mayo, Deborah G. 1996. *Error and the Growth of Experimental Knowledge*. Chicago: University of Chicago Press.

Mercator, Nicholaus. 1676. *Institutionum Astronomicarum Libri Duo, De Motu Astrorum Communi & Proprio*. London: Samuel Simpson.

Meyer, Oskar Emil. 1899. *The Kinetic Theory of Gases*. tr. Robert E. Baynes. London: Longmans, Green; *Die Kinetische Theorie der Gase*. Breslau: Marschke & Berendt, 1899.

Millikan, Robert A. 1911. "The Isolation of an Ion, A Precision Measurement of Its Charge, and the Correction of Stokes's Law," *The Physical Review*, vol. 32, pp. 349–397.

Millikan, Robert A. 1913. "On the Elementary Electrical Charge and the Avogadro Constant," *The Physical Review, Second Series*, vol. II, pp. 109–143.

Naudé, S. Meiring. 1930. "The Isotopes of Nitrogen, Mass 15, and Oxygen, Mass 18 and 17, and their Abundances," *Physical Review*, vol. 36, pp. 333–346.

Nernst, Walther. 1895. *Theoretical Chemistry from the Standpoint of Avogadro's Rule and Thermodynamics*, tr. Charles Skeele Palmer of first German edition. London: Macmillan; 1923. tr. L. W. Cobb of the eighth-tenth German edition. London: Macmillan.

Newton, Isaac. 1999. *The Principia: Mathematical Principles of Natural Philosophy*, tr. I. Bernard Cohen and Anne Whitman. Berkeley: University of California Press.

Perrin, Jean. 1910. *Brownian Movement and Molecular Reality*, tr. F Soddy. London: Taylor and Francis.

Perrin, Jean. 1912. "Les Preuves de la Réalité Moléculaire (Études spécial des émulsions)," in *Théorie du Rayonnement et Les Quanta: Rapports et Discussions de la Réunion Tenue à Bruxelles,du 30 Octobre au 3 Novembre 1911, Sous Les Auspices de M. E. Solvay*, ed. P. Langevin and M. de Broglie, Paris: Gauthier-Villars, pp. 150–250.

Planck, Max. 1900. "On the Theory of the Energy Distribution Law of the Normal Spectrum," tr. D. ter Haar, in D. ter Haar, *The Old Quantum Theory*. Oxford: Pergamon Press, 1967, pp. 82–90.

Psillos, Stathis. 2011a. "Making Contact with Molecules: On Perrin and Achinstein," in *Philosophy of Science Matters: The Philosophy of Peter Achinstein*, ed. Gregory J. Morgan. Oxford: Oxford University Press, pp. 177–190.

Psillos, Stathis. 2011b. "Moving Molecules Above the Scientific Horizon: On Perrin's Case for Realism," *Journal for General Philosophy of Science*, vol. 42, pp. 339–363.

Psillos, Stathis. 2014. "The View from Within and the View from Above: Looking at van Fraassen's Perrin," in *Bas van Fraassen's Approach to Representation and Models in Science*, ed. Wenceslao J. Gonzalez. Dordrecht: Springer, pp. 143–166.

Rutherford, Ernest. and Hans Geiger. 1908. "The Charge and Nature of the α-Particle," *Proceedings of the Royal Society of London, Series A*, vol. 81, No. 546, pp. 162–173.

Salmon, Wesley. 1984. *Scientific Explanation and the Causal Structure of the World*. Princeton: Princeton University Press.

Smith, George E. 2014. "Closing the Loop: Testing Newtonian Gravity, Then and Now," in *Newton and Empiricism*, ed. Zvi Biener and Eric Schliesser. Oxford: Oxford University Press, pp. 262–351.

Stein, Howard. 1989. "Yes, but . . . Some Skeptical Remarks on Realism and Anti-Realism," *Dialectica*, vol. 43, pp. 47–65.

Streete, Thomas. 1661. *Astronomia Carolina, A New Theory of the Celestial Motions*. London: Lodowick Lloyd.

Suppes, Patrick. 1962. "Models of Data," in *Logic, Methodology, and the Philosophy of Science: Proceedings of the 1960 International Conference*, ed. Ernest Nagel, Patrick Suppes, and Alfred Tarski. Stanford: Stanford University Press, pp. 252–261.

Tritton, D. J. 1988. *Physical Fluid Dynamics*, 2nd ed. Oxford: Clarendon Press.

Van Fraassen, Bas C. 2008. *Scientific Representation*. Oxford: Oxford University Press.

Van Fraassen, Bas C. 2009. "The Perils of Perrin, in the Hands of Philosophers," *Philosophical Studies*, vol. 143, No. 1, pp. 5–24.

Van Vleck, J. H. 1936. "On the Isotope Corrections in Molecular Spectra," *Journal of Chemical Physics*, vol. 4, pp. 327–338.

Wing, Vincent. 1669. *Astronomia Britannica*. London: Georgii Sawbridge.

Zewail, Ahmed H. 1994. *Femtochemistry: Ultrafast Dynamics of the Chemical Bond, Volume I*. Singapore: World Scientific.

PART II
CONTEMPORARY SCIENTIFIC REALISM

9

The Historical Challenge to Realism and Essential Deployment

Mario Alai

Department of Pure and Applied Sciences
University of Urbino Carlo Bo
mario.alai@uniurb.it

9.1 Deployment Realism

Scientific realists use the "No Miracle Argument" (NMA): it would be a miracle if theories were false, yet got right so many novel and risky predictions. Hence, predictively successful theories are true. Of course, one could easily make up a theory *with completely false theoretical assumptions which predicted a phenomenon P* (call it a F^P-theory) if she knew P in advance and used it in framing the theory. But how could she think of a F^P-theory, without knowing P? Or knowing P but without using it in building the theory? In fact, it is puzzling how one could have built a F^P-theory even if she used P *inessentially*: suppose Jill built a F^P-theory by knowing and using P, but she *could* have done without it, because, quite independently of her, John built *the same* theory without using P. This I call Jill using P *inessentially*, and it is something hard to explain, because it is understandable how the theory was built by Jill, but not by John (Lipton 1991, 166; Alai 2014c, 301).

So, how could one find a F^P-theory without using P essentially? Well, one could do this if P were very poor in content, i.e., very probable: any false theory could predict the existence of *some* new planet *somewhere* in the universe. But it took Newton's theory to predict *precisely* the existence of a planet of *such and such* mass with *such and such* orbit as Neptune. So, the riskier a prediction is (i.e., the more precise, rich in content, hence improbable it is), the less likely it is that it has been made by a completely false theory.

It might be objected that since a falsity may entail a truth, even false theories might make novel predictions. Granted, in the Hyperuranion of possible theories there are some with completely false theoretical assumptions which

Mario Alai, *The Historical Challenge to Realism and Essential Deployment* In: *Contemporary Scientific Realism.* Edited by: Timothy D. Lyons and Peter Vickers, Oxford University Press. © Oxford University Press 2021. DOI: 10.1093/oso/9780190946814.003.0009 (together with appropriate auxiliary assumptions) make true *novel and improbable* predictions (call them F[NIP]-theories). However, their rate is very small, for it is a logical fact that the less probable a consequence is, the fewer the premises entailing it. Hence, it is extremely improbable that F[NIP]-theories are found *by chance*. But since scientists look for *true* theories, only by chance can they pick a F[NIP]-theory. Therefore, it is extremely unlikely that scientists find a F[NIP]-theory.[1]

For instance, quantum electrodynamics predicted the magnetic moment of the electron to be $1159652359 \times 10^{-12}$, while experiments found $1159652410 \times 10^{-12}$: hence John Wright (2002, 143–144) figured that the probability to get such an accuracy *by chance*, i.e., through a false theory, is as low as 5×10^{-8}.

Here (following Alai 2014c) I will speak of "novel predictions" (NP) when the predicted phenomenon is either unknown (historically novel) or not used (use-novel) or not used essentially (functionally novel) by the theorist. So, it is *practically* impossible that a *novel* prediction which in addition is risky, i.e., a highly improbable, was derived from a completely false theory.

Against the claim that success warrants truth, Laudan (1981) had objected that many false theories in the history of science were successful, so realists replied that however those theories didn't have *novel* predictive success.[2] Unfortunately, it turns out that also many false theories had novel success (see the list in Lyons 2002, 70–72). Selective realists acknowledged the problem and argued that the NMA does not warrant the *complete* truth of theories, but only their *partial* truth. In particular, for Kitcher's (1993) and Psillos's (1999) "Deployment Realism" (DR), we should be committed only to those *particular hypotheses* which were *deployed* in deriving novel predictions: this is Psillos's "*divide et impera*" move against Laudan's "reductio" (Psillos 1999, 102, 108, passim).

A rejoinder came from Timothy Lyons: first, he explained that Laudan's original objection was not "an articulation of the pessimistic *meta-induction* toward the conclusion that *our present theories are probably false*," but a *modus tollens* refutation of the NMA (a "meta-*modus tollens*") (Lyons 2002, 64–65; 2006, 557). Secondly, he criticized DR by listing a number of false *hypotheses* which were actually deployed in the derivation of various novel risky predictions—including Neptune's discovery (2002; 79–83; 2006). Deployment realists, however, can answer this new challenge by arguing that the NMA does not warrant the truth of all hypotheses *actually used* in deriving novel predictions, but only of those which were *essential*, i.e., strictly necessary to those predictions. False hypotheses are no counterexample to the NMA if they were used *de facto*, but actually superfluous, i.e., *non-essential*.

---

[1]   Alai 2014c, §1; 2014d, §5; 2016, §4. More on this objection at §9.7.1.
[2]   Musgrave (1985, 1988); Lipton (1993, 1994); Psillos (1999); Sankey (2001).

Thus, *essentiality* is crucial to two key claims of **DR**: (1) a predicted phenomenon is novel only if it was not used essentially in building the theory which predicts it; and (2) a hypothesis is most probably true only if it was essential in deriving a novel prediction. Here we shall focus on the latter claim.

According to Psillos (1999, 110) a hypothesis **H** is deployed *essentially* in deriving a novel prediction **NP** if

(1)   **NP** follows from **H**, together with some other hypotheses **OHs** and auxiliary assumptions **AA**, but not from **OHs** & **AA** alone;

(2)   no *other* hypothesis **H*** is *available* which can do the same job as **H**, viz. is
   (a)   compatible with **OHs** and **AA**,
   (b)   non-*ad hoc*,
   (c)   potentially explanatory, and
   (d)   together with **OHs** and **AA** predicts **NP**.

Conditions (1) and (2) capture the idea that **H** is essential, i.e., strictly necessary, when (1) **OHs** without **H** could not predict **NP**, and (2) no alternative hypothesis is available which could substitute **H** in the derivation of **NP**.

9.2   Lyons's Criticism of Psillos's Criterion of Essentiality

However Lyons (2006; 2009) criticized this criterion of essentiality, in particular condition (2). First, he argued that it is too vague to be applicable to any historical case (2006, 542). For instance, it is unclear *when* **H*** should not be *available*: when the theory is put forward? when the prediction is derived? when the prediction is confirmed? at some point in the future? Further unclear points are what "potentially explanatory" means; whether each of the elements of **OHs** and **AA** also need to be "essential" by this criterion; whether the replacement hypothesis needs to be consistent with those components of **OHs** and **AA** which are "essential" for other predictions but unnecessary for the prediction of concern; whether **H*** can result in the loss of other confirmed predictions.

Secondly, according to condition (2) a competitor **H*** makes **H** inessential only if **H*** is compatible with **OHs** and **AA**, and it is non-ad hoc (i.e., predicts phenomena not used in constructing **H***: Psillos 1999, 106). But Lyons notices that in actual history "most (if not all) well-respected theories" were either *incompatible* with **OHs** and **AA**, or ad hoc in Psillos's sense, or both (Lyons 2009, 149). So the requirement placed by (2) on admissible competitors is "too strict to permit even some of the most exemplary scientific theories . . . to qualify as competitors" (150), and this means that (2) is too weak (or practically empty), for most (if not all) possible hypotheses **H** will lack admissible competitors. Psillos didn't understand his conditions for essentiality as a definition, but just as a criterion: he writes "H indispensably contributes to the generation of P if . . . [(1) and (2) are met]," not "if *and only if* . . . [(1) and (2) are met]" (1999, 110). Yet, Lyons's criticisms are at least prima facie well-founded, and must be taken seriously.

But Lyons also claims that his criticisms force deployment realists to abandon the essentiality requirement altogether. Besides, he claims that the "crucial" or "fundamental" insight of DR is that we should credit "those and only those constituents that were genuinely responsible for, that actually led scientists to, specific predictions" (2006, 543), no matter whether they were indispensable or not. In fact, he argues that Psillos, after introducing his criterion of essentiality (Psillos 1999, 109–110), "never mentions that rigid criterion again"[3] and fails to show that his case studies comply with it (Lyons 2009, 143). On the contrary, according to Lyons, Psillos often talks of commitment to constituents which simply "generated," or "brought about," or were "responsible for," or "genuinely contributed to," or "really fueled" (Psillos 1999, 108–110) the success of the theory (Lyons 2006, 540). Moreover, Lyons says that the same "fundamental insight" is also embraced, in some form or another, by Kitcher, Niiniluoto, Leplin, and Sankey (538). Thus, he holds, DR should get rid of the essentiality requirement and "return to the crucial insight" of actual deployment (543).

At this point, however, Lyons argues that many assumptions which in history were *actually* employed and led to novel predictions are *false*. In fact, in various papers (2002, 2006, 2016, 2017) he discusses a number of striking novel predictions, toward *each* of which many false assumptions were employed (we shall briefly review some of them later). Hence, he concludes the NMA does not work even when restricted to particular hypotheses, and DR is false: "If we take seriously the no miracles argument [i.e., if novel predictions derived from false assumptions are a miracle] . . . we have witnessed numerous miracles" in history (2006, 557).

However, in reply, a number of points should be noticed: first, by accepting such purported "miracles," one is left with the problem of explaining how they are possible. If one wonders how scientist S made the *novel risky prediction* NP which was then confirmed, a possible answer is that NP followed from true assumptions in S's theory, for all consequences of true assumptions are true. But one cannot answer that NP followed from some false hypothesis in S's theory, because that would entail that S had luckily assumed a false hypothesis entailing NP, and as mentioned, this is extremely improbable.[4] Lyons (2002; 2003) and

---

[3] Personal communication, but see 2006, 542.

[4] More on this at section 9.7.1. One might undercut the NMA and DR also by rejecting the need or possibility of explaining unlikely events, or of explanations in general. But I cannot discuss that radical objection here.

others proposed alternative explanations, not involving the truth of the deployed hypotheses, but I have argued elsewhere that they all fail (Alai 2012, 88–89; 2014a).

Secondly, as concerns which one is the "fundamental insight" of DR that cannot be abandoned vs. what is merely accessory and should be dropped, two questions must be distinguished: (I) the exegetic question whether Psillos understands "deployment" as *merely actual* deployment, or as *essential* deployment; and (II) the theoretical question of how deployment *should* be understood if we are to give DR its best currency in order to correctly assess its tenability.

As to (I), when Psillos claims that we should credit constituents that are "responsible" for success (1999, 108–109), Lyons understands "responsibility" as actual deployment, suggesting that this was the original core intuition of DR (2006, 540, 543). But this is incorrect. As Lyons himself acknowledges (2006, 539) on the next page Psillos specifies that constituents to be credited are those "that made *essential* contributions" to success and that "have an *indispensable* role" in it, and right after this he spells out his conditions (1)–(2) (109–110). So, by responsibility Psillos clearly means *essential* responsibility.

Even apart from this, when an author says two different things, in the exegesis we are allowed to discount one of them only if they are contradictory. But if they are not contradictory we must keep both, and understand how they can be reconciled. In this case, sometimes Psillos talks of responsibility or contribution, or deployment, etc., without further qualifications, sometimes he introduces the essentiality or indispensability qualification. But these two talks are not contradictory, and can be easily reconciled, since the latter is just a specification of the former. In fact, when talking plainly of responsibility or contribution, Psillos never adds "no matter whether essential or not" or the like. Therefore, even his plain talk of responsibility or contribution must be understood as implying "essential" or "indispensable."

Even if in his case studies (1999, ch. 6) Psillos fails to explicitly refer back to his *criterion* of essentiality and to explicitly show how it applies, he does use the essentiality *requirement*. For instance, he argues that we need not be committed to the caloric hypothesis, although it was actually deployed in Laplace's prediction of the speed of sound in air, because that prediction "did not depend on this hypothesis" (121). Hence, he obviously would not give up his essentiality requirement. Moreover, in the absence of any indication to the contrary, and as this quotation suggests, we must assume that he understands the essentiality requirement precisely in terms of his essentiality criterion.

Summing up, from the exegetic point of view, ignoring the essentiality requirement would be misinterpreting Psillos's position. But apart from this, since here we must assess the tenability of DR, what matters more is not the exegetical question, but (II) the theoretical question of how DR *should* be understood in order to give it the most favorable interpretation. Now again, the answer is the essential deployment interpretation, for it immunizes **DR** from Lyons' purported historical counterexamples.

In fact, to begin with, by Occam's razor essentiality is *required* to the cogency of the **NMA**: in abductions we can assume only what is essential, i.e., the weakest hypothesis sufficient to explain a given effect; but if a hypotheses, although deployed, was not essential in deriving **NP**, it is not essential in explaining its derivation either; therefore deployment realists need not (and must not) be committed to its truth.

Moreover, Psillos's conditions (1)–(2) are just *one* way to spell out essentiality. Therefore, even if Lyons succeeds in criticizing them, he has not shown that no other characterization of essentiality is possible, hence he cannot conclude that this condition *must* be abandoned. In fact, he only *suggests* that it be abandoned, so proposing a version of **DR** deprived of the essentiality requirement which he (wrongly, I argued) believes to be Psillos's "fundamental" version. Therefore, although his refutation of *this* latter version by meta-*modus tollens* from historical cases is effective, it does not affect *actual* **DR**, i.e., the essential deployment version.

Granted, Lyons's analysis of Psillos's conditions (1)–(2) is correct, and his criticisms of (2) are convincing. But I will propose a different condition, which performs the task Psillos had in mind, yet is simpler, less troublesome, and escapes Lyons's criticisms. This condition is not altogether new, having been used or presupposed at various places, I suggest, by Psillos himself (e.g., 1999, 119–121) and explicitly proposed by Vickers, who writes:

> H does not merit realist commitment whenever H is doing work in the derivation solely in virtue of the fact that it entails some other proposition which itself is sufficient, when combined with the other assumptions in play, for the relevant derivational step (2016, §3).

9.3 An Improved Criterion of Essentiality Escaping Lyons's Criticisms

I propose to substitute the crucial condition (2) in Psillos's criterion with:

> (2') There is no other hypothesis H* which is *proper part* of H (hence weaker than H) which together with **OHs** and **AA** entails **NP**.

Here in section 9.3 I explain (2'), spelling out the intuitive notion of indispensability more precisely through Yablo's (2014) concept of proper parthood, and arguing that the troublesome requirements of Psillos's criterion can be dropped without loss, thus escaping Lyons's criticisms; in section 9.4 I show how my criterion immunizes DR from Lyons's historical counterexamples; in section 9.5 I hold that essentiality in this sense cannot be detected prospectively, and Vickers's optimism on the identifiability of non-essentiality should be complemented by some less optimistic considerations; but far from causing problems for realists, this frees them from unreasonable obligations; in section 9.6 I explain in more details why my criterion is enough, although apparently weaker than Psillos's; in section 9.7 I answer some actual and potential objections.

Intuitively, a hypothesis H* is a proper part of H iff its content is part of the content of H but does not exhaust it. At first approximation and for most purposes, a part of H is therefore a hypothesis entailed by H but not entailing it. But this is not always enough: for instance consider

(i) It's Spring

and

(ii) It's April.

As (i) is entailed by (ii) without entailing it, (i) is a proper part of (ii). But

(iii) It's April *or* the streets are wet

is implied by (ii) without entailing it, however it is not a (proper) part of it, for it brings in a different subject matter. Therefore, just characterizing parthood in terms of entailment will not always do for our purposes. Instead, we must understand parthood as Yablo (2014):[5] H* is a part of H iff it is entailed by H *and preserves its subject matter.* The subject matter of a proposition consists of two classes: that of its truthmakers (or reasons why it is true) and that of its falsemakers (or reasons why it is false). A proposition H* entailed by H preserves H's subject matter, hence, is *part* of it, iff every truthmaker/falsemaker of H* is entailed by a truthmaker/falsemaker of H. Moreover, H* is a *proper* part of H iff every truthmaker/falsemaker of H* is entailed by a truthmaker/falsemaker of H, *but not vice versa.* This explains the intuitive idea that (iii), although entailed by (ii), is not part of it: because some truthmakers of (iii) (e.g., "it's raining") are not entailed by any truthmaker of (ii).[6]

---

[5] I thank Matteo Plebani for pointing this out to me. See Plebani (2017).

[6] Although Yablo's approach helps us to properly explain essentiality, it was not designed to this purpose, but to answer the general question of "aboutness," i.e., the subject-matter of sentences: why and how sentences with the same truth-conditions say different things, like "here is a sofa" and "here is the front of a currently existing sofa, and behind it is the back." Thus, it offers a natural solution to a wide variety of problems concerning agreement, orders, possibilities, priority, explanation,

Condition (2') expresses precisely the above-mentioned Occam's requirement, namely that H is not redundant, i.e., that one could not explain the derivation of NP by assuming the truth of *something less* than H. For instance, suppose that instead of his actual gravitation theory Newton had advanced the following theory:

(H) inside each massive body there resides a demi-god, which attracts the demi-gods dwelling in each other body by a force $F = Gm_1 m_2/r^2$.

H would predict the same novel phenomena as Newton's actual theory, but we wouldn't need to believe it, because it is redundant, i.e., not *essential* to those predictions. In fact, H consists in the following assumptions:

(H*) each body attracts all other bodies by a force $F = Gm_1 m_2/r^2$
($H^{d\text{-}g}$) F is exerted by a demi-god residing inside each body.

H would violate condition (2'), because there would be another hypothesis H* (i.e., Newton's actual theory) which is a proper part of H and sufficient to derive its novel predictions.

This alternative condition (2') avoids Lyons's vagueness and emptiness objections to Psillos's condition (2). First, (2') is not too vague because of five reasons: (i) there is no question about when the alternative H* is *available*, because (2') does not exclude just that H* is available, but even that it is logically possible (for if it were possible, H would be redundant in the prediction of NP, hence possibly false); (ii) there is no need to specify what "explanatory" means, since (2') has no such requirement; (iii) it is not required that the elements of OHs and AA are also essential;[7] (iv) since (2') excludes only hypotheses H* which are part of H, ipso facto such hypotheses are compatible with all the components of OHs and AA with which H is compatible; (v) alternative hypotheses H* are excluded no matter whether substituting H by H* results in the loss of other confirmed predictions or not: if H allows us to derive two predictions NP_1 and NP_2, and the weaker H* allows us to derive NP_1 but not NP_2, then H is not essential with respect to NP_1, but it may be essential with respect to NP_2, in which case it is most probably true.

Lyons's second criticism was that (2) is practically empty, because it excludes only alternative hypotheses which are compatible with OHs and AA, while in

knowledge, partial truth, and confirmation. For instance, a white marble does not confirm (1) "All crows are black" because (1) does not say the same as (2) "All non-black things are non-crows," and Yablo explains why (2014, pp. xiv, 7–8).

[7] Whether they are or not can be decided in the same way as for H. In section 9.6 I shall notice that different but compatible assumptions may all be essential for a given novel prediction.

most actual cases alternative hypotheses are not compatible with **OHs** and **AA**. My condition (2') escapes this problem, since it excludes, without exceptions, *all* the alternative hypotheses which are part of **H** and can derive **NP**. In fact, in section 9.4 I shall mention some historical examples of hypotheses which violated condition (2'), confirming that it is not empty.

Yet, it might seem that here lurks a further difficult problem: as just said, all the alternative hypotheses excluded by (2') are compatible with **OHs** and **AA**, since they are part of **H**. But how about those alternative hypotheses which are *not* compatible with **OHs** and **AA**, because they are *not* part of **H**? Lyons stresses that they exist in many historical cases, and if so, don't they make the original hypothesis **H** superfluous, i.e., non-essential? Why doesn't my criterion exclude such hypotheses?

In other words, Psillos's condition (2) for essentiality is of medium strength: it excludes all the hypotheses which can derive **NP** when joined with **OHs** and **AA**. Lyons's criticism is that it is too weak, and his implicit suggestion is that it should be strengthened to exclude even the hypotheses which can derive **NP** in conjunction with different collateral assumptions. My condition (2'), instead, is even weaker than Psillos's (for it excludes only a subset of the alternative hypotheses excluded by (2), i.e., only those which are part of **H**), and this is why it escapes Lyons's first criticism: how can it be enough? In particular, shouldn't we require that **H** is *unique*, i.e., that there is no alternative hypothesis **H'** incompatible with **H**, which (no matter whether compatible and joined with **OHs** and **AA**, or incompatible with them and joined with different hypotheses and auxiliary assumptions) can derive **NP**? For if it existed, we couldn't know which of them is true.

I answer that we don't need to explicitly exclude any other **H'** except those which are part of **H** (so, we don't need Psillos's condition (2) with its drawbacks, nor its strengthening implicitly suggested by Lyons), because the required uniqueness of **H** is already ensured by condition (2') together with the *risky* character of the prediction **NP**. In fact, just as it happens for *theories*, it is practically impossible to find a *completely false* hypothesis making risky novel predictions. A false hypothesis **H** may yield a risky novel prediction **NP**, only if it is *partly* true: i.e., if it has a true part **H*** from which **NP** can be derived. But in this case, the essential role in the derivation of **NP** is played by **H***, **H** being inessential by my condition (2').

This much we know a priori from the NMA (as spelled out in the five initial paragraphs of section 9.1), and anti-realists have not been able to spot any inferential fallacy in that argument. However, they challenged it by apparent counterexamples, i.e., by citing historical instances of false hypotheses which were supposedly deployed essentially in deriving novel predictions. But in each of those cases it can be shown that either (a) those hypotheses were not actually false or the predictions not actually true, or (b) the predictions were not actually novel and risky, or (c) they were not deployed essentially (Alai 2014b).

Condition (2') is designed precisely to exclude cases of type (c), and I shall offer six examples in section 9.4 and three in section 9.5. Therefore, if H fulfills (2'), it must be *completely* true. But if so, we don't need *in addition* to exclude that there is any *incompatible* assumption H' from which NP could be essentially derived, because any H' incompatible with H would be (at least partly) false, hence it could not (except by a miracle) be essential in deriving NP: H' could play a role in deriving NP only if it had a completely true part H'* sufficient to derive NP, and obviously H'* would be compatible with H, since both would be completely true.

This is a nutshell account of why my condition (2') is enough to ensure the required uniqueness of H, but I shall explain it in more detail in section 9.6. This account, anyway, entails an empirical claim: in the history of science there have never been a couple of *incompatible* hypotheses both essential in deriving the same novel risky prediction. So, my proposal is empirically testable: it will be falsified if any such couple is found, but confirmed if it is not.

9.4. How the Essentiality Condition Rules out Lyons's Purported Counterexamples

Here is how my condition (2') rules out as inessential five false assumptions which according to Lyons were deployed to derive novel predictions (the first three are more extensively discussed in Alai 2014b, §7). Three further examples will be discussed in section 9.5. Yet another striking case is Arnold Sommerfeld's 1916 prediction of the fine structure energy levels of hydrogen: for decades it was thought to be derived from utterly false assumptions, and so considered "the ultimate historical challenge" to the NMA, for even physicists called Sommerfeld's success a "miracle." But Vickers (2018) shows that the wrong assumptions were not essential in deriving it, since the true part of Sommerfeld's premises was enough.

9.4.1 Dalton

Lyons argues that Dalton derived his true *Law of Multiple Proportions* (LMP) from his false *Principle of Simplicity*:

> PS: "Where two elements A and B form only one compound, its compound atom contains 1 atom of A and 1 of B. If a second compound exists, its atoms

will contain 2 atoms of A and 1 of B, and a third will be composed of 1 of A and 2 of B, etc." (Lyons 2002, 81; Hudson 1992, 81).

But this false **PS** has a *true* part: a weaker principle we might call "*of Multiple Quantities*":

PMQ: The quantity of atoms of B combining with a given number of atoms of A is always a multiple of a given number.

Besides, **PMQ** is enough to derive the *Law of Multiple Proportions*:

LMP: The weights of one element that combine with a fixed weight of the other are in a ratio of small whole numbers.

Therefore, although actually employed in deriving **LMP**, the false **PS** was not essential, since its true part **PMQ** was sufficient to derive **LMP**.

9.4.2 Mendeleev

Mendeleev derived his predictions of new elements from his false *Periodic Law*:

PL: atomic *weights* determine the chemical properties of elements.

PL is false, since chemical properties are rather determined by atomic *numbers* (Lyons 2002, 80, 84). However, **PL** entails that

Q: an atomic quantity *approximately proportional to atomic weight* determines chemical properties (where Mendeleev thought this quantity was atomic weight itself, while we know it is the atomic *number*).

Now, **Q** is a part of **PL**, true, and sufficient to derive Mendeleev' predictions. So, although actually employed, **PL** was not essential, while the essential assumption, **Q**, was true.

9.4.3 Caloric

Lyons (2002, 80) points out that the caloric theory truly predicted that

ER: The rate of expansion is the same for all gases.

But ER was derived from a number of false claims, viz.:

(1) heat is a weightless fluid called caloric;
(2) the greater the amount of caloric in a body, the greater its temperature;
(3) gases have a high degree of caloric;
(4) caloric, being a material itself, is composed of particles;
(5) caloric particles have repulsive properties which, when added to a substance, separate the particles of that substance;
(6) the elasticity of gases is caused by this repulsive property of caloric heat particles (Carrier 1991, 31).

However, these false claims were not *essential* in predicting ER. In fact, each of them includes a true part not referring to caloric, from which ER could still be derived, respectively

(1*) heat is weightless, and it expands like fluids;
(2*) the greater the amount of heat (whatever it consists in) in a body, the greater its temperature;
(3*) gases have a high amount of what heat consists in;
(4*) particles play a crucial role in the constitution of heat;
(5*) heat involves repulsive forces, which separate the particles of substances to which it is added;
(6*) the elasticity of gases is caused by the repulsive forces involved by heat.[8]

So, while the false (1)–(6) were used, they were not essential, since their parts (1*)–(6*) are sufficient to derive ER (and true, as far as we know).

9.4.4 Kepler

In (2006, 545–553) Lyons discusses a number of false premises from which Kepler derived (P1) that the Sun spins, (P2) that a planet's speed is highest at its perihelion and lowest at its aphelion, and (P3) his three laws. Further, since Kepler's laws were used by Newton, those false premises indirectly yielded also the successes of Newton's gravitation theory, including Neptune's discovery. An exhaustive analysis of Lyons's discussion would require a different paper. Here I can only sketch how my essentiality condition rescues the NMA even in this case.

Kepler's false assumptions were:

---

[8] See Psillos (1999, 115–130) for an extensive presentation of this strategy.

(#1) the planets tend to rest, and they move only when forced to move;

(#2) the Sun has an *Anima Motrix* which spins it around its axis;

(#3) this spinning is transmitted to the planets through the Sun's rays, which push the planets in their orbits (thus the Sun exerts on the planets a *directive* rotational force, not an *attractive* one);

(#4) the force of the Sun's rays is inversely proportional to the distance from the Sun, like the intensity of their light.

The model of the universe described by (#1)–(#4) is crazy in the light of our mechanics (#1), metaphysics (#2), and physics (#3), but it was not in Kepler's time. Moreover, it was a natural abductive conclusion from two facts known at that time:

(a) the planets move around the Sun approximately on the same plane and in the same wise;

(b) the order of their velocities is inverse to that of their distances from the Sun.[9]

These facts plausibly suggested a model analogous to mechanisms well known to Kepler, like a sling, where a central hub rotates, and through a strap transmits its motion to a peripheral body, which is then set into circular motion as well. In Kepler's model the hub is the Sun, the peripheral body is a planet, and the strap represents the Sun's rays. The sling has just one projectile, while the planets are many; but there are similar earthly mechanisms which preserve this analogy, like wool winders, wheels to lift water, or clock gears (although the sling model is better from the point of view of fact (b), for (initially) the projectile rotates more slowly than the hand, and the more slowly the longer the strap is).

The predictions (P1) that the Sun spins and (P2) that the velocity of the planets is highest at the perihelion and lowest at the aphelion, are immediate consequences of Kepler's model (#1)–(#4). Of course the model is wrong, for his abduction from the known facts (a) and (b) included some wrong guesses. But it is also unnecessarily strong as an explanation of those facts. Moreover, the wrong part was not *essential* to Kepler's novel predictions or to the discovery of his three laws, since the model had a weaker core which was both true and sufficient to derive them. Kepler might have restricted his assumptions to that weaker core if

---

[9] This was already clear to Copernicus, who knew the times of the orbits and the relative distances from the Sun, hence the relative lengths of the orbits. Kepler also knew that the times grow more than the lengths, hence the velocity diminishes as the distance from the Sun increases. This suggests that the planets be driven by a force emanating from the Sun, so varying inversely with the distance from it (Koyré 1961, II, §1). Kepler thought that the light varies inversely with the distance from its source (rather than with its square), hence he thought the Sun's motrix force did the same.

he had been more cautious in extrapolating from (a) and (b); but unfortunately there were no technological analogies, like the sling, the wool winder, etc., to suggest to him such a model. This true core is the following:

(#5) the solar system moves around the centre of the Sun as a coherent but non-rigid disk;

(#6) its coherence is *mainly* due to *a* force exerted by the Sun on the planets (Kepler thought it was a *directive* rotational force, and the only one in play; while we know it is the *attractive* gravitational force, supplemented by the planets' own attractive forces and the effects of inertia);

(#7) the velocities of the planets are also *proportional* to the same force (Kepler thought this happens because that force is responsible for their motion, by overcoming their tendency to rest; instead we know this happens because their motion is due to their tangential inertia, which must be equal to that (gravitational) force, otherwise the planet would either fly off on its tangent, or collapse on the Sun);

(#8) that force is in *an* inverse relation with the distance from the Sun (for Kepler that relation was $F=1/d$, we know it is $F=1/d^2$).

I am not saying that (#5)–(#8) by themselves constitute a viable model of the universe: we get one only by supplementing them either by the rest of Kepler's assumptions or by the rest of our assumptions[10] (in the first case, of course, we get a viable, but false model, in the second case a true model). For instance, we wouldn't get a viable model just by adding to (#6) that the force is an attractive one, without adding the planets' inertia, for then the planets would collapse on the Sun. This is probably why Kepler himself, after considering the idea of an attractive force, rejected it (Lyons 2006, 547). I am only claiming that (#5)–(#8) are a core shared by two otherwise very different models; that as far as we know they are true; and that they are enough to derive Kepler's novel predictions (P1)–(P3).

In fact, (P1) the spinning of the Sun is already implicit in (#5); (P2) that a planet's speed is highest at its perihelion and lowest at its aphelion is implied by (#7) and (#8); finally, (P3) Kepler's three laws are just kinematic laws, and once given (#5) they were found mainly by making hypotheses on which curves and functions would fit the data and checking them back with the data; but their discovery was helped by Kepler's revolutionary intuition that the orbits must not be found by geometrical models alone, but reasoning on the forces which govern them, as suggested by (#7) and (#8) (Hoskin 1999, ch. V). So, there were no miracles: the true model (#5)–(#8) is part of the false model (1#)–(#4) and

---

[10] Some of which are mentioned within parentheses after (#6), (#7), and (#8) respectively.

enough to generate (P1)–(P3); therefore the model (#1)–(#4) was not essential by condition (2'), hence the NMA does not commit to it.

For the sake of simplicity, in (#6), (#7), and (#8) I spoke of a *force*, as if Newton's gravitational theory were true; but, one could object, General Relativity shows that there are no gravitational *forces*, and the planets' motion is due to the curvature of spacetime caused by the Sun. The objection is fair, but it only shows that we must circumscribe more accurately the true core of Kepler's model, which we can do by generalizing (#6), (#7), and (#8) as follows:

(#5)   the solar system moves around the centre of the Sun as a coherent but not rigid disk;

(#6*)   its coherence is due to an effect (i.e., for us, the curvature of spacetime) which is mainly brought about by the Sun;

(#7*)   the velocities of the planets are also proportional to the same effect;

(#8*)   that effect is in an inverse relation with the distance from the Sun (hence so are the planetary velocities).

This restricted core is shared by Kepler, Newton, and Einstein and is still sufficient to derive Kepler's novel predictions. But what if someday General Relativity is replaced by a better theory and this by another one, etc.?[11] If we constantly weaken our hypotheses to keep them compatible with successive theories, won't we reduce them to mere descriptions of observed phenomena, and scientific realism to empiricism?[12] I don't think this pessimism is warranted, because history shows that each successive theory T', while giving up part of the theoretical content of the earlier theory T, adds some new theoretical content. Even if in turn T' is superseded by T" and some of its content is given up, it is quite possible that (a) some of the original content of T is preserved even in T", and (b) the theoretical content of T' preserved in T" is even larger than that of T preserved in T'.

For instance, in the case discussed here, the directive force has been replaced by the attractive force, and this by a curvature of spacetime, but all of them are unobservable entities or properties. Each theorist while discarding some theoretical assumptions adds some new ones (Newton adds inertia and gravitation force, Einstein spacetime curvature). Tomorrow it might no longer be spacetime curvature, but something else; but whatever it is, the Sun will have a major role in it, and it will have an inverse relation with the distance from the Sun and a direct relation with the velocities of the planets. Besides, each new model is probably closer to the truth, because it has a better fit to the data and greater predictive

---

[11] I deny that our current theories are completely true and will never be rejected: Alai (2017).

[12] I owe this objection to an anonymous reviewer.

power. Thus our theoretical knowledge (our theoretical true justified beliefs) increases over time. While acknowledging that our beliefs are not yet "the whole truth and nothing but the truth," we can trust to be on the right track. I shall further discuss similar worries in §§9.7.3–9.7.5.

9.4.5 Neptune

Finally, Lyons claims that the following false claims were used in the prediction of Neptune (2006, 554):

(1) the sun is a divine being and/or the center of the universe (Kepler);

(2) the natural state of the planets is rest;

(3) a non-attractive emanation coming from the sun pushes the planets in their paths;

(4) the planets have an inclination to rest, which resists the solar push, and contributes to their slowing speed when more distant from the sun;

(5) the planets are pushed by a "directive" magnetic force;

(6) there exists only a single planet and a sun in the universe (Newton);

(7) each body possesses an innate force, which, without impediment, propels it in a straight line infinitely;

(8) between any two bodies there exists an instantaneous action-at-a-distance attractive force;

(9) the planet just beyond Uranus has a mass of 35.7 earth masses (Leverrier)/50 earth masses (Adams);

(10) that planet has an eccentricity 0.10761 (Leverrier)/0.120615 (Adams);

(11) the longitude of that planet's perihelion is 284°, 45' (Leverrier)/299°, 11' (Adams), etc.

However, (1)–(5) were no longer held in the 18th century. They were involved in Neptune's discovery only to the extent that they were involved in the discovery of Kepler's laws, which in turn were used by Newton. But I argued that (1)–(5) were not involved *essentially* in Kepler's discoveries. (9)–(11) concern Neptune, so they could not be premises from which Neptune's existence or location was derived: they are only (partially wrong) parts of a global hypothesis on the planet's existence and behavior. (6) was practically true in this context, given the negligible influence of the other planets. To my knowledge (7) was never held by anybody; anyway, it entails the true claim that inertial motion continues indefinitely in a straight line. Finally, I just argued that (8), gravitational *force*, was not essential in Newton's predictions. So, even Neptune ceases to be a counterexample to DR.

9.5 Essentiality Cannot be Detected Prospectively

For deployment realists the components of discarded theories which were essentially involved in novel predictions are most probably true; but Stanford (2006), Votsis (2011), and Peters (2014) argue that it is not enough to identify these components just *retrospectively*, i.e., as those preserved in current theories. Rather, they must be identified *prospectively*, from the viewpoint of their contemporaries. In fact, deployment realists claim that the hypotheses of past theories still preserved today are true because they were essential. Thus they resist the Pessimistic Meta-Induction (PM-I), maintaining that both past and current theories are at least partly true. But explaining that those hypotheses were essential because they are preserved today, would be explaining that they are true because they are preserved today, so begging the question of the truth of current theories. In fact, since a hypothesis H is preserved today because we believe it is true, saying that it is true because it is essential, and it is essential because it is preserved today, would be saying it is true because we believe that it is true.

Equally, deployment realists claim that the now rejected hypotheses deployed in novel predictions were not essential, hence the NMA did not commit to their truth. In this way they block Laudan and Lyons's meta-*modus tollens* (M-MT). But explaining that they were not essential because they are not preserved today would be stipulating that the NMA warrants all and only the hypotheses accepted today. Thus the realist's defense against both the PM-I and the M-MT would be circular.

This is why it would seem desirable that the components of past theories which are essential for their novel predictions are identified prospectively; moreover, Votsis (2011), Peters (2014), Cordero (2017a, 2017b), and others claim that such prospective identification *is* possible. On the opposite side, Stanford (2006, 167–180; 2009, 385–387) claims this is impossible, therefore the arguments for DR are circular.[13]

I take an intermediate position: any time we believe a hypothesis is essential, it may in fact be, but neither at that time nor later can we ever be certain that it is. Yet, I argue that this does not make the arguments for DR circular. That we cannot ever be certain is shown by many hypotheses which were firmly believed by past scientists because they appeared to be essential to certain novel predictions, but subsequently turned out to be inessential, in fact false.

For instance, Fresnel and Maxwell derived various novel predictions from the hypothesis that

AV: aether vibrates.

---

[13] See also Vickers (2013, 207).

Today we know that **AV** was inessential in these derivations, because it can be substituted by its weaker consequence

   VM: there is a vibrating medium.[14]

Fresnel and Maxwell could easily have seen that **VM** was a proper part of **AV** and enough to derive those predictions. However, they didn't consider **VM** (probably they didn't even think of it), no doubt because they presupposed that

   P_1: all mediums are material,

and/or that

   P_2: all waves are produced by the oscillations of particles.

Hence, given their presuppositions, *any* vibrating medium was a *material* medium *composed of particles* (i.e., either water, or air, or *aether*). Therefore from their viewpoint, **AV** was required by their predictions as much as **VM**.

   Again, Laplace predicted the speed of sound in air starting from the false hypothesis that

   H: the propagation of sound is an adiabatic process, in which some quantity of
      caloric contained by air is released by compression.

Psillos (1999, 119–121) argues that **H** was not essential to Laplace's prediction, which could also be derived from a part of **H**, viz.

   H*: the propagation of sound is an adiabatic process, in which some quantity
      of latent heat contained by air (whatever be the nature of heat) is released by
      compression.

However, at that time adiabatic heating could only be explained as the disengagement of caloric from ordinary matter, caused by mechanical compression (Chang 2003, 904), for it was presupposed that

   P_1: gases can be heated without exchanges with the environment only if they
      contain heat in a latent form,

and

---

[14] Today we call it electromagnetic field.

P_2: only material substances can be contained by material substances in a latent form.

But the material substance of heat was just caloric: therefore, given P_1 and P_2, H^* entailed H, hence H too had to be considered essential.

Generalizing, we might say that at any time in the history of science, when a novel prediction NP is derived from an assumption H, there may be an assumption H^* which is proper part of H and sufficient—from a purely logical viewpoint—to derive NP; however, given certain current explicit or implicit presuppositions PRS, H may "appear to be conceptually or metaphysically entailed by" H^* (Vickers 2016, §4), hence scientists may falsely believe that H is essential to NP. At a later time, however, the advancement of science may show that the PRS are false, so scientists can (retrospectively) recognize that H was inessential, after all. This recognition is facilitated if meanwhile experimental or theoretical doubts have arisen about H, so that scientists started wondering whether H was really necessary. Yet, this recognition is logically independent of those doubts, as it follows already from the falsity of the PRS; hence, although retrospective, it is not circular.

Vickers, however, suggests that although prospectively we cannot ever be certain that H is essential to a prediction, *in some cases* we could (still prospectively) recognize it is *not essential*, although "metaphysically or conceptually" required by other considerations. This may (i) suggest the realist to "restrict her commitments to what is *directly confirmed* by the predictive successes" (i.e., H^*), and (ii) supply "a worthwhile heuristic," i.e., show which commitments should be abandoned first in case of empirical or theoretical refutations (i.e., H).

For instance, Bohr predicted (NP) the spectral lines of ionized helium by assuming that

H: The electron orbits the nucleus only on certain specific orbital trajectories, each characterized by a given quantized energy.

H turns out to be false, but NP could have been derived by the weaker hypothesis that

H^*: The electron can only have certain, specific, quantized energy states.

Moreover, says Vickers, Bohr could have seen (prospectively) that H was inessential, because it entailed H^*, which was enough to derive NP. But he still held H because "it may have been inconceivable at the time to think that electrons could have quantized electron energies *without* having associated quantized orbital trajectories (cf. Stanford 2006, 171)". Nonetheless, Bohr might have distinguished between his reason for believing H* (i.e., the essentiality of H* in predicting NP), and his reason for believing the content of H exceeding H* (i.e., his presupposition that electrons could have quantized energies only by having orbital trajectories). Thus he might have realized that "the latter were not as secure as the former" (Vickers 2016, §4).

I agree, but with the following qualifications:

(α) in many cases prospective recognition of inessentiality will be impossible;
(β) essential hypotheses cannot be identified with certainty, neither prospectively (as argued by Stanford) nor retrospectively;
(γ) however, this is not a problem, because the PM-I can be blocked even without identifying essential hypotheses, and the M-MT can be blocked even by identifying inessential hypotheses *retrospectively*.
(δ) at any rate, recognizing inessentiality (especially prospectively) is not a task for (realist) philosophers, but for scientists;
(ε) even when inessentiality can be recognized prospectively, this does not help the realist defense against the PM-I and M-MT.

Here are my arguments:

(α) Perhaps in Bohr's case the extra-content of H could be distinguished from H*; but in other cases it may consist in principles so obvious or deeply entrenched to pass unnoticed, such as the principles of conservation of energy and mass, isotropy and homogeneity of space, physical causal closure, etc. We usually presuppose so many principles of this kind, that even by paying close attention one cannot be certain to have ruled out all of them. Moreover, it may be extremely difficult to imagine an assumption H* that still entails NP once all of them have been discarded.

Besides, Vickers grants that H may be "conceptually" entailed by H*, i.e., that the contemporaries may lack the conceptual resources needed to distinguish H* from H. For instance, those who used Newtonian *forces* to predict Neptune couldn't even conceive any "effect" (such as the curvature of spacetime) which could cause the planetary motions except a *force*. Therefore, when H *seems* to be essential in deriving NP, it may be impossible (either prospectively or retrospectively) to see that it is not.

(β) A fortiori, we cannot ever be certain that there is *no* weaker component H* from which NP could have been derived. Therefore, we cannot ever be certain that H is essential, neither prospectively (as Stanford claims) nor retrospectively. This is just natural, for if we could identify essential components we would know that they are *completely* true. But if so, we could anticipate scientific progress much more than we actually can: while trusting that current theories are largely true, scientists grant that some of their assumptions are probably wrong, but only future research will tell which ones.

Besides, the PM-I is probably right that no theory or component older than 100 years or so was *completely* true (but not that none was *at least partly* true). Cordero, Peters, and Votsis are also right that we know when a hypothesis H is "true," if this means "*at least partly* true": but even if deployed in novel predictions, H may have been inessential, and even if otherwise well supported it can be partly false, and sooner or later replaced by another more completely true.

The progress of research allows us to drop more and more false presuppositions. Thus we may discover, retrospectively, that H was inessential, for it had some proper part H* which (i) can be true even if H is false and (ii) is sufficient to derive NP. That weaker H* may also be actively sought for, if H encounters empirical or theoretical refutations, which suggest that it cannot possibly be essential (since essential components are completely true). In this case (still retrospectively), one can recognize that H was inessential even before identifying its weaker substitute H*.

(γ) Neither the impossibility of recognizing essentiality, nor the difficulties in prospectively recognizing non-essentiality are a problem for deployment realists, because these recognitions are not necessary to resist the PM-I and the M-MT. In fact, *as soon as* a *risky novel* prediction NP derived from H is confirmed, we know (prospectively, and a fortiori retrospectively) that, short of miracles, there is *some* truth in H. This is enough to refute the PM-I's claims that both past and present theories are completely false.

The M-MT, in turn, is always used *retrospectively*, therefore only a retrospective recognition of inessentiality is needed to resist it: if at time t hypothesis H was firmly believed because considered essential to derive NP, but at t' it is shown to be false, the M-MT uses it as a counterexample to the NMA. But realists can reply by two moves: (i) *arguing* that probably H was inessential, for only by a miracle could NP have been derived from a completely false assumption; (ii) *showing* H was inessential, by identifying a proper part H* sufficient to derive NP, and the false presuppositions PRS that at t prevented us from distinguishing H from H*. Both moves are retrospective (at t' or at a later t"), yet sufficient to block the M-MT. Besides, move (ii) is independent of the refutation of H, therefore the realist's defense is not circular, as Stanford claims.

(δ) Even when it is possible, the recognition of non-essentiality (especially prospectively) is a task for scientists: philosophers just don't have the necessary expertise. The burden of scientific realists is arguing that certain criteria (e.g., essential deployment) can justify our beliefs in the at least partial truth of theories

Figure 9.1

or hypotheses. But Fahrbach (2017) asks too much when he requires that they also apply those criteria to actual research, so teaching practitioners which are the working hypotheses and which are the idle parts in their theories, urging changes, suggesting directions of research, etc.

Therefore, *qua* philosophers, realists need not be committed to any particular theory or hypothesis, not even to the best current ones: as argued by Smart (1963, 36), this is a task for scientists. At most, they may argue that science is convergent, hence, *in general*, current theories are *probably* more largely true than past ones. Therefore, *pace* Stanford (2017), they need not be more conservative than anti-realists. Actually, since they set for hypotheses a higher standard than anti-realists—truth, rather than empirical adequacy or the like—for them it is even more likely that any particular hypothesis fails to reach that standard, hence must be substituted by a better one.

(ε) The discussion at (γ) shows that even when inessentiality can be recognized prospectively, that is superfluous in resisting the PM-I and the M-MT. In fact, the at least partial truth of H can be recognized prospectively, even if H is not essential, and this is enough to block the PM-I. Besides, prospective recognition of inessentiality is superfluous against the M-MT, because the latter is always used retrospectively.

9.6 A More Complete Account of the Dispensability of Psillos's Condition (2)

Here I explain in more detail why my condition (2') is enough and we can do without the greater complexity of Psillos's condition (2) or its possible strengthening implicitly suggested by Lyons. When Psillos excludes any "*other* hypothesis H*," H* may be compatible or incompatible with H, and in different mereological relations to it. The following cases are included:

(a) H* is proper part of H. This is precisely the kind of alternative hypothesis H* excluded by my condition (2') (Figure 9.1)
(b) H is a proper part of H* (Figure 9.2)

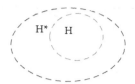

Figure 9.2

Figure 9.3

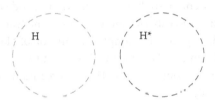

Figure 9.4

(c) H* and H coincide (Figure 9.3)
(d) Neither hypothesis is part of the other, and they have no common content
    (Figure 9.4)
(e) Neither hypothesis is part of the other, since they have some common con-
    tent but two alternative extensions (Figure 9.5)

Now, Psillos's condition (2) excludes alternative hypotheses H* of all these five
kinds, and here the troubles arise, while my condition (2') excludes only alternative hypotheses H* of kind (a), and I claim that this is enough for deployment realists. In cases (b), (d), and (e) H* has some content outside H; so, in
these cases H* might be incompatible with OHs and AA, and Lyons implicitly
suggests they should be excluded along with the cases in which H* is compatible with OHs and AA. But we don't need to worry about any of these cases,
and here is why:

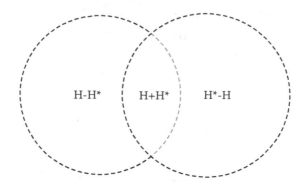

Figure 9.5

In case **(b)** H is weaker than and part of H*. Therefore the existence of H* does not make H inessential, so there is no need to exclude it. In fact, there have always been such unnecessarily rich alternatives to the true hypothesis (like e.g. the ether theory with respect to the simple assumption of Maxwell's equations, or the aforementioned demi-gods theory with respect to Newton's actual theory). From a purely logical viewpoint, for any H there are infinite such H*, but they do not make H inessential.

In case **(c)** H* coincides with H, so again it does not make H inessential and we don't need to exclude it.

In case **(d)** H and H* have no common content, and this may happen in two ways:

(d1) H and H* are compatible, hence, they may be both true. In general it is quite possible that a prediction is reached by independent inferential routes starting from different hypotheses. This is just natural if we conceive the world as a tightly connected whole, in which each phenomenon, at each scale, is causally linked to many other phenomena even at different scales. For instance, Perrin famously derived the same value for Avogadro's number reasoning from tenets of chemistry, thermodynamics, electrical theory, the theory of Brownian motions, and others. In such cases, each hypothesis H_i deployed in a different inferential route to NP is not "essential" in the sense that without H_i one could not have predicted NP (for NP *could* have been predicted through some different inferential routes involving some alternative hypothesis H_j). However, H_i can be *essential* in the sense that NP could not have been predicted along that particular inferential route without H_i, i.e., in the sense that H_i cannot be substituted in the same inferential route by one of its proper parts. This is precisely what my condition (2') requires, and it is all we need from a criterion of essentiality, for it is enough to warrant the complete truth of H_i (for we saw that, short of miracles, a hypothesis deployed in predicting NP cannot be completely false, and it can be partly false only if its false part is dispensable in the derivation). Therefore we need not exclude case (d1), for in this case H is essential in the required sense if it fulfills (2'). By the way, since H entails NP only in conjunction with AA and OHs, also AA and OHs are essential if they satisfy (2').[15]

(d2) Otherwise, H and H* may have no common part and be incompatible. One might think that we need to exclude this case: for if we had two incompatible hypotheses both predicting NP, we couldn't know which of them is true.[16] However, as explained in my nutshell account at section 9.3, this case is already excluded by the *risky* character of NP. In fact, as argued earlier, it is practically impossible to find a *completely* false hypothesis entailing a risky novel prediction. But if H is essential in the sense of (2'), it must be completely true. Therefore, if H* is incompatible with it, we know that (i) H* is at least partly false and (ii) it cannot possibly predict NP, for it contradicts an assumption (i.e., H) which is essential to that prediction.[17] Therefore, my condition (2') (together with the risky character of NP) already excludes that any hypothesis incompatible with H also predicts NP.

Finally, case (e): H and H* partially overlap. Like in case (d), here too we must distinguish:

(e1) If H and H* are compatible, it is possible, though unlikely, that they are both are essential, hence certainly true, as explained in case (d1). Otherwise, the essential assumption is neither H nor H*, but some part of their common content.

(e2) If H and H* are incompatible (like e.g. Rutherford's and Bohr's theories of the atom), neither of their non-overlapping parts is required by the derivation: in fact, NP is predicted by H without using H*-H, and it is predicted by H* without using H-H*. Therefore both

---

[15] Then we know they are true (unless NP was used in building them).

[16] Besides, we would have a counterexample to the NMA, for there would be at least one false hypothesis allowing us to derive a novel prediction.

[17] Instead, if H is not essential in my sense, it consists of an essential part H_e and an inessential part H_i. If H* also has a role in predicting NP, it may consist in an essential part H^*_e and an inessential part H^*_i. Of course H^*_e cannot contradict H_e, but H^*_i may contradict H_i, in which case H and H* are incompatible, yet have both a (non-essential) role in predicting NP, and are both partly true. Instead H_e and H^*_e are two essential but compatible hypotheses as considered in case (d1).

H and H* have a superfluous part, hence neither is essential. If H were essential in my sense, it would be practically impossible that H* predicted NP, for it would contradict an assumption which is needed to predict it. So, even in case (e2) condition (2') is enough to warrant the truth of H.

Summing up, my condition (2') is enough to guarantee that a hypothesis H is *essential* in the derivation of a novel prediction NP, in a sense which warrants its complete truth, and we don't need Psillos's more complex and troublesome condition (2) or its strengthening implicitly suggested by Lyons.

9.7 Objections and Answers

Here I reply to some potential objections.

9.7.1 The "False May Entail True" Objection

Since a falsity may entail a truth, even false theories might make novel predictions.

Answer:

As argued at the beginning of section 9.1, in the Hyperuranion of possible theories there are some with wholly false theoretical assumptions which make true novel and improbable predictions (F^{NIP}-theories). But their rate is exceedingly small, hence it is extremely unlikely that scientists pick one of them (because scientists look for true theories, hence they could find F^{NIP}-theories only by chance). On the opposite, *all* true theories have true consequences, and those sufficiently fecund have true *novel* consequences. Granted, true (and fecund) theories are much fewer than false ones; however, they are not found *by chance*, but *on purpose* and through *reliable methods* (White 2003; Alai 2016, 552–554; 2018, §III).

9.7.2 The Disjunctive Objection

Your condition (2') cannot be fulfilled, hence no hypothesis is essential by your criterion, because any hypothesis H always entails some weaker claim H', like e.g.,

Disjct: H or the Moon is made of cheese.

Answer:

First, condition (2') excludes that NP is entailed by any hypothesis which is *part* of H (in Yablo's sense), not just by any hypothesis entailed by H. In fact, **Disjct** is not *part* of H. Second, disjunctions are weaker than their disjuncts. Therefore, if H entails NP, in most cases H*: «H or H'» does not. Hence, in such cases H fulfills (2'). For instance, Newton's Gravitation Law entailed the existence of Neptune; but the disjunction of this law with 'The Moon is made of cheese' does not. Therefore in certain cases NP cannot be entailed by any hypothesis which is part of H, hence H is essential.

In particular instances the disjunction of H with another hypothesis H' might still be strong enough to entail NP (Vickers 2016, §3). Obviously, however, this can happen only if even H' alone entails NP. But in this case H and H' must be in one of the five mereological relations examined in section 9.6, and I argued that in each of them condition (2') is enough to decide whether H (and H') is essential or not.

9.7.3 The "Theoretician's Dilemma" Objection 1

The author's condition (2') seems to make realism redundant for purely logical reasons. A statement is derivable from a set of statements iff the content of that statement is already included in that set of statements. The only proper part of that set of statements that's required to derive the statement is in fact the statement itself. Thus, a novel prediction NP needs only NP to be entailed. But surely we want to be realists about more things than the content corresponding to NP.[18]

Answer:

Most frequently and typically NP is not part of H, because, as explained by Vickers[19] and others, it is not entailed by H alone, but only in conjunction with certain other hypotheses OHs of T, and various auxiliary assumptions AA. Therefore (2') does not allow us to dispense with H in favor of NP.

But the objection might become that since H&OHs&AA entails NP, we can dispense with H&OHs&AA in favor of NP. This is reminiscent of Hempel's (1958) "theoretician's dilemma": theoretical hypotheses H are required to connect the observable initial conditions IC to the observable final conditions FC, and they are justified only if they succeed in this, i.e., if IC→H→FC. But then why not drop H and just keep the empirical laws IC→FC? Hempel answered that this would work from a purely logical point of view, but not from an epistemological point of view: H does not predict only IC→FC, but many other phenomena as well, most of which we would never have discovered without assuming H.

---

[18] Objection raised by an anonymous referee.
[19] Vickers (2013, p. 202 footnote 9; 2016, footnote 8).

The answer here is similar: we would never have found NP, and many other predictions, unless H&OHs&AA were true. In fact, NP is novel and risky, hence it cannot have been found just by chance. Besides, the auxiliary assumptions AA are the consequences of many independent theories T', T''. . . T^n, typically dealing with matters different from T. Therefore, it would be a miraculous coincidence if even one of them were completely false, yet the conjunction of their consequences entailed NP. The only plausible explanation is that T, T', T''. . . T^n are all at least partially true (Alai 2014c, §4). Besides, each of T, T', T''. . . T^n, in conjunction with still different assumptions, issues many other predictions, which are only explainable in the same way.

One might reply that guessing n true theories is even less likely than guessing n false theories with one true joint consequence. This is right, but guessing n false theories with one true, novel, and risky joint consequence is still too improbable to be a minimally plausible explanation of our frequent predictive successes. So, how was it possible to derive NP and so many other novel predictions? The only plausible explanation is that theories are not *guessed*, but found by a method which is reliable in tracking truth. Scientific realists have a reasonable account of why the scientific method is reliable, while anti-realists have never been successful in providing an account of how predictive success might be achieved by pure chance or "miracles" (Alai 2014a). Notice, here the truth-conduciveness of the scientific method is not a *petitio principii*, but the conclusion of my inference to the only plausible explanation in this paragraph (the second in this section).[20]

9.7.4 The "Theoretician's Dilemma" Objection 2

By your criterion no hypothesis will ever be essential, because whenever H together with OHs and AA entails prediction P, there is always a weaker hypothesis doing the same, viz.

H*: If OHs&AA, then P.[21]

Answer:
Again, here H* is not part of H, hence its existence does not make H inessential. For instance, suppose the following:

H: The atoll is loaded with deadly radioactive material;

---

[20] White (2003); Alai (2016, 552–554; 2018 §III).
[21] Objection raised in discussion by John Worrall, who cited a *topos* from the debates on confirmation.

OHs&AA: Carl lands on the atoll;
P: Carl will die;
H*: If Carl lands on the atoll he will die.

Of course H* together with OHs&AA entails P. However H* is not part of H, because there are truthmakers of H* (e.g., «There will be a disastrous *tsunami* on the atoll») not implied by any truthmaker of H. One relevant consequence is that while H explains P, H* does not. Even more importantly, in cases like this H* is a purely "empirical" conditional, connecting the initial conditions with an empirical consequence. Therefore, unless it were inferred from the "theoretical" hypothesis H, H* could be discovered only a posteriori, by witnessing P. Therefore P would not be novel, and H* would not have been deployed in a *novel* prediction.

9.7.5 The "Ramsey Sentence" Objection

Given that whatever follows from H also follows its Ramsey sentence (RS^H), and that H implies RS^H, wouldn't it follow that every hypothesis is dispensable in favor of its Ramsey sentence?[22]

Answer:

No: in general, the essential part H* of H does not coincide with RS^H, because it contains both something more and something less than RS^H. It must contain something more because RS^H would not allow to derive NP. For instance, assuming in Newton's gravitation law (N) the terms "force" and "mass" are theoretical and "distance" is observational, its Ramsey sentence is

$$\mathbf{RS}^N : \exists x, \exists y \left(x = G y_1 y_2 / D^2 \right).$$

But RS^N is not enough to predict Neptune. We also need to say what kind of properties or relations are x and y, how they behave, etc.: we need an expanded Ramsey sentence like

$$\mathbf{RS}^{NEx} : \exists x, \exists y \left(....x.... \ \& \ ....y.... \ \& \ x = G y_1 y_2 / D^2 \right),$$

where '....x....' and '...y...' are various laws concerning x and y (which, in practice, characterize them respectively as force and mass).

---

[22] Objection raised by an anonymous referee.

But typically the essential part H* contains also something less than RS^H or RS^{HEx}. For instance, Newton's Gravitation Law is partly false (since there exists no gravitation force). Consequently, also RS^N and RS^{NEx} are partly false, hence, inessential. Therefore, in general, the essential H* is different from RS^H. It might still be representable by an expanded Ramsey sentence RS^{H*Ex} (a proper part of RS^{HEx}), but this wouldn't be a problem for realism, because:

(i) Like any Ramsey sentence, RS^{H*Ex} is not a merely empirical claim, it states the existence of unobservable entities, even without naming them.

(ii) In general, RS^{HEx} will have a very complex structure, hence it might be practically impossible to write it out completely, so to *dispense* with H. Besides, as argued in section 9.5, identifying the essential part of hypotheses is very difficult, so even if we could spell out RS^{HEx}, often we would be unable to dispense with it in favor of its essential part RS^{H*Ex}.

(iii) At any rate, my claim is that the NMA commits only to essential hypotheses, not that only essential hypotheses can be true. So, if the essential hypothesis turns out to be something like RS^{H*E}, since Ramsey sentences characterize only partially the entities to which they refer, realists may hold that something stronger than RS^{H*E} is true.

9.7.6 The "Modus Ponens" Objection

By your criterion no hypothesis will ever be essential, because whenever H together with OHs and AA entails NP, numberless but intuitively irrelevant hypotheses do the same, for instance,

H**: (If God exists, then H) & God exists.[23]

Answer:
As explained above, H** is neither part of H nor implied by it. Therefore H** does not make H inessential.

Acknowledgments

I am grateful to the organizers of the "Quo Vadis Selective Scientific Realism?" conference in Durham (2017) for generously helping my participation with

---

[23] Objection raised in discussion by John Worrall, referring to another *topos* from the debates on confirmation.

a travel bursary. I received helpful comments and suggestions from Gustavo Cevolani, José Díez, Vincenzo Fano, Ludwig Fahrbach, Greg Frost-Arnold, Pablo Lorenzano, Giovanni Macchia, Flavia Marcacci, Matteo Plebani, Andrea Sereni, John Worrall, and especially Peter Vickers and Timothy Lyons. Thanks to Maria Grazia Severi and Peter Vickers for checking my English grammar.

This research was supported by the University of Urbino Carlo Bo through the *Contributo d'Ateneo progetti PRIN2015* and the DiSPeA Research Projects 2017 and by the Italian Ministry of Education, University and Research through the PRIN 2017 project "The Manifest Image and the Scientific Image" prot. 2017ZNWW7F_004. For this work, I profited from a leave of absence granted by the University of Urbino (Department of Pure and Applied Sciences) in 2020, during which I was affiliated with the International Academy for Philosophy of Science.

References

Alai, M. (2012) "Levin and Ghins on the 'No Miracle' Argument and Naturalism," *European Journal for Philosophy of Science* vol. 2, n. 1, 85–110. http://rdcu.be/mTcs.

Alai, M. (2014a) "Why Antirealists Can't Explain Success," in F. Bacchini, S. Caputo and M. Dell'Utri (eds.) *Metaphysics and Ontology Without Myths*, Newcastle upon Tyne: Cambridge Scholars Publishing, pp. 48–66.

Alai, M. (2014b) "Defending Deployment Realism against Alleged Counterexamples," in G. Bonino, G. Jesson, J. Cumpa (eds.) *Defending Realism. Ontological and Epistemological Investigations*, Boston-Berlin-Munich: De Gruyter, pp. 265–290.

Alai, M. (2014c) "Novel Predictions and the No Miracle Argument," *Erkenntnis* vol. 79, n. 2, 297–326. http://rdcu.be/mSra.

Alai, M. (2014d) "Explanatory Realism," in E. Agazzi (ed.) *Science, Metaphysics, Religion: Proceedings of the Conference of the International Academy of Philosophy of Science, Siroki Brijeg 24–24 July 2013* Milano, Franco Angeli, pp. 99–116.

Alai, M. (2016) "The No Miracle Argument and Strong Predictivism vs. Barnes," in L. Magnani and C. Casadio (eds.) *Model-Based Reasoning in Science and Technology. Logical, Epistemological, and Cognitive Issues*, Cham: Springer, pp. 541–556.

Alai, M. (2017) "Resisting the Historical Objections to Realism: Is Doppelt's a Viable Solution?" *Synthese* vol. 194, n. 9, 3267–3290. http://rdcu.be/mSrG.

Alai, M. (2018) "How Deployment Realism Withstands Doppelt's Criticisms," *Spontaneous Generations* vol. 9, n. 1, 122–135.

Carrier, M. (1991) "What is Wrong with the Miracle Argument?" *Studies in History and Philosophy of Science* vol. 22, 23–36.

Chang, H. (2003) "Preservative Realism and Its Discontents: Revisiting Caloric," *Philosophy of Science* vol. 70, 902–912.

Clarke, Steve, and Timothy D. Lyons (eds.) (2002) *Recent Themes in the Philosophy of Science. Scientific Realism and Commonsense.* Dordrecht: Kluwer.

Cordero, A. (2017a). *Retention, truth-content and selective realism*, in E. Agazzi (ed.) *Varieties of Scientific Realism*, Cham: Springer, pp. 245–256.

Cordero, A. (2017b) "Making Content Selective Realism the Best Realist Game in Town," *Quo Vadis Selective Scientific Realism?* Conference, Durham, UK 5–7 August 2017.

Fahrbach, L. (2017) "How Philosophy Could Save Science," *Quo Vadis Selective Scientific Realism?* Conference, Durham, UK 5–7 August 2017.

Hempel, Carl G. (1958) "The Theoretician's Dilemma: A Study in the Logic of Theory Construction," *Minnesota Studies in the Philosophy of Science* vol. 2, 173–226.

Hoskin, M. (ed.) (1999) *The Cambridge Concise History of Astronomy*, Cambridge: University Press.

Hudson, J. (1992) *The History of Chemistry.* London: Macmillan Press.

Kitcher, P. (1993) *The Advancement of Science.* New York: Oxford University Press.

Koyré, A. (1961) *La révolution astronomique.* Paris: Hermann.

Laudan, L. (1981) "A Confutation of Convergent Realism," *Philosophy of Science* vol. 48, 19–49.

Lipton, P. (1991) *Inference to the Best Explanation*, London: Routledge.

Lipton, P. (1993) "VI—Is the Best Good Enough?," *Proceedings of the Aristotelian Society* vol. 93, 89–104.

Lipton, P. (1994) "Truth, Existence, and the Best Explanation," in A. A. Derksen (ed.), *The Scientific Realism of Rom Harré.* Tilburg: Tilburg University Press, pp. 89–11.

Lyons, T. D. (2002) "The Pessimistic Meta-Modus Tollens," in S. Clarke and T. D. Lyons (eds.) *Recent Themes in the Philosophy Science. Australasian Studies in History and Philosophy of Science*, vol 17, Dordrecht: Springer, pp. 63–90. doi:10.1007/978-94-017-2862-1_4.

Lyons, T. D. (2003) "Explaining the Success of a Scientific Theory," *Philosophy of Science* vol. 70, 891–901.

Lyons, T.D. (2006) "Scientific Realism and the Stratagema de Divide et Impera," *The British Journal for the Philosophy of Science*, vol. 57, 537–560.

Lyons, T.D. (2009) "Criteria for Attributing Predictive Responsibility in the Scientific Realism debate: Deployment, Essentiality, Belief, Retention," *Human Affairs* vol. 19, 138–152.

Lyons, T.D. (2016) "Structural Realism versus Deployment Realism: A Comparative Evaluation," *Studies in History and Philosophy of Science* Part A vol. 59, 95–105.

Lyons, T.D. (2017) "Epistemic Selectivity, Historical Threats, and the Non-Epistemic Tenets of Scientific Realism," *Synthese* vol. 194 n. 9, 3203–3219.

Musgrave, A. (1985) "Realism versus Constructive Empiricism," in P. Churchland and C. Hooker (eds.) *Images of Science*, Chicago, Chicago University Press, pp. 197–221.

Musgrave, A. (1988) "The Ultimate Argument for Scientific Realism," in Robert Nola (ed.) *Relativism and Realism in Science*, Dordrecht: Springer, pp. 229–252.

Peters, D. (2014) "What Elements of Successful Scientific Theories are the Correct Targets for "Selective" Scientific Realism?" *Philosophy of Science* vol. 81, 377–397.

Plebani, M. (2017) "Aboutness for Impatients," *Academia.edu*, https://www.academia.edu/34295884/Aboutness_for_the_impatient.

Psillos, S. (1999) *Scientific Realism: How Science Tracks Truth.* London-New York: Routledge.

Sankey, H. (2001) "Scientific Realism: An Elaboration And A Defense," *Theoria*, vol. 98, 35–54.

Smart, J.J.C. (1963) *Philosophy and Scientific Realism.* London: Routledge.

Stanford, P. K. (2006). *Exceeding Our Grasp: Science, History, and the Problem of Unconceived Alternatives.* Oxford: Oxford University Press.

Stanford, K. P. (2009) "Author's Response," in "Grasping at Realist Straws," a Review Symposium of Stanford (2006), *Metascience* vol. 18, 379–390.

Stanford, P. K. (2017) "A Difference That Makes a Difference: Stein on Realism, Instrumentalism, and Intellectually Nourishing Snacks," *Quo Vadis Selective Scientific Realism?* Conference, Durham, UK 5–7 August 2017.

Vickers, P. (2013) "A Confrontation of Convergent Realism," *Philosophy of Science* vol. 80, n. 2, 189–211.

Vickers, P. (2016) "Understanding the Selective Realist Defence against the PMI," *Synthese* vol. 194, 3221–3232. https://doi:10.1007/s11229-016-1082-4.

Vickers, P. (2018) "Disarming the Ultimate Historical Challenge to Scientific Realism," *The British Journal for the Philosophy of Science* axy035, https://doi.org/10.1093/bjps/axy035

Votsis, I. (2011) "The Prospective Stance in Realism," *Philosophy of Science* vol. 78, 1223–1234.

White, R. (2003) "The Epistemic Advantage of Prediction over Accommodation," *Mind* vol. 112 n. 448, 653–683.

Wright, J. (2002) "Some Surprising Phenomena and Some Unsatisfactory Explanations of Them," in Clarke and Lyons (eds.), *Recent Themes in the Philosophy of Science.* Australasian Studies in History and Philosophy of Science, vol. 17. Dordrecht: Springer, pp. 139–153. https://doi:10.1007/978-94-017-2862-1_4.

Yablo, S. (2014) *Aboutness.* Princeton: Princeton University Press.

10

Realism, Instrumentalism, Particularism

A Middle Path Forward in the Scientific Realism Debate

P. Kyle Stanford

Department of Logic and Philosophy of Science
University of California, Irvine
stanford@uci.edu

10.1 Introduction

Here I propose a particular conception of both the current state of play and what remains at stake in the ongoing debate concerning scientific realism. Recent decades have witnessed considerable evolution in this debate, including a welcome moderation of both realist and instrumentalist positions in response to evidence gathered from the historical record of scientific inquiry itself. In fact, I will suggest that a set of commitments has gradually emerged that are now embraced by many (though by no means all) who call themselves scientific realists and also by many (though by no means all) who would instead characterize themselves as instrumentalists or (perhaps less helpfully) "antirealists." I will go on to suggest that these commitments collectively constitute a shared "Middle Path" on which many historically sophisticated realists and instrumentalists have already made substantial progress together, perhaps without realizing how much closer their respective views have thereby become to one another than either is to the classical forms of realism and instrumentalism whose labels and slogans they nonetheless conspicuously retain. I will then go on to suggest, however, that at least one crucial disagreement still remains even between realists and instrumentalists who walk this Middle Path together, and that this remaining difference actually *makes* a difference to how we should go about conducting scientific inquiry itself. Recognizing this difference poses a further challenge for instrumentalists both on and off the Middle Path, however, and responding to it will illuminate a further dimension of the realism debate itself: namely, the level(s) of abstraction or generality at which we should seek and expect to find useful epistemic guidance concerning scientific theories or beliefs. Along this dimension of the debate,

P. Kyle Stanford, *Realism, Instrumentalism, Particularism* In: *Contemporary Scientific Realism.*
Edited by: Timothy D. Lyons and Peter Vickers, Oxford University Press. © Oxford University Press 2021.
DOI: 10.1093/oso/9780190946814.003.0010

I suggest, Middle Path realists and instrumentalists are once again united, but this time against both the shared presuppositions of their classical predecessors on the one hand and radical or extreme forms of particularism on the other.

10.2 Finding a Middle Path

The first of the shared commitments that I suggest constitute a Middle Path between classical forms of both realism and instrumentalism is to what I've elsewhere (Stanford 2015) called "Uniformitarianism," in parallel with the great battle in 19th century geology between Catastrophists and Uniformitarians concerning the causes and pattern of changes to the Earth. Famously, Uniformitarians (like Charles Lyell) argued that the broad topographic and geographic features of the Earth were the product of natural causes like floods, volcanoes, and earthquakes operating over immense stretches of time at roughly the same frequencies and magnitudes at which we find them acting today. Their Catastrophist opponents (like Georges Cuvier) held instead that such natural causes had operated in the past with considerably greater frequency and/or magnitude than those we now observe, on the order of the difference between a contemporary flood and the great Noachian deluge reported in the Christian Bible. Catastrophists thus held that the Earth had steadily quieted down over the course of its history, that truly fundamental and wide-ranging changes to its topography and geography are now confined to the distant past, and that contemporary natural causes will further modify that topography and geography only in comparatively marginal and limited ways. By contrast, Uniformitarians held that if given enough time to operate present-day natural causes would continue to transform the existing topography and geography of the Earth just as profoundly as it was transformed in the past.

Likewise in the case of the realism dispute, Uniformitarians take the view that the future of the scientific enterprise will continue to be characterized by theoretical revolutions and conceptual transformations just as profound and fundamental as those we have witnessed throughout the history of that enterprise. By contrast, a textbook description of what I will call classical or Catastrophist realism holds that "Our mature scientific theories, the ones used to underwrite our scientific projects and experiments, are mostly correct" and "[w]hat errors our mature theories contain are minor errors of detail" (Klee 1999 313–4). Such Catastrophists take the view that truly profound and fundamental revisions to our scientific understanding of the world are either largely or completely confined to the past and that the future of scientific inquiry will not be characterized by the sorts of fundamental revolutions or transformations by which theories like Newton's mechanics, Dalton's atomism, and Weismann's theory of the germ-plasm ultimately came to be profoundly modified, qualified, amended, or simply replaced. That is, Catastrophist realists hold that the theoretical orthodoxy embraced by future scientific communities will include what seem both to them and us to be simply expanded, amended, and more sophisticated versions of the most successful theories we ourselves have already adopted. Thus, while Uniformitarians see us as being in the midst of an ongoing historical process of fundamental revolution or transformation in our scientific beliefs, rather than as having the enviable good fortune of living at or near the end of that process, Catastrophist scientific realists instead seem forced to adopt some form of exceptionalism concerning at least the most successful scientific theories of the present day.

Such classical or Catastrophist realism has not only been influentially championed by philosophical luminaries like Smart (1963) and Putnam (1975) but is also often claimed to be the view of the matter favored by common sense itself. It is therefore striking that in recent years many prominent historically sophisticated scientific realists have abandoned such Catastrophism in favor of the Uniformitarian alternative while insisting that this concession does not undermine the more modest realist epistemic entitlements that they themselves defend. So-called "selective" scientific realists have argued, for example, that although we should indeed anticipate further radical and fundamental changes in our theoretical conception of the natural world, we can nonetheless identify particular elements, aspects, or features of our best scientific theories that we can justifiably expect to find preserved throughout the course of such further changes, whether those privileged elements are held to be their claims about the "structure" of nature (Worrall 1989), their "working posits" (Kitcher 1993), their "core causal descriptions" (Psillos 1999), their "detection properties" (Chakravartty 2007, Egg 2016), the posits that "unify the accurate empirical claims" of a theory (Peters 2014), the verae causae they identify (Novick and Scholl 2020), or something else altogether. This commitment to Uniformitarianism would seem to *unite* such selective realists with historically motivated defenders of instrumentalism or anti-realism and against classical, commonsense, or Catastrophist varieties of scientific realism itself.

But realists and instrumentalists on the Middle Path are similarly united in rejecting central commitments of many classical forms of instrumentalism or anti-realism as well, including the idea, perhaps most familiar from Thomas Kuhn in his most exuberant moods, that the changes still to come in our scientific conception of the world will periodically be *so* radical and profound that they will render it impossible to impartially, neutrally, or fairly compare theories on either side of such a revolutionary divide either to one another or to the available body of empirical evidence. Instead, Middle Path instrumentalists allow not only that we can perfectly well articulate the competing claims of two scientific theories about a single shared world using language that privileges neither theory, but also that we can characterize the available evidence in support of each theoretical alternative in ways that are at least neutral *between those two competing theories* (if not independent of any and all "theorizing" whatsoever) in reaching an impartial judgment that one is better or worse confirmed by the existing evidence, or even that one or both theories are definitively refuted (within the bounds of fallibilism) by that evidence. They similarly reject the further consequence drawn by various interest-driven theorists of science that the outcomes of such theoretical competitions are therefore typically determined less by the available evidence in support of each theoretical alternative than by the comparative power, standing, resourcefulness, and determination of the social groups who advance and defend them. Where Uniformitarianism assures us that further profound and fundamental changes are still to come for our scientific picture of the world, a further commitment to what we might naturally call *commensurability* assures us that such changes will neither prevent us from fully understanding the competing conceptions of nature thereby proposed nor from impartially adjudicating the character and strength of the evidence in support of each of those competing conceptions. This is certainly not to suggest that resolving such competitions is simple or straightforward, or that an individual scientist's own theoretical (and other) sympathies do not influence her evaluation of the evidence, or even that such comparisons can always be convincingly resolved by the particular body of evidence available at a given time. But it is to claim that their ultimate resolution typically depends on the accumulation of evidence that the relevant scientific community rightly sees as objectively and impartially favoring one proposal over the other. Such a commitment to commensurability would seem to similarly unite instrumentalists and realists on the Middle Path together with one another and against such radically Kuhnian varieties of instrumentalism or anti-realism.

Realists and instrumentalists who travel this Middle Path together are also united against other classical varieties of instrumentalism or anti-realism by their commitment to what I will call the Maddy/Wilson Principle, which I first encountered when my colleague Penelope Maddy skeptically responded to my own instrumentalist sympathies with this especially pithy formulation: "well, nothing works by *accident* or for *no* reason." That is, she seemed to regard instrumentalism as committed not simply to the view that our best scientific theories are powerful cognitive instruments that guide our predictions, interventions, and other forms of practical engagement with nature successfully, but also that it is misguided or somehow illegitimate even to ask, much less try to discover or explain, how they manage to achieve that success when and where they do. Having recognized the ability of our best theories to guide our practical engagement with nature successfully, such instrumentalism rejects even the demand for any explanation of how or why they are able to do so.

When I reminded her of this conversation some years later, she was quick to credit the influence of Mark Wilson's *Wandering Significance* as having inspired this particular expression of her disquiet. And indeed, Wilson does articulate something very like this same concern in his own inimitable way:

> However, I regard this [instrumentalist] terminology as misleading because successful instrumentalities, whether they be of a mechanical or a symbolic nature, always work for *reasons*, even if we often cannot correctly diagnose the nature of these operations until long after we have learned to work profitably with the instruments themselves. (2006 220, original emphasis)

Both Maddy and Wilson, then, endorse the broad principle that when a scientific theory enjoys a track record of fine-grained and wide-ranging success in guiding our practical engagement with the world, it typically does so in virtue of *some* systematic connection or relationship between the description of the world offered by that theory and how things actually stand in the world itself.

It is no accident that Maddy and Wilson both regard instrumentalism as committed to the idea that there is simply nothing more to say or know about how and why our instrumentally powerful scientific theories work as well as they do when and where they do. Scientific instrumentalism has had a wide variety of incarnations over the last several centuries, and some of the most influential are indeed characterized by explicit or implicit versions of this commitment. Perhaps most famously, some logical positivist and logical empiricist thinkers held that the claims of our best scientific theories are not even assertions about the world in the first place, nor, therefore, even *candidates* for truth or falsity, but instead simply "inference tickets" allowing us to predict or infer some observable states of affairs from others. More recently, van Fraassen's Constructive Empiricism argues that it is illegitimate to demand anything more than the empirical adequacy of a theory as an explanation of its success, insisting instead "that the observable phenomena exhibit these regularities, because of which they fit the theory, is merely a brute fact, and may or may not have an explanation in terms of unobservable facts 'behind the phenomena'" (1980 24). Thus, there is indeed a long and distinguished philosophical tradition advocating forms of instrumentalism that either reject or remain agnostic concerning the Maddy/Wilson Principle, insisting instead that once we have recognized the instrumental utility of our scientific theories it is somehow misguided or illegitimate even to ask how or why those theories manage to be so instrumentally useful when and where they do.

The influential legacy of such classical instrumentalism makes it critical to recognize not just the plausibility but also the foundational importance of the Maddy/Wilson Principle. In science as in ordinary life, it is generally true that when things work, they work for reasons, and when scientific theories are able to achieve robust empirical and practical success, it is surely at least reasonable to think that the reasons for that success will consist in *some* systematic relationship or connection between how the theory represents (some part of) the world as being and how things actually stand there—otherwise that empirical success really would be miraculous! Indeed, the Middle Path not only embraces the Maddy/Wilson Principle's insistence that there must *be* some reason why a cognitive instrument that works well does so, but also goes so far as to insist that the historical record itself provides us with abundant exemplars of at least the broad *sorts* of such systematic relationships that presumably also constitute the reasons for the successes of various contemporary scientific theories as well, whether or not we are ever in a position to specify those reasons more precisely in particular cases.

To see how, suppose for a moment that Einstein's relativistic mechanics represents a true and complete account of how things stand in the otherwise inaccessible domain of nature that it seeks to describe. On that assumption, we can explain how and why Newton's gravitational mechanics is so spectacularly successful when and where it is by describing in detail how aspects or elements of Newtonian mechanics are systematically related to features of the actual world (as described by General Relativity) and how these systematic relationships permit the theoretical apparatus of Newtonian mechanics to make accurate predictions and guide interventions successfully across a wide though not unlimited range of circumstances. *If* contemporary theoretical orthodoxy in mechanics simply describes how things really stand in nature, the details of those systematic relationships themselves constitute the "reasons" why Newtonian mechanics works as well as it does. Assuming the truth of contemporary theoretical orthodoxy more broadly would similarly allow us to identify the actual reasons for the systematic instrumental utility of many other successful but ultimately rejected scientific theories of the past such as the caloric theory of heat, phlogistic chemistry, the wave theory of light, Weismann's theory of the germ-plasm, and many other familiar examples. And *even if we think all contemporary theories are fundamentally mistaken*, we should fully expect that highly successful contemporary scientific theories stand in one or more systematic relationships to the truth of the matter regarding various otherwise inaccessible domains of nature that are *like* the relationship between Newtonian and relativistic mechanics, or *like* that between caloric theory and contemporary thermodynamics, or like that between Weismann's theory of the germ-plasm and contemporary molecular genetics, and so on. Of course, even this small collection of familiar examples makes clear that there is no *single* such relationship holding between all earlier successful theories and their contemporary successors: realists would need to appeal to a wide and heterogeneous array of different systematic relationships or "reasons" of this sort in explaining the diverse particular empirical successes of the various instrumentally powerful, past scientific theories that have since been abandoned or replaced.

I suggest that many realists and instrumentalists alike, especially those who rely on evidence drawn from the historical record, have gradually come to embrace these shared commitments to Uniformitarianism, commensurability, and the Maddy/Wilson Principle, perhaps without fully realizing how much more they have thereby come to agree with one another than either of them does with the more classical doctrines whose labels and slogans they nonetheless retain. Their broadly shared conception of the past, present, and future of scientific inquiry anticipates many substantial continuities between future scientific orthodoxy and that of the present day (just as there are between present theoretical orthodoxy and that of the past), but also many substantial discontinuities just as profound and significant as those we now find in past historical episodes of fundamental revolution or upheaval. Moreover, I suggest that realists and instrumentalists who travel this Middle Path have already made considerable progress in using further historical evidence to refine and elaborate that broadly shared conception. Important recent work conducted by self-described realists and antirealists alike, for example, has revealed that the mistaken (i.e., ultimately rejected) claims or components of successful past theories have often played crucial and ineliminable roles in generating those theories' empirical successes (e.g., Lyons (2002, 2006, 2016, 2017), Saatsi and Vickers (2011); for a fairly comprehensive listing, see Vickers (2013) and the references contained therein). Likewise, Peter Vickers (2013, 2017) has recently argued that we can at least identify posits (including some that do genuine work in generating a theory's successful implications) that are nonetheless inessential or eliminable from those theories in a sense that renders them poor candidates for realist commitment. Such increases in the nuance and sophistication of their broadly shared vision of the past, present, and future of scientific inquiry should be recognizable as important forms of progress by both realists and instrumentalists alike on the Middle Path.

10.3 Trouble in Paradise: A Remaining Difference that Makes a Difference

I do not mean to suggest, however, that these shared commitments of realists and instrumentalists on the Middle Path leave no room for significant disagreement between them. As noted earlier, "selective" scientific realists are largely motivated by the idea that the historical record itself (perhaps together with other considerations) puts us in a position to reliably distinguish the *particular* elements, aspects, or features of our own best scientific theories (e.g., working posits, core causal descriptions, structural claims, etc.) genuinely responsible for their empirical successes from those that are instead "idle" or otherwise not required for those successes (e.g., the electromagnetic ether). Even if we expect further profound and dramatic changes in theoretical orthodoxy as scientific inquiry proceeds into the future, Middle Path realists insist that we can nonetheless justifiably expect these privileged elements, aspects, or features to be retained and ratified in some recognizable form throughout the further course of scientific inquiry itself. Note that without a commitment to at least this minimal form of stability or persistence, the selective realist loses not only her grounds for claiming to have identified secure epistemic possessions on which we can safely rely as inquiry proceeds, but also her grounds for claiming that these particular elements, aspects, or features of our best scientific theories were those genuinely *responsible* for their various empirical successes (for which they have turned out not to be essential or required after all).

By contrast, those on the Middle Path who embrace the "instrumentalist" label certainly accept the Maddy/Wilson Principle's insistence that there must *be* such reasons for the successes of our best scientific theories, but they nonetheless deny that we ourselves are generally in a position to know what those reasons are in particular cases, or to reliably identify which particular elements or features of our best scientific theories we should therefore expect to find persisting throughout the course of further inquiry. Such instrumentalists will see the wide *variety* of such reasons that (as we noted earlier) contemporary scientific orthodoxy would need to invoke in explaining the various successes of various past scientific theories as piling up counterexamples rather than confirmation for the idea that any *single* aspect, element, or feature of a successful scientific theory is invariably or even just reliably preserved in its successors: *sometimes* equations or claims about the "structure" of nature are preserved from one successful scientific theory to its successors, but other times it is the fundamental entities posited by that theory, or the "core causal descriptions" of those entities, or something else altogether—no *one* of these forms of continuity (or corresponding versions of selective realism itself) seems to capture even an especially wide range of central historical examples.

Rather than seeking to adjudicate this remaining central point of disagreement dividing even realists and instrumentalists traveling together on the Middle Path, I will instead argue for its importance. It truly matters, I suggest, because it actually *makes* a difference to how we should go about conducting further scientific inquiry itself. To see why, recall first the case of classical, commonsense, or Catastrophist realists, who are confident that the theoretical orthodoxy embraced by future scientific communities will include what seem both to us and to the members of those communities to be simply updated, expanded, and more sophisticated versions of at least the most successful theories that we ourselves have already embraced. At least with respect to those successful theories, then, the classical or Catastrophist realist simply does not see any real need for what the National Science Foundation and other granting agencies call "transformative science" and characterize explicitly as "revolutionizing entire disciplines; creating entirely new fields; or disrupting accepted theories and perspectives" (Bement 2007). The realist should be perfectly happy (or at least systematically happier than her instrumentalist counterpart, as will become clear below) for review boards to reject lines of research or theoretical proposals that fundamentally contradict or violate existing successful scientific theories, as she thinks it quite *un*likely that any such alternative will ever become part of the theoretical orthodoxy we come to embrace in the future. Indeed, the more conflict there is between a given theoretical proposal and the central claims of existing successful theoretical orthodoxy, the more confident she will be that it is misguided in some fundamental and fatal way.

Of course, Middle Path realists instead hold only that we can justifiably predict which specific elements, aspects, or features of our best scientific theories are responsible for their successes, but they similarly expect those same elements, aspects, or features of our own theories to persist and be preserved in some recognizable form in any future scientific theory we ultimately adopt. (Again, if that expectation is defeated, the realist loses her claim to have picked out either the trustworthy elements of contemporary scientific theories or those actually responsible for their empirical successes.) Accordingly, it seems that they too should be systematically skeptical of theoretical proposals or avenues of research that contradict or fail to preserve *whatever* elements, aspects, and/or features of our best scientific theories engender those realist commitments. In both cases, realists seem to have systematic grounds for *prioritizing* investments in finding, exploring, and testing theoretical alternatives that preserve *whatever it is that they are realists about* over the pursuit of theoretical alternatives that fail to do so.

The realist may, of course, have *instrumental* or *strategic* reasons for taking seriously or exploring particular theoretical possibilities that conflict with what she takes herself to already know: as Ronald Fisher famously suggested, for example, "No practical biologist interested in sexual reproduction would be led to work out the detailed consequences experienced by organisms having three or more sexes, yet what else should he do if he wishes to understand why the sexes are, in fact, always two?" (1930 ix). She might even be convinced (for whatever reason) that the best way to make incremental progress in improving existing theories is by trying to articulate radically different alternative theoretical proposals with which to compare them. But the instrumentalist shares any such strategic or instrumental reasons the realist may have for exploring or developing a theoretical alternative conflicting with (some privileged part of) existing theoretical orthodoxy. The instrumentalist also has another that is ultimately far more important: she believes that even more instrumentally powerful alternatives radically distinct from contemporary scientific theories are actually out there still waiting to be discovered.

Evaluating the promise, interest, or appeal of any particular theoretical proposal is, of course, a complex and multi-dimensional affair, so the suggestion here is certainly not that realists must always favor investing in the pursuit of any theoretical proposal (about anything) consistent with existing orthodoxy over the pursuit of any theoretical proposal (about anything) contradicting that orthodoxy. Nor is there some threshold degree of theoretical conservatism that realists must meet or exceed, either in general or in any particular case. The point instead is that those who are realists regarding some particular scientific theory (or privileged part thereof) should be prepared to treat the inconsistency of a given alternative theoretical proposal with the central claims (or otherwise privileged elements) of that theory as a reason to doubt that the alternative in question is even a viable candidate for representing the truth about the domain of nature it seeks to describe. All else being equal, this in turn constitutes a reason (for the realist) to discount or disfavor that alternative in competition for funding or support against otherwise equally attractive alternatives that do not similarly contradict (the relevant parts of) existing orthodoxy and therefore remain more plausible avenues for successfully extending, expanding, sophisticating, or supplementing our existing scientific conception of ourselves and the world around us. But this same reason for increased skepticism remains in force for the realist even when the alternatives being compared are anything *but* otherwise equally attractive, that is, even when it is simply one among a wide range of considerations bearing on the comparative attractions of investing in the pursuit of one theoretical proposal rather than another. Thus, holding all such further considerations *fixed* in any particular case reveals that realists have reasons instrumentalists lack for skepticism about just those theoretical proposals that violate (the relevant parts of) existing theoretical orthodoxy.

Of course, the instrumentalist no less than the realist will need to make difficult choices about how to invest and distribute the scarce resources available to support scientific inquiry itself. Many of the considerations she will weigh in making such judgments or decisions will operate for her in precisely the same way that they do for the realist: both, for example, will see alternative theoretical proposals as more attractive or promising targets for investment the better able they are to recover and/or explain whatever empirical consequences or implications of existing theories have already been independently confirmed by experiment and observation. Realists and instrumentalists will likewise appeal in much the same way to a wide array of further considerations to which granting agencies already direct the attention of their referees and review boards, such as the extent to which a proposal is well-reasoned, well-organized, and based on a sound rationale; the extent to which it promises to benefit society or advance desired societal outcomes; and the extent to which the proposers are well qualified and have access to the resources needed to carry out the proposed activities. And both realists and instrumentalists are entitled to make nuanced, discriminating judgments driven by the details of the theory and evidence in question concerning the extent to which the exploration and development of any particular theoretical alternative or line of investigation is comparatively more or less likely to help us make progress either in refining and extending existing theoretical orthodoxy or in finding and developing even more empirically successful theories that will ultimately supplant those we now embrace. But for those who are realists about a given scientific theory (or particular elements, aspects, and features of that theory) that calculation *should* reflect what she *already* takes herself to know about the claims, elements, aspects, or features of contemporary theoretical orthodoxy that will be retained and ratified in some recognizable form as part of any theoretical orthodoxy we come to embrace in the future. This same consideration simply does not arise for those who are instead instrumentalists about that theory.

10.3.1 Theoretical Conservatism: A Double-Edged Sword

Although realists may bristle at the suggestion that they defend a systematically more theoretically conservative form of scientific inquiry than instrumentalists do, this difference seems to generate an even more direct and immediate challenge to instrumentalism instead. After all, if such commitments to various parts of contemporary theoretical orthodoxy are what enable realists to dismiss alternative theoretical possibilities out of hand when they conflict with too much of what we think we already know about the world, it seems natural to worry that the instrumentalist will, by contrast, wind up forced into an absurd permissiveness with respect to the alternative theoretical possibilities that she is prepared to take seriously and/or consider potentially deserving investments of the time, energy, money, and other scarce resources available for pursuing scientific inquiry itself. Surely even the instrumentalist thinks we should take a dim view of investing scarce resources in finding and/or developing alternative theoretical proposals contradicting claims like "fossils are the remains of once-living organisms," "many diseases are caused by bacterial or viral infections," or "the world around us is filled with microscopic organisms" as well as many others that seem put beyond serious question or reasonable doubt by the evidence we already have. Although the instrumentalist remains open to the possibility that future scientific communities may ultimately *express* these firmly established facts using a very different theoretical vocabulary than our own,[1] she should nonetheless take a skeptical view of alternative theoretical proposals asserting or implying the *falsity* of such claims (even as we ourselves express them). (Here and throughout I assume that the theoretical alternatives in question *simply* contradict, explicitly or implicitly, the relevant claims concerning fossils, infections, microscopic organisms, etc., and do not offer convincing *alternative* explanations of the evidence we presently take to support those claims.) But if the instrumentalist restricts her beliefs to only the empirical consequences or observable implications of her best scientific theories, she might seem to lose any ground for thinking that theoretical alternatives contradicting such claims should be taken any less seriously or regarded with any more suspicion than those which instead simply contradict, say, the far more speculative claims of our best scientific theories concerning the nature of dark matter or dark energy.

Recall, however, that the instrumentalist's view is not that all claims of contemporary theoretical orthodoxy in science will ultimately be abandoned, but instead simply that many central and fundamental claims will be and that we are not *generally* in a position to predict in advance just which claims these will be. What she rejects is the realist's claim to have identified *general* or *categorical* features of scientific theories and/or their supporting evidence from which their approximate truth (or some analogue) can be reliably inferred. That is, the realism debate itself has been most fundamentally concerned with whether there is any *general* or *categorical* variety of empirical success or evidential support that serves as a reliable indicator that a theory (or its privileged parts) will be retained and ratified throughout the course of further inquiry. But realists and instrumentalists alike can recognize exceptions to their general or generic expectations in particular cases based on evidence or other considerations specific to the case in question. Many realists, for example, are prepared to make such an exception in the case of quantum mechanics, whose empirical success is extraordinary but whose very *intelligibility* to us as a description of how things actually stand in some otherwise inaccessible domain of nature remains controversial. Moreover, many realists are inclined to see the theory of evolution by natural selection as extremely well confirmed despite the fact that the evidence supporting that theory includes little in the way of the "novel predictive success" that they argue is *generally* required to justify the claim that a theory (or its privileged

---

[1] For example, when Joseph Priestly reported that after breathing "dephlogisticated air . . . my breast felt peculiarly light and easy for some time afterwards" we judge that he made a true claim about the effects of breathing oxygen using flawed or dated theoretical vocabulary, not a false or empty claim about a substance that does not exist (see Kitcher 1993 100).

parts) will persist throughout the course of further inquiry. While the instrumentalist holds instead that no *generic* characteristic or category of theories or the evidence available in support of them entitles us to any more than the expectation that the theory in question is a useful conceptual tool or instrument for guiding our practical interaction with nature, she nonetheless remains just as free as the realist to recognize particular cases as exceptions to this generic expectation on the basis of considerations specific to those cases. What she cannot do is respond to the challenge by specifying generic epistemic characteristics or categories that distinguish trustworthy from untrustworthy scientific beliefs, for such generic characteristics and categories are just what she thinks we have yet to identify successfully.

Accordingly, her judgment that a particular belief or claim is established beyond a reasonable doubt will have to be a function of her evaluation of the details of the *specific* evidence she has in support of that particular belief. I have argued at length elsewhere (Stanford 2010), for example, that the details of the evidence we now have supporting the (once highly contentious) hypothesis that fossils are the remains of previously living organisms should lead us to conclude that this hypothesis is not merely a useful cognitive instrument but is in addition an accurate description of how things stand in nature itself. Of particular importance in that case, I suggested, was an abundance of what I called "projective" evidence in support of this hypothesis (especially from the field of experimental taphonomy) in addition to merely eliminative and abductive forms of evidence. In a similar fashion, it will be the details of the evidence available in particular cases which convince the instrumentalist that any particular belief is established beyond a reasonable doubt. (Indeed, this same response might also appeal to selective or Middle Path realists, who face their own version of the problem insofar as proposed theoretical alternatives might well *preserve* the "working posits," "structural claims," or other privileged elements of contemporary theoretical orthodoxy while *nonetheless* disqualifying themselves from serious consideration by contradicting claims like "many diseases are caused by bacterial infections" or "fossils are the remains of once-living organisms.")

Here an important strategic difference emerges between the realist's and the instrumentalist's respective engagements with the historical record. The realist begins from the attempt to explain the success of science and seeks to defend the credibility of one particular explanation against potentially undermining historical counterevidence, in the process refining whatever criterion of epistemic security she proposes for identifying just which categories of scientific claims and commitments she thinks can be trusted to persist throughout further scientific inquiry. By contrast, the instrumentalist begins from the demonstrably serious threat to the persistence of our scientific beliefs posed by the historical record and tries to find ways to restrict our beliefs so as to obviate or sufficiently mitigate that threat. Because she knows of no general way to distinguish contemporary scientific claims or beliefs that will be abandoned or overturned from those that will instead be ratified and retained in future theoretical orthodoxy, she will proceed cautiously, presuming that a *generically* successful scientific theory is simply a useful conceptual tool or instrument (like Newtonian mechanics or caloric thermodynamics) unless and until presented with compelling specific reasons to regard a particular theory, belief, or commitment as an accurate description of some otherwise inaccessible part of nature (like the organic origins of fossils). That is, she starts with strict constraints regarding the beliefs to which a generically successful theory entitles us (such as the theory's "empirical implications" or its claims about "observable" matters of fact, though cf. Stanford 2006 Ch. 8) and then looks for reasons to relax these restrictions in particular cases. In the meantime, to oversimplify, given a motley collection of suspects with checkered pasts and conflicting evidence, realists are generously presuming our successful theories (or privileged parts thereof) innocent unless and until proven guilty, while instrumentalists are cynically presuming guilt unless innocence can be convincingly established.

This freedom of both realists and instrumentalists on the Middle Path to countenance exceptions to their respective general expectations or inferential entitlements concerning the fates of generically successful theories does, however, represent a further departure from their classical predecessors, who instead (as we noted earlier) typically saw themselves as articulating competing proposals concerning the appropriate epistemic attitude to take toward "successful scientific theories" as such. In fact, recognizing this further divergence of Middle Path realism and instrumentalism from their classical counterparts invites renewed attention to a much less widely recognized further dimension of disagreement in the modern realism debate concerning the *level of abstraction or generality* at which we should seek or expect to find useful epistemic guidance concerning our scientific beliefs. Along this dimension, I will now suggest, Middle Path realists and instrumentalists are once again largely united with one another, not only in contrast to their classical predecessors' fully general ambitions and expectations, but also to the diametrically opposed expectations of those who defend radical forms of what we might call "particularism" regarding the scientific realism debate.

10.3.2 Particularism and Generality in the Realism Debate

Since its inception in the writings of Smart (1963), Putnam (1975), van Fraassen (1980), Boyd (1984), Laudan (1981), and others, the modern realism debate has been predicated on the assumption that there is some *point* to ascending to the levels of abstraction at which we generalize about our "mature scientific theories" and their "empirical successes" or "approximate truth." For realists, the point of that ascent was to explain the successes of such theories in a way that revealed general or abstract epistemic categories we might use to reliably pick out those particular theories or otherwise privileged elements of contemporary scientific orthodoxy that represent secure epistemic possessions we can justifiably expect to persist in some recognizable form throughout the remaining course of scientific inquiry itself. The instrumentalist thinks we learn quite different lessons from considering matters at this level of abstraction and generality: she is convinced by the historical evidence, for example, that we should expect many of even the most fundamental commitments of our most successful contemporary theoretical orthodoxy to be eventually overturned and that there are no such general epistemic features or categories we might use to form reliable expectations regarding which of those commitments will or will not be preserved in some recognizable form throughout the course of our further scientific investigation of the world. Of course, if the instrumentalist goes on to claim that we should believe only a successful theory's empirical implications, or what it says about observable matters of fact, or some such (cf. Stanford 2006 Ch. 8), she too is offering a (competing) abstract and general criterion of epistemic security for our scientific beliefs.

In contrast to both realism and instrumentalism, particularism holds that there is not now nor was there ever any point in ascending to these heights of abstraction and generality in the first place and suggests that no useful guidance for evaluating the epistemic status of scientific claims can be gleaned from doing so. The particularist thinks the very best we can do in deciding whether some particular scientific claim or commitment is true and/or will be retained and ratified throughout the course of further inquiry is to carefully evaluate the details of the specific evidence we have for and against that particular claim or commitment. That is, she thinks that the delicate and painstaking *scientific* work of evaluating particular claims and commitments *already* represents our most sophisticated efforts to determine the appropriate level of confidence we should have in particular claims about what things exist and how they behave, and she has more confidence in the outcome of those efforts than in any far less specific guidance we might hope to find by seeking broad patterns in the historical record or from any generic conception of how science works. That is, she thinks that the evidential import of any purported general relationship between empirical success and truth, or facts about how often past successful theories have turned out to be not even approximately true, or how reliably we've failed to conceive of well-confirmed theoretical alternatives when they existed, or how frequently past scientists themselves have held spectacularly mistaken beliefs about which parts of their own theories were conclusively confirmed by the available evidence, simply pales into comparative insignificance when confronted with the ordinary first-order evidence we have in favor of or against any particular scientific claim. In this way the particularist sees the existing debate between scientific realists and instrumentalists as simply *superfluous* to our real efforts to find out anything about the world: broad reflections on the scientific enterprise as a whole or patterns in the historical record simply add nothing of substance to the outcomes of those investigations in particular cases. Moreover, such generalities and abstractions are simply insensitive to precisely the sorts of variation in the *details* of the evidence we have in different cases that the particularist thinks really should generate varying degrees of confidence concerning various claims about the existence and character of fossils, or dark matter, or electrical charge, or bacterial infections.

Particularist sentiments of this sort, I suggest, constitute an important part of Arthur Fine's motivation for embracing what he calls the Natural Ontological Attitude (NOA). Fine insists that realists and antirealists alike make a profound mistake when they seek to provide universalizing interpretations that purport to characterize the general aim of science, what our scientific claims really mean or say, and/or the epistemic accomplishments of science as a whole. "What binds realism and antirealism together," he says, is that "[t]hey see science as a set of practices in need of an interpretation, and they see themselves as providing just the right interpretation. But science is not needy in this way" (1986a 147–8). Instead, he suggests, the first-order scientific claims and counterclaims of ordinary scientific practice already represent our most careful efforts to decide what entities exist and what claims are true (in the humble, quotidian, philosophically unanalyzed senses of those terms), and we must treat such claims and beliefs as standing on their own bottom rather than standing in need of any further, distinctively philosophical analysis of what they really mean, or which ones we should actually believe, or even the point of making such claims in the first place. In fact, Fine rejects the idea that the scientific enterprise has any general aim or goal. Although scientific activity is replete with important goals and purposes, these are specific to particular contexts and practices, such as "For what purpose is this particular instrument being used, or why use a tungsten filament here rather than a copper one?" But he notes that it is simply a gross fallacy in quantifier logic to move from "They all have aims" to "There is an aim they all have" (1986b 173). When we go on to try to identify the general goal, aim, or purpose of science itself, he says, "we find ourselves in a quandary, just as we do when asked 'What is the purpose of life?' or indeed the corresponding sort of question for any sufficiently rich and varied practice or institution" (1984 148): "the quest for a general aim [for science], like the quest for the meaning of life, is just hermeneuticism run amok" (1986b 174).

Fine insists that it is similarly misguided to appeal to any global philosophical analysis or interpretation of science in order to decide which scientific claims to believe (e.g., those concerning "observable" states of affairs) or how much confidence to invest in them. Instead,

> NOA's attitude makes it wonder whether any *theory* of evidence is called for. The result is to open up the question of whether in particular contexts the evidence can reasonably be held to support belief (regardless of the character of the objects of belief). Thus NOA, as such, has no specific ontological commitments. It has only an attitude to recommend: namely, to look and see as openly as one can what it is reasonable to believe in and then to go with the belief and commitment that emerges. Different NOAers could, therefore, disagree about what exists, just as different, knowledgeable scientists disagree. (1986b 176–7)

What binds these NOAers together, it seems, is simply their skepticism about whether ascending to some more general or abstract philosophical level of investigation, analysis, or reflection on science itself will help them make any progress in deciding what to believe about the world. Scraping away or refusing to indulge in the universalizing interpretations offered by both realist and anti-realist philosophers of science simply leaves us with the first order practices of advancing, challenging, and defending particular scientific claims in particular scientific contexts. "The general lesson," Fine suggests, "is that, in the context of science, adopting an attitude of belief has as warrant precisely that which science itself grants, nothing more but certainly nothing less" (Fine 1986a 147).

In a similar fashion, Penelope Maddy sees the trouble at the root of both realism and instrumentalism as their shared inclination to try to ascend to some higher philosophical court of epistemic evaluation in which we leave behind the ordinary sorts of evidence and evaluative standards that are characteristic of science itself. She sees van Fraassen, for example, as conceding that the evidence in favor of the existence of atoms is perfectly adequate and convincing for scientific purposes but insisting that this does not settle whether or not we have sufficient epistemological or philosophical grounds for holding such beliefs:

> As far as methodology goes, the actual practice of science, it is perfectly reasonable for our scientist to take the Einstein/Perrin evidence as establishing the real existence of atoms. But for the proper "interpretation" of atomic theory, we must adopt a point of view other than that of the practicing scientist: "stepping back for a moment," we adopt an "epistemic attitude" towards the theory ([van Fraassen 1980] 82). Only then, answering the question as epistemologists, do we determine that the Einstein/Perrin evidence is not enough, and indeed, that no evidence can be enough to establish the existence of entities that cannot

be perceived by unaided human senses. Here we have yet another two-level theory: at the ordinary scientific level, we have good evidence that atoms are real; at the interpretive, epistemic level, we do not. (Maddy 2001 43–4)

She suggests that realists like Boyd mistakenly take the bait here, trying to rise to the challenge of defending our scientific beliefs in this higher, philosophical court of inquiry where ordinary scientific evidence is disallowed, and thereby being pushed "away from the details of the local debate over atoms and towards global debates over such questions as whether or not the theoretical terms of mature scientific theories typically refer" (2001 46).

By contrast, Maddy's "Second Philosopher" (2007) simply declines the invitation to leave ordinary scientific evidence and methods of evaluation behind and ascend to any such extrascientific or philosophical court of evaluation for scientific beliefs. Indeed, she feels no temptation to follow realists and antirealists down this shared rabbit hole unless and until someone can convincingly explain what this further distinctively epistemological or philosophical inquiry is supposed to achieve or accomplish and why the ordinary scientific evidence that actually convinces her of the reality of atoms should be treated as inadequate or irrelevant to that inquiry. She insists instead that scientific inquiry itself already represents our best efforts to decide (for any purposes we do or should actually care about) whether or not particular scientific claims are true and/or how much confidence in any particular claim is warranted, and that such questions can only be convincingly answered in the case-by-case or piecemeal manner that science itself employs:

> where the constructive empiricist issues a blanket rejection of all unobservable posits, the Realist issues an equally blanket endorsement; the Second Philosopher faults both for passing over the details of the evidence for each particular posit, for shirking the responsibility to evaluate each case individually. (2007 310n)

Although both Fine and Maddy deny that there are any useful answers to questions about science at the level of generality at which philosophers have traditionally sought them, they are both careful to leave room for at least the bare possibility that we might secure claims of somewhat greater generality (regarding confirmation, explanation, and the like) "bottom up," as it were, by generalizing over and abstracting away from the details of particular investigations (Fine 1986a, 179–80; Maddy 2007 403n). But at least some particularists are explicitly skeptical regarding this possibility. Magnus and Callender, for example, have argued influentially that "profitable" realism debates must be conducted at the "retail" level of individual, particular scientific inquiries rather than "wholesale." And although they concede that it might be "logically possible" to abstract away from or generalize over the results of such retail investigations, they doubt that the resulting guidance will be "either interesting or useful," suggesting instead that "[w]e should pay attention to particular cases for their own sake and not as proxies for something else" and that "the great hope for realism and anti-realism lies in retail arguments that attend to the details of particular cases" (2004 336). And Maddy herself has argued that when any such general epistemic guidance we embrace conflicts with the more detailed, informed, and contextual judgments of confirmation we make regarding particular cases, we typically prioritize the normative force of the particular judgments over that of the more general guidance. This is in part because such general guidance must itself remain sufficiently vague, indeterminate, and open-ended to permit further specification or adjustment in response to new developments: even if we generalized and abstracted our way bottom-up from historical or other evidence to the conclusion that we should only believe in entities we can "detect," for instance, she suggests that we will subsequently adjust the boundaries of what we count as "detection" (thereby including or excluding some new method) so as to preserve the normative force of our reflective confirmational judgments about individual scientific cases (including those making use of the new method). In this way, as Maddy says it, "the ordinary science, not the [general] criterion, is doing the work" (2007 402).

Moreover, even many of those who do still seek "interesting and useful" general epistemic guidance concerning our scientific beliefs have in recent decades done so in ways that reflect growing particularist sympathies, suggesting that such guidance may well consist in the accumulation and synthesis of principles each of which applies to only a limited range of cases rather than the sort of fully general criterion of epistemic security sought by their classical predecessors. Richard Miller (1987), for example, argues that convincing realist commitments must be grounded in "topic-specific truisms" whose force is restricted to particular contexts or domains of inquiry, while Jamin Asay (2019) suggests that realism debates should be conducted at the level of particular sciences rather than science as a whole. More generally, Larry Sklar argues that "by far the most interesting issues will be found in the detail of how the global, skeptical claims . . . become particularized and concrete when they appear as specific problems about theories within the context of ongoing physical theorizing" (2000 78). And more recently, Juha Saatsi has advocated a similarly modest particularism, arguing first that it is "a manifestation of philosophical arrogance to think that as a realist philosopher one commits oneself to providing a global recipe—largely independently of the science steeped in relevant details—for revealing what aspects of theory-world correspondence makes any given theory in mature science tick," but explicitly contending nonetheless that we should seek to generalize from the way particular exemplars latch onto the world to other cases "relevantly similar to those exemplars" (2017 3240–1; see also Saatsi and Vickers 2011).

Those who harbor modest particularist sympathies of this sort will presumably welcome my earlier suggestion that Middle Path realists and instrumentalists alike remain free to recognize exceptions to the general expectations they defend concerning the fates of generically successful scientific theories. Like such modest particularists, Middle Path realists and instrumentalists are prepared to abandon their classical predecessors' aspirations to fully general criteria of epistemic security for our scientific beliefs (what Saatsi calls a "global recipe"), but they regard the radical particularist's contrary conviction that there is little or no useful epistemic guidance to be found at any level of generality or abstraction higher than that of the individual case (what Asay calls "hyperlocalism") as a counsel of despair. After all, "to look and see as openly as one can what it is reasonable to believe in and then to go with the belief and commitment that emerges" (as Fine and NOA recommend) is presumably what scientists themselves have been doing all along, and many of their resulting sincere and carefully considered judgments have been among those ultimately overturned and abandoned in the course of further inquiry. Moreover, scientists' own explicit judgments of conclusive confirmation for particular scientific theories, particular parts or aspects of those theories, and particular scientific claims have repeatedly turned out to be spectacularly mistaken (Stanford 2006 Ch. 7). Our hope is to make *more* reliable judgments of this sort, and Middle Path realists and instrumentalists alike think that we can and do find considerable useful epistemic guidance informing those judgments at levels of abstraction and generality higher than that of the individual scientific investigation.

My own earlier examination of the case of organic fossil origins (2010), for example, suggested a tentative general moral: the greater the extent to which the supporting evidence for a given scientific theory or belief is eliminative or abductive in character, the greater its vulnerability to the problem of unconceived alternatives and the more cautious we should be about endorsing its truth. The historical record might also seem to support the view that highly successful scientific theories are more likely to be subsequently overturned in favor of previously unconceived alternatives when they concern questions of fundamental ontology in domains of nature far removed from ordinary human experience (like particle physics and cosmology as opposed to ecology or geology). And earlier we noted the detailed case made by Lyons and by Vickers for the claim that the subsequently abandoned components of successful past theories have regularly played crucial and ineliminable roles in generating the empirical successes (including the novel predictive successes) of those theories. None of these claims amounts to anything like the fully general criteria of epistemic security sought by classical realists and antirealists, but each is nonetheless a clear example of epistemic guidance intermediate in generality between those classical ambitions and the radical particularist's competing conviction that useful epistemic guidance for science can only be found in the details of the evidence we have in support of particular scientific claims. Middle Path realists and instrumentalists are thus once again united, this time charting a course between the fully general commitments and expectations of their classical predecessors on the one hand and the sharply opposed expectations of radical or extreme forms of particularism on the other.

10.4 Conclusion

I've suggested here that the modern scientific realism debate is most centrally concerned with whether or not we have yet discovered some particular form(s) of empirical success or evidential support allowing us to reliably predict whether particular theories (or privileged parts thereof) will be retained and ratified throughout the course of further scientific inquiry. As I've tried to emphasize, realists and instrumentalists fundamentally disagree on the answer to this question (and thus on the *predictability* of further changes in our theoretical conception of the world around us) in ways that actually make a difference to how we should pursue our further scientific investigation of the world. But I have also argued that in recent decades many contemporary realists and instrumentalists have come to share a set of fundamental commitments that unite them more closely with one another than either is to their classical predecessors, including Uniformitarianism, commensurability, and the Maddy/Wilson principle, as well as to seeking useful guidance concerning the epistemic security of our scientific theories and beliefs at a level of abstraction and generality intermediate between that anticipated by radical or extreme forms of particularism and the perfectly general criteria of epistemic security envisioned by classical forms of realism and instrumentalism alike. For such Middle Path realists and instrumentalists, arguably more important than any remaining point of debate or disagreement between them is the shared epistemic *project* in which they are jointly engaged: realists and instrumentalists alike on the Middle Path are already actively seeking to identify, evaluate, and refine candidate indicators of epistemic security for our scientific beliefs, and both see the historical record of scientific inquiry itself as the most important source of evidence we have available to us for pursuing that joint project together.

References

Asay, J. (2019) Going local: A defense of methodological localism about scientific realism. *Synthese* 196: 587–609.

Bement, A. L., Jr. (2007) Important notice 130: Transformative research, National Science Foundation, Office of the Director. https://www.nsf.gov/pubs/issuances/in130.pdf.

Boyd, R. (1984) The current status of scientific realism, in Jarrett Leplin (ed.) *Scientific Realism*. Berkeley: University of California Press, 41–82.

Chakravartty, A. (2007) *A Metaphysics for Scientific Realism: Knowing the Unobservable*. Cambridge: Cambridge University Press.

Egg, M. (2016) Expanding our grasp: Causal knowledge and the problem of unconceived alternatives. *British Journal for the Philosophy of Science* 67: 115–141.

Fine, A. (1986a) *The Shaky Game: Einstein, Realism, and the Quantum Theory*. Chicago: University of Chicago Press.

Fine, A. (1986b) Unnatural attitudes: Realist and instrumentalist attachments to science. *Mind* 95: 149–179.

Fischer, R. (1930) *The Genetical Theory of Natural Selection*. Oxford: Clarendon Press.

Kitcher, P.S. (1993) *The Advancement of Science*. New York: Oxford University Press.

Klee, R. (1999) *Scientific Inquiry: Readings in the Philosophy of Science*. New York: Oxford University Press.

Laudan, L. (1981) A confutation of convergent realism. *Philosophy of Science* 48: 19–48.

Lyons, T. (2002) Scientific realism and pessimistic meta-modus tollens, in S. Clarke and T.D. Lyons (eds.) *Recent Themes in the Philosophy of Science: Scientific Realism and Commonsense*. Dordrecht: Springer, 63–90.

Lyons, T. (2006) Scientific realism and the stratagema de divide et impera. *British Journal for the Philosophy of Science* 57: 537–560.

Lyons, T. (2016) Structural realism versus deployment realism: A comparative evaluation. *Studies in History and Philosophy of Science Part A* 59: 95–105.

Lyons, T. (2017) Epistemic selectivity, historical threats, and the non-epistemic tenets of scientific realism. *Synthese* 194: 3203–3219.

Maddy, P. (2001) Naturalism: Friends and foes. *Philosophical Perspectives* 15: 37–67.

Maddy, P. (2007) *Second Philosophy: A Naturalistic Method*. Oxford: Oxford UP.

Magnus, P.D., and Callender, C. (2004) Realist ennui and the base rate fallacy. *Philosophy of Science* 71: 320–338.

Miller, R. (1987) *Fact and Method: Explanation, Confirmation, and Reality in the Natural and the Social Sciences*. Princeton: Princeton University Press.

Novick, A., and Scholl, R. (2020) Presume it not: True causes in the search for the basis of heredity. *British Journal for the Philosophy of Science* 71: 59–86.

Peters, D. (2014) What elements of successful scientific theories are the correct targets for "selective" scientific realism? *Philosophy of Science* 81: 377–397.

Putnam, H. (1975) *Mathematics, Matter, and Method (Philosophical Papers Vol. 1)*. London: Cambridge University Press.

Psillos, S. (1999) *Scientific Realism: How Science Tracks Truth*. London: Routledge.

Saatsi, J.T. (2017) Replacing recipe realism. *Synthese* 194: 3233–3244.

Saatsi, J.T., and Vickers, P. (2011) Miraculous success? Inconsistency and untruth in Kirchhoff's diffraction theory. *British Journal for the Philosophy of Science* 62: 29–46.

Sklar, L. (2002) *Theory and Truth: Philosophical Critique within Foundational Science*. Oxford: Clarendon Press.

Smart, J.J.C. (1968) *Between Science and Philosophy*. New York: Random House.

Stanford, P.K. (2006) *Exceeding Our Grasp: Science, History, and the Problem of Unconceived Alternatives*. New York: Oxford University Press.

Stanford, P.K. (2010) Getting real: The hypothesis of organic fossil origins. *Modern Schoolman* 87: 219–243.

Stanford, P.K. (2015) Catastrophism, Uniformitarianism, and a Scientific Realism Debate That Makes a Difference. *Philosophy of Science* 82: 867–878. doi:10.1086/683325

Van Fraassen, B.C. (1980) *The Scientific Image*. Oxford: Clarendon Press.

Vickers, P. (2013) A confrontation of convergent realism. *Philosophy of Science* 80: 189–211.

Vickers, P. (2017) Understanding the selective realist defence against the PMI. *Synthese* 194: 3221–3232.

Wilson, M. (2006) *Wandering Significance: An Essay on Conceptual Behaviour*. Oxford: Clarendon Press.

Worrall, J. (1989) Structural realism: The best of both worlds? *Dialectica* 43: 99–124.

11

Structure not Selection

James Ladyman

Department of Philosophy
University of Bristol
James.Ladyman@bristol.ac.uk

11.1 Introduction

Structural realism is the best of both worlds, according to John Worrall (1989), because it takes account of the most powerful argument against scientific realism, as well as the motivations for it. The problem for scientific realism that he takes most seriously is that posed by the actual historical record of scientific theories in physics and chemistry, because it shows, all philosophical argument aside, that not everything that is supposed by the highly empirically successful theories of the past is real by the lights of current science. He was not troubled by other arguments against scientific realism (such as the argument from the underdetermination of theory by evidence), and the problem of theory change is the only problem for standard scientific realism that Worrall sought to solve by adopting structural realism. It is the problem that others, notably Stathis Psillos (1999), seek to solve by what David Papineau dubbed "selective realism" (1996), which involves analyzing case studies from the history of science to find a formula that restricts the epistemic commitment of scientific realists to parts of theories that will be retained. The idea is that close examination of the history of science will reveal what it is about abandoned theoretical constituents that distinguishes them from those that are retained, so that a selective commitment to current theories can be then be applied in the confidence that the relevant ontology will not subsequently be abandoned. (Psillos's formula is roughly "only believe in the reference of the central theoretical terms of theories that play an essential role in generating novel predictive success.") While selective realism is motivated and tested by studying the history of science, it is a form of realism that involves criteria that can be applied to our best current theories to select in advance what of their ontology will be preserved. If selective realism is understood as selecting part of science which we can be confident will be retained on

James Ladyman, *Structure not Selection* In: *Contemporary Scientific Realism*. Edited by: Timothy D. Lyons and Peter Vickers, Oxford University Press. © Oxford University Press 2021. DOI: 10.1093/oso/9780190946814.003.0011 theory change, it is simple to think of structural realism as the kind of selective realism that selects structure. Hence, naturally enough structural realism is often interpreted as a form of selective realism (as indeed it was by Papineau).

It might therefore be surprising that ontic structural realism (OSR) as I proposed and developed it (including with Don Ross in Ladyman and Ross 2007) is not intended as a form of selective realism in the sense just outlined. Furthermore, Steven French (whose recent work (2014) defends a somewhat different version of OSR) also makes it clear that he does not regard OSR as a form of selective realism, and nor does John Worrall understand his original structural realism that way. For Worrall, structural realism is a general kind of epistemic humility about what we know on the basis of our best scientific theories, not a criterion of epistemic commitment that can be applied within theories past or present. While OSR is supposed to deal with the problem of theory change, it is also motivated by the need to solve problems with the ontology of standard scientific realism as applied to physics. For French, Ladyman, and Ross, OSR is not an epistemological modification of standard scientific realism wholly, primarily or even partly (see also, for example, Esfeld and Lam (2008), where a moderate version of OSR is proposed solely for the metaphysical interpretation of spacetime). OSR is an epistemic thesis to the extent that it incorporates the epistemic commitment to our best science that all forms of scientific realism involve, but it is distinctive in proposing a metaphysics along with it.

Of course, authors have no ultimate authority over the readings of their texts, but, in any case, the assimilation of structural realism to selective realism creates pseudo-problems and is otherwise unhelpful. The next section (which forms the bulk of the paper) explains that structural realism does not respond to the problem of theory change by offering a way of demarcating in advance the parts of theories that are likely to be retained in future as a response. Rather structural realism in both its epistemic and ontic guises solves the problem of theory change with general departures from standard scientific realism.[1] (Epistemic structural realism is also associated with Russell's structuralism and epistemological and logical issues not addressed here.)

There are several ontological problems for standard scientific realism that do not figure in the canonical debate between realists and antirealists, which was the context for the development of both Worrall's structural realism and the current literature on selective realism, and which are not discussed in early work on OSR. The form of OSR developed by Ladyman and Ross (2007) addresses them (as does French 2014). Hence, there are two reasons why their OSR is not a

---

[1] A referee claims that epistemic structural realism involves looking at our current best theories and trying to identify the structure to which we then make a realist commitment. However, nobody has ever actually done this with our best science in the attempt to tell what will be retained on theory change.

kind of selective realism. The first is that it does not offer a criterion for selective realist commitment to current theories (as argued in the next section), and the second is that it addresses ontological problems other than the problem of theory change.

Section 11.3 argues that the construction of a positive realist metaphysics that can deal with the problem of theory change, must also take account of ontological issues, most importantly the lack of a fundamental level and scale-relativity (and explains how these problems are related to the problem of theory change). Section 11.4 reviews the form of OSR involving the theory of real patterns developed by Ladyman and Ross, and shows how it addresses the ontological problems for scientific realism, emphasizing the centrality of the idea of objective modal structure (where this is also retained on theory change even when ontology changes radically). It is the latter that makes OSR a form of realism, and differentiates it from van Fraassen's structural empiricism, as well as allowing OSR to avoid collapsing the distinction between abstract and concrete structure (van Fraassen's problem of "pure" structuralism, 2006), and making an account of causation and other modal aspects of scientific knowledge possible. The paper concludes with some remarks about the relationship between structural realism and the historiography of science.

11.2 Structural Realism is not Selective Realism

The problem of theory change is an empirical challenge to scientific realism pressed in Larry Laudan's work (1977 126, 1984), following Henri Poincaré (1905/1952 160), Ernst Mach (1911 17), and Hilary Putnam (1978 25). In its most basic form the problem is that the history of science is littered with laws, propositions, and theories that are now regarded as only approximately true and/ or true in a restricted domain, or outright false theories. This provides a reason for skepticism about the first-order methods of science that recommend belief in entities such as black holes and electrons. The problem of theory change is not addressed much in the debate about scientific realism centered on the work of Bas van Fraassen. In that context the big issue is the underdetermination of theory by evidence. Realists argue that beliefs about observables that have not been observed and inductive generalizations in everyday life are in general just as underdetermined as beliefs about unobservables. The fundamental argument of many scientific realists is that the observable/unobservable distinction is of no philosophical significance (see Churchland and Hooker 1985). In particular, it is of no epistemic significance and does not demarcate the knowable from the unknowable, and it is of no ontological significance in the sense that it has nothing to do with what things, or kinds of things, exist. Hence, realists argue that the fallible methods, in particular inference to the best explanation, used in everyday life to arrive at beliefs about the unobserved are just as legitimate when extended and refined to arrive at beliefs involving unobservables in science. On this way of thinking, skepticism about unobservables based on underdetermination is analogous to skepticism about the external world or other minds (and may be disregarded by the scientifically minded philosopher as "merely philosophical" (Worrall 1989)). On the other hand, the problem of theory change is based on what we know about the actual history of science.[2]

The simplest form of argument from theory change against scientific realism is the Pessimistic Meta-Induction that can be rendered as follows:

(i) There have been many empirically successful theories in the history of science that have subsequently been rejected, and whose theoretical terms do not refer according to our best current theories.

(ii) Our best current theories are no different in kind from those discarded theories, and so we have no reason to think they will not ultimately be replaced as well.

So, by induction we have positive reason to expect that our best current theories will be replaced by new theories according to which some of the central theoretical terms of our best current theories do not refer, and hence, we should not believe in the approximate truth or the successful reference of the theoretical terms of our best current theories.

Both (i) and (ii) are dubious. The number of theories that can be listed in support of (i) is drastically reduced if the notion of empirical success is restricted to making successful novel predictions (Psillos 1999). (ii) is undermined by the unprecedented degree of quantitative accuracy and practical and predictive success of current science. Furthermore, it may be argued that only since the twentieth century have the physical sciences been fully integrated because of two developments, namely the atomic valence theory of the periodic table and electronic theory of the chemical bond. Science is now very mature in the sense that testing hypotheses at the cutting edge in one field relies upon background theories from the basics of other sciences that are fully integrated with each other. The common system of quantities and units, and many of the core background theories used across all the sciences have been stable for a long time. There is no precedent in history for the current extent of overlap and reinforcement among

---

[2] Of course, they can be related by asking whether knowledge of the observable phenomena is the limit of what is retained on theory change. Obviously observable phenomena are lost on theory change, in so far as new theories are more empirically adequate than their predecessors.

theories in the natural sciences, nor for the commonality of statistical and other methods.

However, there is another form of the problem of theory change that does not have the form of an induction based on Laudan's much-discussed list. Rather it is an argument against the main argument for scientific realism namely the no miracles argument. Laudan's paper was also intended to show that the successful reference of its theoretical terms is *not* a necessary condition for the novel predictive success of a theory (1981 45), and so there are counterexamples to the no miracles argument. No attempt at producing a large inductive base need be made; rather, one or two cases are argued to be counter-arguments to the realist thesis that novel predictive success can only be explained by successful reference of key theoretical terms. As Psillos (1999 108) concedes, even if there are only a couple of examples of false and non-referring, but mature and strongly successful theories, then the "explanatory connection between empirical success and truth-likeness is still undermined." Hence, Laudan's ultimate argument from theory change against scientific realism is not really an induction of any kind, but a reductio.

The Argument from Theory Change:
  (I) Successful reference of its central theoretical terms is a necessary condition for the approximate truth of a theory.
  (II) There are examples of theories that were mature and had novel predictive success but whose central theoretical terms do not refer.

So there are examples of theories that were mature and had novel predictive success but which are not approximately true.

Approximate truth and successful reference of central theoretical terms is not a necessary condition for the novel-predictive success of scientific theories. So, the no miracles argument is undermined since, if approximate truth and successful reference are not available to be part of the explanation of some theories' novel predictive success, there is no reason to think that the novel predictive success of other theories has to be explained by realism (Ladyman and Ross 2007 84–85). Recall that the realist claims that novel predictive success is only intelligible on the realist view so once it is established that it is possible where realism is untenable then the realist's argument is much less if at all compelling (assuming it was to some extent compelling in the first place).

The ether theory of light and the caloric theory of heat seem to have enjoyed novel predictive success by anyone's standards.

If their central theoretical terms do not refer, the realist's claim that approximate truth explains empirical success will no longer be enough to establish realism, because we will need some other explanation for success of the caloric and ether theories. If this will do for these theories then it ought to do for others where we happened to have retained the central theoretical terms, and then we do not need the realist's preferred explanation that such theories are true and successfully refer to unobservable entities. (Ladyman and Ross 2007 84, see also Ladyman 2002)

Stathis Psillos (1999) first argues that these are the only two truly problematic cases for the scientific realist. He then adopts his "divide and conquer strategy," which is to adopt a different solution for each though both are ways of denying (II). In the case of the ether, he argues that according to his causal-descriptivist theory of reference, the term "ether" refers, while in the case of caloric he argues that there was never any reason why a realist of the time had to commit to it being a material substance and that it was not after all a "central" theoretical term in the sense that demands successful reference. Hence, he argues that we can be confident that all the central theoretical terms of our current science do in fact refer to things in the world.

Psillos argues that Fresnel's use of the term ether referred to the electromagnetic field and that continuity of reference being assured the problem posed by that case is solved. This is contentious but anyway other cases are also problematic. For example, consider the relationship between classical and quantum physics. Classical physics describes phenomena in diverse domains to very high accuracy and made predictions of qualitatively new phenomena. It had a vast amount of positive evidence in its favor. The rationalists and Kant had in various ways attempted to derive its principles a priori. However, it turned out to be completely wrong for very fast relative velocities and for very small particles, as was shown by relativistic and quantum physics respectively. This is surely the problem of theory change in its most stark form, yet the problem is nothing to do with the abandonment of theoretical terms, or difficulty in securing reference.[3] Clearly the reference of the term "energy" at least overlaps in classical and quantum theories because quantities in each coincide in the limit as the number of particles increases. This does not solve the problem that the radical ontological and metaphysical differences between the theories pose for the realist. Different theories often use the same terms, to say very different things about what the world is like. The question of which terms refer does not settle whether it is

---

[3] A referee objects that this may be thought to be a less serious form of theory change because of the precise mathematical relationships of approximation between the laws of the respective theories but these can only be interpreted in terms of approximate truth if the ontological discontinuities in question are not taken to be significant, while the laws are taken to have more than the purely empirical content van Fraassen and other antirealists take them to have. This is just what ontic structural realism recommends.

plausible to say that the metaphysics and ontology of the theory is even approximately true, because in many cases the same theoretical term is still used despite radical changes.

For this reason, Worrall (1989) argues that the project of defending realism by securing reference for abandoned theoretical terms is missing the point of the problem as well as the key to its solution.[4] Arguably "field" is just the new word for what the ancients and early moderns termed "aether" or "ether." Fields like the ether permeate all of space and are posited to explain certain phenomena. Anaxagoras is said to have introduced the idea of the ether to propel the planets, and subsequently there were many ethers to account for various things, notably the propagation of electricity and light through empty space. Fresnel's optical ether was supposed to be a solid and Maxwell's field behaves very differently and is said to be immaterial. However, from our current perspective the ether has much more in common with the field than atoms do with the indivisible particles of antiquity (Stein 1989).

Structural realism is not motivated by the need to secure reference across theory change but by Heinz Post's (1971) General Correspondence Principle, according to which the well-confirmed laws of old theories are retained by their successors as approximations within certain domains. Even comparing the well-confirmed laws of theories that differ wildly in their metaphysical dimensions such as quantum mechanics and classical mechanics, or general relativity and Newtonian gravitation, the past theories are limiting cases of the successor theories. This confirms Poincaré's idea of physics as ruins built on ruins in the sense that the old theories form the foundations of the new ones rather than the ground being cleared before work begins. The key point is that more than the empirical content and phenomenological laws of past theories is retained. For example, it is not just Kepler's laws and Galileo's kinematical laws that are preserved as limiting cases of general relativity, but Newton's inverse square force law that unifies and corrects those laws is retained in the form of the Poisson equation as a low-energy limit of Einstein's field equation. Similarly, the classical limit of quantum physics gives rise to the approximate truth of the laws of Newtonian mechanics for macroscopic bodies.[5]

The laws take mathematical form, and there are special cases, such as that of Fresnel's equations, where the very same equations are reinterpreted in terms of different entities. However, of course this is not the norm, and in many cases mathematical structure is lost and radically modified on theory change.

---

[4] Whether or not Worrall is right there is less much emphasis on reference in the subsequent literature on scientific realism.

[5] The structure of the world described by science and preserved on theory change can be taken as just the occurrent regularities between actual events. This is structural empiricism as defended by van Fraassen.

Structural realism does not require that *all* mathematical (or any other kind of) structure is preserved on theory change. If it did it would be refuted by the fact that the mathematical form of the physics of Maxwell is different from that of the physics of Einstein. As French and Ladyman (2011) put it, "[t]he advocate of OSR is not claiming that the structure of our current theories will be preserved simpliciter but rather that the well-confirmed relations between the phenomena will be preserved in at least approximate form and that the modal structure of the theories that underlies them, and plays the appropriate explanatory role, will also be preserved in approximate form" (32–33). However, this does not enable any kind of selective realism because it says nothing at all about how the laws of the theory are only approximate, and so does not enable the selection of what will be retained in advance. French and Ladyman (2003) emphasized that they understood OSR as the view that the mathematical structure of theories represents, to some extent, the modal structure of the world. Hence, when the mathematical structure changes this often reflects the fact that some of how the world's modal structure was represented to be was wrong. For example, the modal structure of the world is different according to Special Relativity than it is according to Newtonian mechanics because, for example, the velocity addition law of the latter is not even approximately true in case of frames moving relative to each other at velocities close to that of light, and this is reflected in the mathematical change to the Lorentz transformations from the Galilean transformations.

The point of OSR in the context of theory change is that it inflates the ontological importance of relational structure to take account of the fact that the ontological status of the relevant entities may be very different in the different theories, yet the relationship between the way they represent the modal structure of the world can nonetheless be studied in depth by investigating the relevant mathematical structures (even in the case of geocentrism and heliocentrism, as shown by Saunders 1993). Ladyman (1997) argues that mathematical representation is ineliminable in much of science and takes this to be key to OSR (and work by French, Ross, and Wallace on OSR is also predicated on it). However, this does not mean that this kind of structural realism only applies to mathematicised theories, as shown by the following case.

Kuhn and Feyerabend popularized the view that theory change in the history of science disrupts narratives of continuous progress and realism. The response of many philosophers of science to this challenge, particularly in Popper's school, was to examine episodes in the history of science in detail to see if the historically inspired critiques were really supported by the evidence. Noretta Koertege (1968) considered phlogiston theory as an example supporting the general correspondence principle because the well-confirmed empirical regularities stated in terms of phlogiston theory (such as that air saturated with phlogiston by combustion does not support respiration) are true when translated into the language of oxygen (de-oxygenated air does not support respiration). George Gale (1968) argued that phlogiston theory was also a good explanatory theory that explained the loss of weight of wood, coal, and ordinary substances when burnt, among many other things. Furthermore, and crucially for structural realism, the fact that combustion, respiration, and calcination of metals are all the same kind of reaction and there is an inverse kind of reaction too is an example of a theoretical relation that is retained from phlogiston theory. Even though there is no such thing as phlogiston, the tables of affinity and antipathy of phlogistic chemistry express real patterns (of which more in the next section) that we now express in terms of reducing and oxidizing power.

Indeed as also discussed by Gerhard Schurz (2009), the terms "phlogistication" and "dephlogistication," meaning the assimilation and release of phlogiston respectively when they were used in chemistry, can now be regarded as referring to the processes of oxidation and reduction (where oxidation of X = the formation of an ionic bond with an electronegative substance and reduction of X is the regaining of electrons). If the oxidizing agent is oxygen, and the oxidized compound is a source of carbon then the product is carbon dioxide ("fixed air"), and if the oxidizing agent is an acid, then hydrogen ("inflammable air") is emitted, in keeping with phlogiston theory. One could go further and allow that "phlogiston rich" and "phlogiston deficient" refer too, namely to strongly electro-negative and electro-positive molecules respectively. One could even argue that "phlogiston" refers to electrons in the outer orbital of an atom (as suggested by Andrew Pyle, see Ladyman 2011). However, none of this changes the fact that phlogiston theory is wrong and that there is nothing contained in all flammable materials that leaves them on combustion, and nothing that metals lose when they become calces. No form of selective realism could have told us in advance that phlogiston would be abandoned, because it was essential to much empirical success.

The examples from mathematical physics cited earlier and the case of phlogiston theory show that even though the ontology of science may change quite radically at the level of objects and properties (for example, there is no elastic solid ether, no principle of combustion), there can be continuity of the modal structure that is attributed to the world. OSR is not selective realism; rather, it is the generic modification of standard scientific realism as exemplified by Psillos (1999). The point is that "[t]heories, like Newtonian mechanics, can be literally false as fundamental physics, but still capture important modal structure and relations" (Ladyman and Ross 2007 118). Likewise important features of the modal structure of the world can be represented in terms of phlogiston, as explained earlier. When Worrall made his dialectical move in the realism-anti-realism debate, philosophers of science had given a lot of attention to the problem of the reference of theoretical terms, because it had been realized that they could be discarded from even very successful science. In the light of this discussion it should be clear that the real problem is not the reference of theoretical terms but that the notion of approximate truth is hard to apply to the ontology and metaphysics of scientific theories. Atoms are simply not indivisible in metaphysical terms, though they are of course in practical terms without a great deal of very sophisticated engineering. OSR does not give an account of metaphysical approximation, but it begins from the insight that more than the empirical content of theories is preserved. The claim that modal structure is preserved is compatible with radical ontological discontinuity, even in the extreme case of phlogiston, which Psillos did not seek to redeem.

As mentioned in section 11.1, OSR was also introduced to address ontological problems other than the problem of theory change. The next section outlines several that are not discussed in the literature about selective realism. Among the most important are the issues of fundamentality and scale-relativity.

11.3 Ontological Problems for Scientific Realism

The problems explained next are only specifically ontological problems for scientific realism, because scientific realism involves ontological commitment on the basis of science, but they are problems more generally for any philosophy of science. There are a number of ontological problems for scientific realism that are not discussed in the context of selective realism or epistemic structural realism. Like the problem of theory change, these problems with standard scientific realism arise from how science actually is.[6]

(a) Scientific realism often incorporates metaphysical ideas of fundamentality and reduction and an epistemologically and methodologically impoverished view of scientific practice.[7] Nancy Cartwright (1983) criticizes ideas of fundamentality and reduction by showing how scientific practice involves approximation and idealization in ways that do not fit with the standard scientific realist picture of how theoretical laws are related to empirical content.

(b) Much, most, or even all of science is not about what is ontologically fundamental. There has always been a tension within scientific realism

---

[6] As mentioned earlier OSR is also supposed to take account of the problems with the standard scientific realist interpretation of physics, in particular, to take account of the irreducibility of relational structure and the problems of identity and individuality in quantum mechanics and spacetime physics. These issues are not discussed in the present paper but are central for Ladyman (1998), French and Ladyman (2003), Ladyman and Ross (2007), and French (2014). For a recent discussion see Ladyman (2015, 2016).

[7] Of course, many scientific realists are not fundamentalists and reject the idea that ontology should be based on fundamental physics, including Cartwright herself.

because: on the one hand it is defended as the extension to science of common sense forms of reasoning that are used to arrive at ontological commitments (such as inference to the best explanation); and on the other hand, many take scientific realism to motivate or even require eliminativism about common sense ontology. More generally, there is the problem of how the ontologies of the sciences relate, and the question of whether there is a fundamental level, and if so whether anything else exists. This is equivalent, for most philosophers, to the question as to how the special sciences relate to fundamental physics.

(c) The scale-relativity of ontology is the thesis defended by Ladyman and Ross (2007) according to which entities may exist at some energy, length, or time scales and not at others. The scientific realist who does not embrace the scale relativity of ontology must otherwise accommodate the fact that within science in general there are entities that are recognized at one scale of description but not at others. In physics this is formalized by the mathematical relationships among the "effective field" theories that form the Standard Model of particle physics. Entities such as mesons and hadrons are not part of the high-energy description of quantum chromodynamics, but they can be described by a lower-energy description, and the way that energy cut-offs can be used to relate such theories at different scales is relatively well-understood (see Ladyman 2015; Wallace 2011; Franklin forthcoming). Beyond physics, entities are similarly bound to scale. For example, in economics prices only exist at long time scales relative to individual exchanges of goods and services, and in geography a torrent of water is not a river if it does not persist.[8]

These problems are related. (a) suggests that a straightforward realist interpretation of scientific theories and models is naïve. For example, the equator, the orbitals of atoms, and temperature of a gas are all abstractions and idealizations to a degree. (a) relates to (b) because something discrete may be represented at a less fundamental level as something continuous (as with ordinary fluids), and something heterogeneous at one scale may be idealized as something homogeneous at another, as with any polyatomic molecular substance. (b) relates to (c) because the special sciences are often associated with restricted scales of energy, length, and time. These problems must be faced by any scientific realist. These problems are also related to the problem of theory change, because past theories are also often those that we now regard as working at particular scales and as describing the emergent entities that are now described by a part of

---

[8] Entities like rivers and waves are arguably necessarily extended in time but nothing here depends on that claim.

physics that has the status of a special science compared to fundamental physics. For example, atoms are now part of atomic physics, which is not fundamental and describes reality at a restricted scale. The ontology of such theories remains "effective," in the sense that the entities it posits are part of empirically successful descriptions and models.

Defenders of scientific realism such as Psillos (1999) and Anjan Chakravartty (2017) are primarily concerned with epistemological issues rather than the problems just outlined. However, in other contexts scientific realism is argued to lead to the elimination of all but the most fundamental physical entities (see, for example, Rosenberg 2011). In this way, scientific realism becomes anti-naturalistic and cannot be claimed to take the ontological commitments of science at face value. On the other hand, those who reject eliminativism or reductionism need a version of realism that takes account of (a)–(c). The next section reviews how the real patterns account of ontology incorporated into OSR by Ladyman and Ross makes for a unified solution to the ontological problems for scientific realism.

11.4 A Realistic Metaphysics

For French and Ladyman (2003) and in their later publications jointly and separately, OSR is to be understood as the claim that science represents the objective modal structure of the world. Ladyman and Ross (2007) and French (2014) argue that incorporation of the idea that the world has a modal structure that is described by science into OSR makes it a realist philosophy of science, and distinct from the structural empiricism of van Fraassen (2006). They also agree that OSR so construed can accommodate the modal aspects of scientific knowledge including causation, law, and symmetry.[9] Modal structure is to be understood roughly as an abstraction of causal and nomological structure. For example, it is part of the modal structure of the world that metals expand when heated, that mass is conserved in chemical reactions, and that heat cannot be converted into work with perfect efficiency. These are all modal claims that support counterfactuals and explanations, and they have all survived very different ideas about the ontology in terms of which they are stated.

However, the forms of OSR that they defend also differ significantly especially in respect of how they relate to problems (a)–(c) of the last section. French (2014) is an eliminativist about all ontology except the structures of physics, and

---

[9] Berenstain and Ladyman (2012) also argue that the arguments for scientific realism are undercut without it. Note that this makes OSR an immodest and to some extent speculative philosophical position, unlike epistemic structural realism which is quietist about metaphysics and very modest about epistemology.

in this way avoids problems (b) and (c) by denying that anything that is not fundamental exists. This is a big price to pay for a naturalist because it departs so radically from taking science at face value.[10]

Ladyman and Ross (2007) develop a version of OSR involving the theory of real patterns in order to address (a)–(c) and develop a position that takes the special sciences to give us irreducible knowledge of the modal structure of the world.[11] The commitment to objective modal structure is essential to making the resulting naturalized metaphysics a form of realism, since the theory of real patterns can also be developed as a form of instrumentalism. We take it that all our scientific knowledge is about effective ontology in the sense of the previous section, and accordingly seek a unified way of thinking about the entities of physics and the special sciences. We propose the idea of real patterns to this end. Real patterns are the features that are used in everyday life and in science to describe the world in terms of projectible regularities. As such they are not confined to any ontological category and may be events, objects, processes, or properties. For example, Mount Everest is a real pattern that features in a host of projectible regularities about the world. To be genuine entities they need to be non-redundant, so that, for example, the disjunction of real patterns is not a new real pattern. Redundancy can be captured in different ways, the most obvious being information theory. Real patterns compress information (though at the cost of loss of precision). Much can be said about Everest that would be much more costly to state otherwise. Describing the world without talking about Mount Everest would be possible, but it would be harder. Real patterns simplify the description of the world relative to some background already established language (of more basic real patterns). Alternatively, redundancy can be understood dynamically. Some coarse-grained variables feature in a much simpler dynamics than that of the underlying variables; think of a rolling ball describable by the position and momentum of its center of mass and its angular velocity, instead of all the positions and momenta of its parts. However, some such coarse-grained variables do not feature in any simpler dynamics and they are redundant and anything defined in terms of them is not a real pattern.[12]

[10] See Ladyman (2019) for discussion of French (2014).
[11] David Wallace (2015) also argues for the real patterns account of effective emergent entities such as quasi-particles. Ladyman (2017) provides a summary of the main theses of Ladyman and Ross (2007), some of which concern topics that are not addressed at all here, including the unity of science and the primacy of physics. Ladyman (2017) argues that the further ontological problems of vagueness of composition and identity over time, and the problems of generation and corruption, apply to both special science objects and everyday ones and that the real patterns account of ontology offers a unified solution for them.
[12] Of course, this is a grey area. A cloud is a real pattern to a very limited extent because it is so ephemeral. A rainbow is a real pattern but only with respect to vision perception. Real patterns can also include events such as parties or revolutions which are of course notoriously difficult to individuate. The theory of real patterns does not solve those problems of individuation but takes them to be practical problems in some cases and pseudo-problems otherwise.

The real patterns account of ontology is compatible with the scale relativity of ontology, because it requires only that there be some regime in which the real pattern in question is projectible (and non-redundant) for it to exist. Different emergent structures are found at different scales, and accordingly there are real patterns that exist at some scales and not at others. For example, Mount Everest does not exist at atomic scale because there are no projectible regularities about entities at that scale that involve Mount Everest. The real patterns account is compatible with taking the ontologies of the special sciences at face value because insofar as the science in question captures the modal structure of the world, its ontology involves real patterns just as much as that of physics. Hence, problems (b) and (c) can be solved without adopting eliminativism about everything except fundamental physics.

As Dennett (1991) makes clear in his original discussion, being a real pattern is not an all or nothing matter. Mountains don't have exact boundaries. As many philosophers find mystifying, almost all events, objects, and properties commonly recognized as real involve some if not many kinds of vagueness. The real patterns account accommodates (a) because it is not based on unrealistic ideas of a pristine scientific ontology to accompany exact and fundamental laws. Representations can be more or less good at capturing the modal structure of the phenomena, and real patterns are indispensable to such successful representations, even though approximation and idealization are essential to the way the representation works.

The real patterns account fits well with the history of theory change in science. The well-confirmed laws of past theories describe aspects of the modal structure of the world, however, the ontologies in terms of which they are stated are not apt for describing other aspects of the modal structure of the world. In so far as past theories are empirically adequate, the entities they involve are real patterns and it may be appropriate to continue to refer to them in some regimes. For example, the force of gravity, the light waves of Fresnel, the electric and magnetic fields of classical electrodynamics, and the heat flows of Carnot are all real patterns and part of an effective ontology in the respective regime. However, sometimes the real patterns in question are best not described in terms of the old ontology, as with the case of phlogiston, because it does not get enough of the modal structure right. However, even in such cases we can say that there were some real patterns to the old theory, in particular, the commonality of different forms of dephlogistication is a real part of the modal structure of the world, as explained earlier. The real patterns account allows that continuity of reference across theory change can always be secured to some extent whenever there are real patterns that are carried over as approximations.

11.5 Realism and the Historiography of Science

Despite Kuhn's subsequent clarifications of his own view of science (1977), his *Structure of Scientific Revolutions* (1962) inspired schools of history and sociology of science that demur explanations of theory change in terms of evidence or experimental results (internalism). Rather, it became standard for explanations of theory choice in science, in so far as they are offered, to appeal to economic, psychological, and social (external) factors rather than the tribunal of experiment (a classic of the genre being Shapin and Shaffer 1985). In the history of science, celebrated recent studies have emphasized the rationality of the losers and the psychological and social influences on the victors, sometimes going as far as claiming that abandoned theories should have been retained contrary to the theory-choice made by the scientific community (for example, Chang (2012) argues that phlogiston theory should have been retained despite its many failings; see Blumenthal and Ladyman (2017, 2018). In this and other cases, notably that of quantum mechanics (see Cushing 1995), it is claimed on the basis of the general considerations of underdetermination that an alternative history could have delivered the same or more scientific success.

The orthodoxy in current historiography is that actors and their social networks in the history of science should always be represented sympathetically, and that their perspective be adopted in describing the relevant evidence and theories. The opening lines of Harvard's STS website describe the field as an "approach to historical and social studies of science, in which scientific facts were seen as products of scientists' socially conditioned investigations rather than as objective representations of nature".[13] Realist historiography is often said to be Whiggish and triumphalist. However, there is no reason why realists must be Whiggish in the relevant sense and no reason why celebrating the success of science is not compatible with recognizing the bias, error, and missteps that are also part of its history, and the role of external factors where they are relevant.

While Whiggism is much derided it is less often clearly defined (see Alvargonzález (2013) for a judicious and informative discussion). In political history it is associated with an implausible teleology. In the history of science it is similarly associated with the idea that the development of theories is an inevitable progression toward the truth. However, the idea that scientific methodology should track the truth is compatible with a degree of contingency. Furthermore, it is not unreasonable to think that there is a degree of inevitability to some scientific discoveries, because so many people have come close to the same ideas around the same time. For example, the inverse square law of gravity may not have been known to Hooke, but he at least came near to it (Westfall

---

[13] http://sts.hks.harvard.edu/about/whatissts.html

1967). In any case, acknowledging the actuality of scientific progress does not require believing it is inexorable, so the teleological element of Whiggism is not required for a realist historiography of science.

Triumphalism is associated with judging past theories by the lights of present ones and with dismissing the successes of abandoned theories and ridiculing the reasoning of the historical actors who defended them. However, celebrating the success of theory-choice and development in particular cases, and considering the history of science in the light of what we know now (presentism), does not imply denigrating or ignoring the success of abandoned or rival theories or shallow judgment of historical actors. As discussed earlier, Gale (1968) and Koertege (1969) are very clear about the successes of phlogiston theory in the course of their explanations of why ultimately it had to go. It is question-begging against realism to insist that the history of science should be explained without reference to what we now know about the world, and that internalist and presentist explanations of theory-choice are ruled out. From the realist perspective, historiography that does not take account of current scientific knowledge is likely to mislead us about the history of science and to beg the question against all forms of realism. What is so compelling about the problem of theory change is that it works from within realist historiography. Solving the problem requires being realistic about the history of science and the limitations of scientific progress, while celebrating its success.

Acknowledgments

Many thanks to Alexander Bird, Geoff Blumenthal, Steven French, Stephan Lewandowsky, Tim Lyons, Naomi Oreskes, Stathis Psillos, Don Ross, Peter Vickers, and the participants of the Durham Selective Realism conference.

References

Alvargonzález, D. (2013), "Is the History of Science Essentially Whiggish?" *History of Science* 51 (1), pp. 85–99.

Berenstain, N, and Ladyman, J. (2012), "Ontic Structural Realism and Modality" in Landry, E. & Rickles, D. (eds.) *Structural Realism: Structure, Object and Causality, The Western Ontario Series in Philosophy of Science* 77, Dordrecht: Springer, pp. 149–168.

Blumenthal, G., and Ladyman J. (2017), "The Development of Problems within the Phlogiston Theories, 1766–1791." *Foundations of Chemistry* 19 (3), pp. 241–280.

Blumenthal, G., and Ladyman, J. (2018), "Theory Comparison and Choice in Chemistry, 1766–1791." *Foundations of Chemistry* 20 (3), pp. 169–189.

Cartwright, N. (1983), *How the Laws of Physics Lie*, Oxford: Oxford University Press.

Chakravartty, A. (2017), *Scientific Ontology: Integrating Naturalised Metaphysics and Voluntarist Epistemology*, New York: Oxford University Press.

Chang, H. (2012), *Is Water H2O? Evidence, Realism and Pluralism*, Boston Studies in the Philosophy and History of Science, Dordrecht: Springer.

Churchland, P., and Hooker, C. (eds.). (1985), *Images of Science: Essays on Realism and Empiricism, (with a Reply from Bas C. van Fraassen)*, Chicago: University of Chicago Press.

Dennett, D. (1991), "Real Patterns." *Journal of Philosophy* 88 (1), pp. 27–51. doi:10.2307/2027085

Esfeld, M., and Lam, V. (2008), "Moderate Structural Realism about Space-Time." *Synthese* 160, pp. 27–46.

French, S. (2014), *The Structure of the World: Metaphysics and Representation*, Oxford: Oxford University Press.

French, S., and Ladyman, J. (2003), "Remodelling Structural Realism: Quantum Physics and the Metaphysics of Structure." *Synthese* 136 (1), pp. 31–56

French, S., and Ladyman, J. (2011), "In Defence of Ontic Structural Realism" ' in Bokulich, A. & Bokulich, P. (eds.) *Boston Studies in the Philosophy of Science* 281, *Scientific Structuralism*, Dordrecht: Springer, pp. 25–42.

Gale, G. (1968), "Phlogiston Revisited: Explanatory Models and Conceptual Change." *Chemistry* 41, pp. 16–20.

Koertege, N. (1969), *The General Correspondence Principle: A Study of Relations Between Scientific Theories*, Doctoral Thesis, University of London.

Kuhn, T.S. (1962), *The Structure of Scientific Revolutions*, Chicago: University of Chicago Press.

Kuhn, T.S. (1977), *The Essential Tension: Selected Studies in Scientific Tradition and Change*, Chicago: University of Chicago Press.

Ladyman, J. (1998), "What is Structural Realism?," *Studies in History and Philosophy of Science* 29 (3), pp. 409–424.

Ladyman, J. (2002), *Understanding Philosophy of Science*, London: Routledge.

Ladyman, J. (2011), "Structural Realism versus Standard Scientific Realism: The Case of Phlogiston and Dephlogisticated Air." *Synthese* 180 (2), pp. 87–101.

Ladyman, J. (2015), "Are There Individuals in Physics, and If So, What Are They?" in Guay, A. and Pradeu, T. (eds.) *Individuals across the Sciences*, Oxford: Oxford University Press, pp. 193–206.

Ladyman, J. (2016), "The Foundations of Structuralism and the Metaphysics of Relations" in Marmodoro, A. and Yates, D. (eds.) *The Metaphysics of Relations*, Oxford: Oxford University Press, pp. 177–197.

Ladyman, J. (2017), "An Apology for Naturalised Metaphysics" in Slater, M. and Yudell, Z. (eds.) *Metaphysics and the Philosophy of Science: New Essays*, New York: Oxford University Press, pp, 141–162.

Ladyman, J. (2019), "Structuralists of the World Unite." *Studies in History and Philosophy of Science* 74, pp. 1–3.

Ladyman J. and Ross, D., Spurrett, D., and Collier, J.G. (2007), *Every Thing Must Go: Metaphysics Naturalised*, New York: Oxford University Press.

Laudan, L. (1977), *Progress and its Problems: Toward a Theory of Scientific Growth*, Berkeley: University of California Press.

Laudan, L. (1984), *Science and Values: The Aims of Science and their Role in Scientific Debate*, Berkeley: University of California Press.

Mach, E. (1911), *History and Root of the Principle of Conservation of Energy*, Chicago: Open Court.

Papineau, D. (ed.). (1996), *The Philosophy of Science*, Oxford: Oxford University Press.

Poincaré, H. (1905/1952), *Science and Hypothesis*, New York: Dover.

Post, H.R. (1971), "Correspondence, Invariance and Heuristics: In Praise of Conservative Induction." *Studies in History and Philosophy of Science* 2, pp. 213–255.

Psillos, S. (1999), *How Science Tracks Truth*, London: Routledge.

Putnam, H. (1978), *Meaning and the Moral Sciences*, London: Routledge.

Rosenberg, A. (2011), *The Atheist's Guide to Reality: Enjoying Life Without Illusions*, New York: W.W. Norton.

Saunders, S. (1993), "To What Physics Corresponds" in French, S. and Kamminga, H. (eds.), *Correspondence, Invariance and Heuristics*, Dordrecht: Kluwer Academic Publishers, pp. 295–325.

Schurtz, G. (2009), "When Empirical Success Implies Theoretical Reference: A Structural Correspondence Theorem." *British Journal for the Philosophy of Science* 60, pp. 101–133.

Shapin, S., and Shaffer, S. (1985), *Leviathan and the Air-Pump: Hobbes, Boyle and the Experimental Life*, Princeton: Princeton University Press.

Stein, H. (1989), "Yes, but . . . Some Skeptical Remarks on Realism and Anti-Realism." Dialectica 43, pp. 47–65.

Van Fraassen, B.C. (2006), "Structure: Its Shadow and Substance." *The British Journal for the Philosophy of Science* 57, pp. 275–307.

Wallace, David. (2011), "Taking Particle Physics Seriously: A Critique of the Algebraic Approach to Quantum Field Theory." *Studies in History and Philosophy of Science Part B: Studies in History and Philosophy of Modern Physics* 42 (2), pp. 116–125.

Wallace, D. (2015), *The Emergent Multiverse*, Oxford: Oxford University Press.

Westfall, R. (1967), "Hooke and the Law of Universal Gravitation: A Reappraisal of a Reappraisal." *British Journal for the History of Science* 3 (3), pp. 245–261.

Worrall, J. (1989), "Structural Realism: The Best of Both Worlds?" *Dialectica* 43 (1–2), pp. 99–124.

12

The Case of the Consumption Function

Structural Realism in Macroeconomics

Jennifer S. Jhun

Department of Philosophy
Duke University
jennifer.jhun@duke.edu

12.1 Introduction

Consider the aggregate consumption function. In its barest form:

$$C_t = f(Y_t)$$

C_t is current aggregate consumption and Y_t is current aggregate income. In the United States, final consumption expenditure of gross domestic product (GDP) is somewhere around 70%. It is one of the simplest macroeconomic relationships. But what does the consumption function represent, if it represents anything at all?

The traditional realist claims that our best theories (and models) say (approximately) true things about the world and refer to real entities that exist. But closer inspection of the conceptual and historical background of the consumption function will cast doubt on whether economics is the kind of discipline apt for a realist interpretation. After all, it's a truism that the central posits of economic theory are often idealizations.[1]

This paper is an attempt to make space for realism, though perhaps not the traditional kind. I argue that the underpinnings of the modern consumption function—namely the Euler equation—are not meant to correspond to

---

[1] This is actually true of other sciences as well, including physics.

Jennifer S. Jhun, *The Case of the Consumption Function* In: *Contemporary Scientific Realism.*
Edited by: Timothy D. Lyons and Peter Vickers, Oxford University Press. © Oxford University Press 2021.
DOI: 10.1093/oso/9780190946814.003.0012

any particular things in the real world. Rather, they are methodological posits that, in and of themselves, do not state truths; but they help the scientist uncover the truth about the way the world is. In particular, I'll argue that such posits help us discover scale-dependent structural features of the economy that are real.

We begin in section 12.2 with a (very!) stylized reconstruction of the history of the consumption function, starting with Keynes (1936) and ending with Hall's (1978) response to the so-called Lucas Critique. Hall's insight was to explicitly incorporate agent expectations into the formulation of the consumption function; in this way, he could avoid the criticism that agents would alter their behavior in anticipation of policy changes and therefore change the very structure of the economy that the modeler attempted to capture (rendering the whole project moot). This explicit incorporation of intertemporal choice is what Chao (2003), inspired by Worrall's structural realism, identifies as *real* structure.

However, a number of authors have expressed skepticism about its empirical adequacy. It seems as if actual consumer behavior suggests that the Euler equation, or the hypotheses it helps scaffold, generally *don't* hold. Some more investigative work must be done if we want to account for actual economic behavior. Yet, the Euler equation approach persists as a cornerstone of consumption modeling. In order to make sense of what role it is playing in economists' reasoning strategies, section 12.3 considers the empirical challenges that the Euler equation faces—both by itself and embedded in, for instance, the rational expectations permanent income hypothesis proposed by Hall. The suggestion is that the consumption function—and indeed, the underlying Euler equation itself—may not be the right object of the realist attitude. We offer a proposal: the Euler equation partakes in a methodology that aims to find out about real structure. That is, we can embed the Euler equation in a larger realist project.

These considerations will help us identify a modest structural realism in section 12.4 that respects that theoretical fixtures such as the Euler equation in themselves do not represent real structure, and yet recognizes that they are still used to uncover real structure. To hash out our position, we look to Lyons's (2005, 2011, 2017) axiological realism—which maintains that the aim of theories is to find truth (whether or not we actually do, and whether or not we are justified in thinking we do) in order to formulate an epistemic counterpart. This epistemic counterpart will depart, however, fundamentally from Lyons's vision in that it takes a *pragmatic, perspectival* view of truth. Finally, in section 12.5, we argue that this version of realism has the benefit of sidestepping some of Reiss's (2012) criticisms of standard realist defenses against charges of instrumentalism.

12.2 A Brief History of the Consumption Function

The Keynesian formulation of the consumption function was conceived as a linear (fairly invariant) relationship.[2]

$$C_t = \alpha + \beta Y_t$$

Here, α is a constant, β is the marginal propensity to consume, and Y_t is disposable income. This equation is supposed to represent, according to Keynes, "The fundamental psychological law . . . that men are disposed, as a rule and on the average, to increase their consumption as their income increases, but not by as much as the increase in their income" (1936, 96). Known as the *absolute income hypothesis*, this formula implies that changes in consumption depend on changes in income. That is, given increases in GDP, we expect some increased consumption and increased savings.

According to this theory, average propensity to consume (APC) would be greater than marginal propensity to consume (MPC), which is less than one per the fundamental psychological law. This implies that average propensity to consume would decrease and average propensity to save would increase as income increased. So, for instance, it predicts that post-war savings would increase and that consumption would decrease as government spending fell and the economy headed toward a recession.

However, empirical issues plagued the Keynesian consumption function. As demonstrated by a series of papers by Kuznets and his colleagues, it became clear that the function could not hold in the long run.[3] For instance, there was a contradiction between cross-section (declining) and time-series (constant average) data of the savings-income ratio, which reported different amounts for the average propensity to save. In 1942, Kuznets, who was then associate director of the Bureau of Planning and Statistics of the War Production Board, showed that even though there was a significant increase in per income capita from 1879 to 1928, the savings-income ratio remained constant. Keynesian theory had predicted that it would decrease.[4] A 1946 publication demonstrated that

---

[2] Technically, it is not Keynesian as in *Keynes* himself did not advocate for this (and certainly not in the sense of the *General Theory*). People who considered themselves Keynesians presented it in this way, but even for them it was a convenient tool rather than something to be unshakably committed to.

[3] Notably, in the long run it also looks linear through the origin. APC is more or less constant over time, but not over the business cycle. That fact gave rise to an entire literature on the consumption function from the 1940s to the 1970s. Friedman was working with Kuznets during this time, and the original permanent income hypothesis can actually be found in a joint publication (1945).

[4] He would document this long-run constancy again in a 1952 paper.

the time series ran contrary to "Keynesian" theory, and instead (as Chao 2003 documents):[5]

> (1) MPC is less than APC in budget data and short-run time-series data but is equal to APC in the long run; (2) APS [average propensity to save] and APC did not rise secularly, and (3) private demand increased sharply and APS was sharply lower than in the interwar period level. (85)[6]

Alongside Duesenberry's (1949) relative income hypothesis and Modigliani and Brumberg's (1954) life-cycle hypothesis models, Friedman (1957) too would offer an alternative in response to these difficulties: the *permanent income hypothesis*. Though he himself did not present a formalized version, the intuition was as follows. Consumption would depend not just on the amount of my current after-tax income, but my estimation of the overall income I believe I'll earn in the future. In particular it's going to depend on the permanent (long-run) rather than transitory income that I receive. As a forward-looking agent, I try to assess what future income might look like by forming adaptive expectations based on my past experiences.[7] And I should react more strongly to permanent shocks than transitory ones. So most of my consumption is going to be based on permanent income—the average amount I expect to receive over the years in the future—rather than transitory income, like lottery winnings. Along the way I learn adaptively—sometimes my estimates will be off. A standard economics textbook may present the general form of my consumption function in the following way.[8] Let permanent income be:

---

[5] And here I put Keynesian in scare-quotes; Keynes himself in the second chapter of the *General Theory* seems to gesture at something that looks like the permanent income hypothesis.

[6] A number of other features cast doubt on the Keynesian formulation. For example, Molana (1993) documents the following empirical challenges (assuming the consumption function takes the form of $C_t = \alpha + \beta Y_t$):

> For instance: (i) Kuznets' (1946) study illustrated that the APC did not contain a significant trend, (ii) the cross section budget investigations by Brady and Friedman (1947) showed that α was in fact positive with a tendency to shift upwards, (iii) Smithies' (1945) time series study confirmed that the presence of a positive deterministic trend in a could not be ruled out, and finally, (iv) estimates of β usually turned out to be lower than expected (see Haavelmo, 1947) and the model persistently under predicted consumption. (338)

Cate (2013) further notes that the Keynesian account neglected the fact that consumption is not only affected by changes in income but is also by wealth (572).

[7] Modigliani's life-cycle hypothesis of consumption is similarly based on forward-looking expectations—this is why such models are now called permanent income-life cycle hypothesis models. Both Keynes and Modigliani argued that consumers tend to smooth their consumption over time, so consumption doesn't depend so much on current income.

[8] I note this because—as is unsurprising with any stylized history—this is not Friedman's formulation.

$$Y_t^P = Y_{t-1} + q \, (Y_{t-1}^P - Y_{t-1})$$

q is the fraction of last year's error subtracted from Y_{t-1} in order to estimate this period's permanent income, Y_t^P. It is the amount by which my estimated income differs from my estimate last year. The consumption function tells us how permanent consumption depends on permanent income (rather than current income), i.e.

$$C_t^P = kY_t^P$$

Therefore,

$$C_t^P = k(1-q)Y_{t-1} + kqY_{t-1}^P$$

k is the marginal propensity to consume out of my permanent income. Subtract my permanent income from my actual income, and the remainder is transitory income. The idea is that people like to smooth their consumption behavior over time, rather than letting it fluctuate in response to short-term changes. The permanent income hypothesis even resolves the earlier empirical puzzle that long-run and short-run consumption habits seem to run in different directions—wealthier households save more, yet the savings rate stays constant over time. We might, for instance, attribute those wealthier households with a large amount of transitory income—such as bonuses from work—but poorer households tend to have negative transitory income.

However, even with this adjustment, there remained a worry that we call the Lucas Critique. Lucas (1976) had noted a more general problem. Even when Keynesian models worked well as forecasting models, they were inappropriate for assessing alternative policies—i.e. they could not accommodate policy change because "any change in policy will systematically alter the structure of econometric models" (41). Lucas's proposed solution was that economists ought to seek "deep" parameters invariant to policy changes, though this really meant explicitly modeling consumer expectations, tastes, and technology. This involved modeling fixed psychological features. Lucas and Sargent (1979) emphatically criticize those models that did not as "incapable of providing reliable guidance in formulating monetary, fiscal and other types of policy" (69).[9]

---

[9] And even this is somewhat overdramatic. These models were only just being used when Lucas criticized them, so they didn't at this time have a historical record of being the spectacular failures they were made out to be.

Robert Hall (1990) would try to get around this problem by explicitly incorporating optimization behavior into his formalism, locating structure in the underlying decision-making mechanism.[10]

> Although Lucas was scornful of existing econometric policy evaluation models, his message was not completely destructive of all model-building or empirical research. There are structural relationships in the economy, but the consumption function is not among them. For consumption, the structural relation, invariant to policy interventions and other shifts elsewhere in the economy, is the intertemporal preference ordering. (135)

Aligning Friedman's insights with Lucas's, Hall proposed extending the permanent income hypothesis account to incorporate rational expectations, i.e. assumptions about belief formation, directly into the model. The permanent income hypothesis tells us that if permanent income were known (deterministic), consumption would remain the same over time. Yet, it doesn't—we have uncertainty with respect to our income. Agents, according to rational expectations, take into account all information in order to forecast the future. Supposing that our agents are utility maximizers, the Euler equation encodes rational expectations as governing intertemporal choice. In the case of income uncertainty, it manifests as the following first order condition on the agent's decision-making problem.

$$E_t u'(c_{t+1}) = \left[(1 + d)/(1 + r) \right] u'(c_t)$$

Here, c is the agent's consumption (indexed to a particular time), d is the subjective time discount rate, r is the real interest rate, and E represents mathematical expectations. According to this formula, there are no reallocations of goods such that the agent's utility at the margin can be improved. In the case the utility function is quadratic and d = r, then

$$Ec_{t+1} = c_t$$

Then by mathematical (rational) expectations:

$$c_{t+1} = c_t + e_{t+1}$$

---

[10] As a note, Friedman did consider uncertainty, but not formally.

That is, consumption tomorrow will be the same as today with some error. Consumption follows a random walk, and any actual changes are unpredictable and random. If I'm forward looking, I'll anticipate changes in my income by smoothing consumption over time (but given that actual changes are random, I do not anticipate substantial changes in income).

12.3 The Question of Realism

12.3.1 Reductionism and Microfoundations

One can read the Lucas Critique as an urge to pursue a reductionist project, and interpret the ubiquitous representative agent as a first-step maneuver in such a project. As economics progresses, we will get ever closer to a model of the individual agents in the economy rather than aggregates.[11] So, as a first approximation, we treat the consumption function as reflecting the behavior of a macroeconomic-sized representative agent. Perhaps we can move next to (aggregate) heterogeneous agents, and so on. According to this picture, aggregates are just a shorthand way of talking about groups of individuals. Macroeconomics, thus, talks about real things insofar as the entities in question are really inhabitants of a microeconomic ontology. Ultimately, what we really need to do is find and then put all household consumption functions together in some appropriate way.

There's something suspect about this particular way Lucas's criticism has been borne out. This is because it strikes me as an unpromising strategy to pursue a reductionist project along these lines. If macroeconomic aggregates are reducible to talk of microeconomic entities—the *microfoundations* project—as the Lucas Critique seems to require, then it seems like macroeconomic features are not really autonomous from microeconomic ones. So, if we take the Lucas Critique seriously, then to be a realist about macroeconomics implies that we are realist only insofar as there are microfoundations for our theory.

The difficulty is that there is often no straightforward relationship between the microscale and the macroscale of the economy. In fact, formal theorems such as the 1970s Sonneschein-Mantel-Debreu (SMD) theorem—also known as the "anything goes" theorem—emphasize that there is no such correspondence between the configurations of individual agents and behavior at the macroscale. In the context of excess demand functions, the SMD theorem itself states that under standard assumptions on agents, no interesting macroscopic properties are guaranteed to emerge from the population. That is, they don't give rise to any

---

[11] For a critical overview, see Hoover (2015).

particular equilibrium vector, meaning that properties at the microscale do not transfer to the macroscale and there can be an arbitrary number of equilibria. We could assign all agents a downward sloping demand curve, and yet we may not see that behavior in the aggregate.

We should worry that similar difficulties will arise in the case of household versus aggregate consumption. This was exactly the problem that arose during the years after World War II; the usual practice in economic forecasting was to put together several different consumption functions for different classes of goods and simply add them together (Bronfenbrenner 1948, 318). Using these as major parts of econometric models, economists would offer policy advice with disappointing results.[12] So facts about the macroscale cannot simply be deduced from facts about goings-on at the more microscale, indicating that at least former kinds of facts are in some sense autonomous from the latter kind—facts about aggregate features may be independent of the sum of facts about individual ones, though of course the former depend on the latter's existing in the first place.[13]

But, the Federal Reserve thinks about the national economy in aggregate terms; for instance, expansionary money policy, which increases the money supply, is thought to be accompanied by an increase in GDP as well as an increase in spending. Policy is aimed at the aggregate and is expected to bring about certain effects. Without a realist reading of economics, interventions of this kind simply don't make much sense. But what kind of realist reading is the appropriate one?

12.3.2 Worrall's Structural Realism and Chao's Realism about Structure

One of the most popular—if not the most popular—account of realism is *structural* realism. The paradigmatic defense of structural realism comes from Worrall (1989), who argues that:

> It would be a miracle, a coincidence on a near cosmic scale, if a theory made as many correct empirical predictions as, say, the general theory of relativity or the photon theory of light *without* what that theory says about the fundamental structure of the universe being correct or "essentially" or "basically" correct. But we shouldn't accept miracles, not at any rate if there is a non-miraculous alternative. If what these theories say is going on "behind" the phenomena is

---

[12] See Hart (1946).

[13] Note that even aggregation is not uniformly problematic, nor is it always problematic. Nominal GDP is an unproblematic aggregate that consists in just exact aggregation of the values of final goods.

indeed true or "approximately true" then it is no wonder that they get the phenomena right. So it is plausible to conclude that presently accepted theories are indeed "essentially" correct. (101)

Known as the No Miracles Argument, it insists that theories would not be as successful as they are unless something had gone right—and this something, says Worrall, is structure. One way of interpreting the position (and the usual way to interpret Worrall) is in an *epistemic* sense, i.e. a claim about what we can have knowledge of, where this is typically "the (preserved) mathematical structure of our theories" (Morganti 2011, 1166).[14]

In what follows I mean structure quite loosely. For one, I am here interested in *causal structure*, where structure consists of modal relationships that satisfy an interventionist conception of causation á la Hoover (2001, 2011) or Woodward (2005). Toggle one thing, and something else changes as a result. That relationship must be (to some extent) invariant: "When a relationship is invariant under at least some interventions . . . it is potentially usable in the sense that . . . if an intervention on X were to occur, this would be a way of manipulating or controlling the value of Y" (Woodward 2005, 16).[15] Something that partakes in a causal relationship is part of the causal structure of a system, and articulating a bit of causal structure will involve articulating the relata in a difference-making relationship (that is apt for counterfactual analysis). One important disclaimer, however, is that the way I am thinking about structure will generalize from the way that structure is typically understood by economists. Economists, at least post–Lucas Critique, consider "deep parameters"—those that are invariant under policy interventions—as structural features, and structure itself is defined as a system of equations.[16] Unlike the Cowles era economists, I am *not* using it

[14] For the most part, the distinction doesn't bear much on the suggestion I'm trying to make in this paper, because (spoilers) the account I turn to is pragmatist in nature; for more on the distinction, see Ladyman (1998). Ross's (2008) ontic structural realism claims that "The basic objects of economic theory are optimization problems" (741). Objects, on this view, are "heuristics, bookkeeping devices that help investigators manipulate partial models of reality so as to stay focused on common regions of measurement from one probe to the next." What really matters, say, in the case of economic games, are that they are "mathematical structures, networks of relationships whose relata are distinguished as such by the mathematical representations of the structures in question—just like quarks and bosons."

[15] One implication is that insofar as I am thinking about structure, I am really only concerned with *causal structure*. I am here agnostic as to whether and what other kinds of structure there are. I should note here that I am not using causation in the strictly Woodwardian sense either, who identifies causal relationships with a particular kind of relationship between *variables*. Despite this, causal relationships we think of as holding in the world may very well on our account be the kinds of things we try to capture with structural equations.

[16] And even *how* and *what* exactly gets considered structure is going to go hand-in-hand with the particular kind of methodology a school uses—for example, London School of Economics style econometrics and Cowles Commission style econometric methodology proceed in different directions, disagreeing on how theory and data bear on modeling. And the New Classical approach that takes rational expectations as foundational, too, is relying upon a different methodology.

in order to draw a contrast with what is known in econometrics as *reduced-form* equations.

Chao (2007), inspired by the semantic view of theories, proposes what he calls *realism about structure*. This account supplements Worrall's account by distinguishing explicitly between structure and non-structure in theories. The test case is the consumption function; Chao identifies intertemporal choice (encapsulated by the Euler equation) as the underlying structure. The structure that we regard as real, Chao tells us, consists in the mathematical equations that are preserved as theory develops. The mathematical equations that represent the abstract structure of a system related to their concrete empirical counterpart models up to isomorphism. The empirical models and the mathematical equations are related via a representation theorem, which assures us that we have singled out relevant structure in our model. Empirical structure remains invariant to disturbances. Analogously, the mathematical equations that capture its structure are invariant to transformations (such as when there is theory development).

> [W]hen Worrall observes that in the history of science the mathematical equations of the new and old theories are invariant (isomorphism among theoretical models), given that in the successful or mature scientific theories, the theoretical model flourishingly represents a feature of the world (isomorphism between theoretical and empirical models), we can conclude that the represented empirical models are also invariant under the same type of transformation as the theoretical one . . .
>
> In this interpretation the represented empirical model is invariant and therefore can be regarded as structure. (239)

These "invariant relations are structural and real." (240)

The story is an optimistic one; it is possible to think about economics in realist terms. But then the story takes an odd turn: "we accept a theory if it yields a model containing the right structure. If a model is not supported by empirical data when the model is considered as containing the true structure, we do not reject the model but instead construct a new model with the same true structure" (240). That is, in the face of recalcitrant experience, we stay committed to the Euler equation as if it were part of a Lakatosian hard core or Kuhnian paradigm; we are wedded to a strong a priori belief in it.

I do not disagree that the Fisherian intertemporal framework has survived over time as consumption theory has evolved. But I want to suggest that we can diverge from or at least reinterpret Chao with respect to the claim that the Euler equation is real. For us, the Euler equation is not what is left once we have cleaned up our mistaken suppositions in a model or a theory. It serves a different methodological purpose.

Now, the Euler equation itself doesn't determine consumption. While it doesn't refute the Euler condition itself, from a practical standpoint, we might be a bit unsettled if it systematically turned out that the Euler equation approach—such as that embodied in the permanent income hypothesis—fails. And it does seem that the permanent income hypothesis does not always describe consumer behavior very well. As Blinder et al. (1985) remark:

> ... the research done to date has not supported the econometric restrictions implied by the Euler equation approach. Nor has further investigation validated the hypothesis that the response of consumption to income (henceforth, Y) reflects only the usefulness of current Y in predicting future Y. Instead, research typically finds "excess sensitivity" to current income. (467)

Flavin's (1985) noteworthy study found that modeling income time series follows an autoregressive–moving-average (ARMA) process, and also that consumption exhibits excess sensitivity: "The empirical results indicate that the observed sensitivity of consumption to current income is greater than is warranted by the permanent income-life cycle hypothesis" (976). Because of the statistical discrepancy between the null hypothesis and the data, she proposes liquidity constraints on household consumption to explain why it is that consumption is excessively sensitive to changes in income.[17]

There are ways to use posits like the random walk equation, or even the underlying Euler equation, without assuming that there must be some straightforward correspondence (or failure as such) to something in the real world in order to help us get a grip on the economy. Consider Campbell and Mankiw's (1989, 1990, 1991) papers that aimed to show that rational expectations did not correctly capture macroeconomic aggregate behavior. Given the aggregate macroeconomic data, they "propose a simple, alternative characterization of the time series data ... [that] the data are best viewed as generated not by a single forward-looking consumer but by two types of consumers" (1989, 185). About half of these consumers behave like rational agents, but the other half follow a rule of thumb, consuming their current income. (Both, as a note, optimize their utility preferences.) This alternative hypothesis is "an economically important deviation from the permanent income hypothesis" (187).

Campbell and Mankiw embed the permanent income hypothesis model within a generalized model that allows for a fraction of income to accrue to forward-looking agents and the rest to those who consume their permanent

---

[17] Later in 1985 she notes that "The null hypothesis in this empirical literature typically consists of the joint hypothesis that 1) agents' expectations are formed rationally, 2) desired consumption is determined by permanent income, and 3) capital markets are 'perfect'" (117–118). There she diagnoses excess sensitivity as a failure of the third assumption.

income. While according to the traditional approach, the change in consumption should simply be an error, the generalized model is:

$$\Delta C_t = \Delta C_{1t} + \Delta C_{2t} = \lambda \Delta Y_t + (1 - \lambda)\varepsilon_t$$

They can directly test the permanent income hypothesis by setting λ to zero. And in fact, their estimate of λ—that portion of the population that consumes current income—is about 0.5 (195). The macroscopic phenomenon of smoothness is explained by the fact that aggregate consumption "is a 'diversified portfolio' of the consumption of two groups of agents" (210).

The point of this exercise, which again *does not discard the original consumption formalism* associated with the random walk hypothesis, is not a straightforward case of starting with an idealized model and gradually de-idealizing it to make it more realistic. If real behavior deviates from the idealized benchmark (like one that assumes a single representative agent), however, we should seek out what factors explain that deviation. The Euler construction is a tool that helps us erect and navigate a larger, more complex framework, making space for potentially relevant structural information (though not every and all pieces of information) that makes a difference to a system's behavior. And for Mankiw and Campbell in particular, the structural information about their population of interest is how much of that income accrues to people who abide by rational expectations as opposed to those who consume their current income. They choose to look at the distribution of agent kinds in the population. That is, theirs is a project that tries to capture what we might call scale-dominant behavior.[18] This is behavior that is characteristic of a system at a particular scale, where particular structural characteristics will be salient there.[19]

---

[18] Wilson (2017), in the context of representative volume element (RVE) methods in physics, points out that complex behavior can be managed by parsing it out into separate sub-models, each "assigned the comparatively circumscribed duty of capturing only the central physical processes normally witnessed at its characteristic scale length . . . each localized model renders descriptive justice only to the dominant behaviors it normally encounters" (19). And indeed, as Batterman (2013) emphasizes, "Many systems, say a steel girder, manifest radically different, dominant behaviors at different length scales. At the scale of meters, we are interested in its bending properties, its buckling strength, etc. At the scale of nanometers or smaller, it is composed of many atoms, and features of interest include lattice properties, ionic bonding strengths, etc." (255).

[19] In order to check whether the results were robust, the authors carry out a battery of tests to supplement their instrumental variables method. These include Monte Carlo methods to examine the small-sample distribution of the test statistics in order to check for the small-sample bias. The Monte Carlo experiment deploys their framework with adjustable parameters for population distribution and aims to produce artificial data with similar structural—in this case statistical—properties (such as that log-income and consumption are integrated processes and consumption and income are cointegrated). In addition, they explore various generalizations of their model to see if there are other candidate explanations.

Again, however (and as Chao points out), one might object that it is only the random walk hypothesis that is in danger, rather than the Euler equation approach itself, as Chao notes: "Anomalies may reject the permanent income hypothesis or the life-cycle hypothesis, but they only motivate modelers to modify the Euler equations instead of abandoning them" (243). For example, instead of discarding the Euler equation, discrepancies in behavior are used to modify it.[20] One might suggest that this only indicates that one or more of the assumptions going into the random walk hypothesis itself must be problematic, but it need not also imply that the Euler equation (the intertemporal preference ordering) is problematic. We have something of an underdetermination problem; the consumption equation's failure might be due to any number of things, such as wrong assumptions about preferences (like that of constant relative risk aversion, which is quite common).

But there are reasons to question the viability of the Euler equation itself. Carroll (2001) questions whether we are able to estimate the (log-linearized) Euler equations at all. Canzoneri et al (2007) compare the interest rates predicted by the Euler equation against the money market interest rate to find out that they are negatively correlated. Though they start out with preferences that satisfy constant relative risk aversion, they do try out different utility functions. Estrella and Fuhrer (1999) claim that "evidence that shows that some forward-looking models from the recent literature may be less stable—more susceptible to the Lucas critique—than their better-fitting backward-looking counterparts," (4) a count against the Euler approach. Blinder and Deaton (1985) have their "doubts about the wisdom of modeling aggregate consumption as the interior solution to a single individual's optimization problem in adjacent periods, but in any case think it fair to say that the research done to date has not supported the econometric restrictions implied by the Euler equation approach" (467). There are at least two reasons for this: corner solutions due to liquidity constraints and aggregation problems. For instance, at least at the aggregate level, the Euler equation seems suspicious.[21] Attanasio (1999) finds that "The Euler equations for (nondurable) consumption . . . are, for most specifications of preferences, non-linear. As they refer to individual households, their aggregation is problematic" (781). Elsewhere, Hvranek's (2015) meta-analysis finds that on average, the estimated

<hr/>

[20] For another interesting case study, see Zeldes (1989) and Runkle (1991), who estimated versions of the Euler equation with different results. Attanasio (1999) reports that Zeldes (1989) "splits the sample according to the wealth held and finds that the rate of growth of consumption is related with the lagged level of income for the low wealth sample. The same result does not hold for the high wealth sample. Zeldes interprets this result as evidence of binding liquidity constraints for a large fraction of the population" (790). That is, the low-wealth sample violates the standard Euler equation while the high-wealth sample does not. On the other hand, Runkle (1991) finds no such evidence.

[21] Deaton (1992) specifies that "if tastes are the same, the Euler conditions will aggregate perfectly if three conditions hold: (a) people live for ever, (b) felicity functions are quadratic (or the time interval is very short), and (c) individuals know all the aggregate information" (167).

elasticity of intertemporal substitution is zero—too low, given the way the Euler equation is usually specified (consumption should be sensitive to variations in the real interest rate).[22],[23]

It's doubtful that even individual agents optimize in this way, and when they do it may be limited in scope—after all, agents are often thought to be "myopic" in nature. (Of course, we could always restrict the behavior to a finite number of periods, rather than an infinite horizon, which is entirely consistent with the Euler approach.) Even if we were only considering individuals as the kinds of agents that intertemporally optimize, aggregation problems notwithstanding, this very well may be false except for fairly short time horizons. One substantial task would then be to assess over what temporal scale intertemporal optimization holds. So it seems a bit odd to say that the Euler equation is the product of our uncovering some deep structure that is actually out there in the world.

Work since Hall (1978) has focused on extending the baseline Euler model in various directions. For instance, one might incorporate habit-formation in consumption behavior, which might explain the lagged behavior of consumption in response to interest rate or follow Campbell and Mankiw and incorporate hand-to-mouth consumers. Throughout all this, as Chao notes, the Euler construction still stands. Now, rather than taking the Euler construction itself to be real, however, I'm more inclined to say that it helps find out what is real. I'd argue that stipulating the Euler equation is a special case of a story about equilibrium reasoning that I have told elsewhere.[24] Equilibrium conditions, constraints on behavior, do not in themselves state truths about the way the world is—but they do help yield interesting, informative insights about the world. That is, equilibrium conditions are methodological tools rather than straightforward claims about or corresponding to states of affairs. For instance, equilibrium conditions like the Euler equation function more like the *ceteris paribus* laws we see commonly in economics. And while it may hold somewhere, and across some range of circumstances, it need not hold universally.

Here's what I mean. For example, we typically do not think "An increase in the supply of a good, *ceteris paribus*, leads to an increase in price of that good" is invariant in the sense that it holds everywhere. We do, however, use it to figure out in which pockets of the economy in which it does hold, so it will have to be invariant over some range of circumstances for some market over some time scale. If our local widget economy is somewhat (causally) isolated from the rest of the nationwide economy, we could apply *ceteris paribus* reasoning in order to think about what would happen under certain kinds of shocks or interventions. I think

---

[22] A finding consistent with Yogo (20045).

[23] Readers should also note that there is work on sensitivity at the micro rather than at the aggregate level. See Shea (1995), Parker (1999), and Hall and Miskin (1982).

[24] See Jhun (2018).

the Euler equation works in somewhat the same way—it's not that it, itself, is invariant—but rather, we use it to locate where relationships can be treated as invariant, which means picking out a relevant scale of interest. When it fails, we seek reasons as to why actual behavior diverges. And these relationships of interest are often causal, in that they are apt for an interventionist difference-making interpretation.

In particular, using the Euler equation means trying to see where, and how well, it fits. To take an example, when Mumtaz and Surico (2011) adjust parameters to fit the data over time, what they are noting is where it is and in what form that the Euler equation holds and doesn't hold, as "the parameters of the aggregate consumption Euler equation may vary over the business cycle" (5). One of their self-imposed tasks is to show that "periods of conditionally low (high) consumption correspond to periods of below-trend (above-trend) consumption" (12). What this helps highlight is the particular scale of the system we should be paying attention to. So one thing that Mumtaz and Surico's account does is make salient particular time scales (and salient behavioral properties at those scales) as relevant to the investigation. For instance, they identify recession periods in the 1970s, 1981, 1992, and 2008 as periods during which agents tend to be more backward-looking and consumption tends to be less sensitive to the real interest rate (13). Similarly, we can use the Euler equation to investigate whether agents are subject to borrowing constraints; strictly speaking this doesn't mean that the Euler equation should be thrown out, but that it applies within a certain scope. Rather than merely pointing out that agents don't optimize over time, we say something about the extent to which they do. That is, I can tell an analogous story here about the role of the Euler equation as I did with that of the random walk hypothesis in the Campbell and Mankiw case. These equations are fixtures that help us identify particular scales of interest and the behaviors that appear dominant at those scales—scale-dominant behaviors.

Let me use an analogy from materials engineering to concretize what I mean by scale-dominant behavior. Though this reductionist attitude is now less popular, one might think that the right way to determine the behavior of a system is first to accumulate all the details at a micro-level. After doing so, we will be able to infer or deduce information about what goes on at the macro-level. So a bottom-up methodology would aim to achieve a picture of the whole system (be it physical or social) first by gleaning information about its constituents and then somehow putting all that information together.

But this approach is unsatisfactory in engineering contexts. For instance, one everyday kind of problem we might worry about is how to manufacture materials in order to construct sturdy buildings that can sustain some wear and tear. Consider a steel beam:

If we engaged in a purely bottom-up lattice view about steel, paying attention only to the structure for the pure crystal lattice, then we would get completely wrong estimates for its total energy, for its average density, and for its elastic properties. The relevant Hamiltonians require terms that simply do not appear at the smallest scales. (Batterman 2013, 268)

What's *also* important in this case is structural detail at the *mesoscale*, where steel's granular structure becomes apparent—say at 30 nanometers. These boundaries between grains affect the way molecular bonds break and reform when external force is exerted on the beam. So, this structure that manifests at the mesoscale contributes to the system's macroscale behavior.

We can even tell a similar story about the permanent income hypothesis. Campbell and Mankiw's search for relevant population parameters (or Mumtaz and Surico looking to see patterns of forward-looking behavior) is akin to looking for details about grain in the steel beam. And like granularity, population distribution is a mesoscale detail that we can treat as real as goings-on at any other scale. But we had to do a little bit of work first—figuring out where the Euler equation could plausibly fit in as successfully describing behavior. Rather than describing the structure of the economy-at-large, the Euler consumption equation might characterize the behavior of some members of the economy over some (but not the infinite) time horizon. The particular configuration of these members might be something that matters. Distributional features capture additional relevant structure that affects the macro-behavior of a system; this is additional information we need to understand this system's behavior. This information also explains why it is that actual population behavior departs from the idealized case with a homogeneous population where everybody consumes current income.

Explaining the behavior of complex systems seems to require treating features that appear as salient at particular scales as real parts of a real structure, because they are the kinds of things that figure into causal relationships. But our treating economic structure as real is not the same as Worrall's structural realism; we are not realists about the random walk hypothesis, or even about the Euler equation. The Euler equation plays a different role for us. It does not figure as a theoretical posit that we understand as true, and thus is preserved over time as the theory progresses. Rather, the Euler equation helps us orient ourselves in a larger framework; it allows us to establish a vantage point from which to look at a particular system of interest and get a grip on the relevant causal relationships involved. We can think of investigation at a particular scale to be investigation from a particular perspective of that system. To accommodate this, we need to be perspectival realists, which, according to Hoover (2011), is "predicated on the belief that models are used to assert true general claims about causal relationships . . . Such a realism is compatible with models viewing the world from different perspectives . . . The truth that we seek is what the world is actually like when seen this way" (6).[25]

For the most part, economic (causal) structure that accounts for deviations in a system's behavior from, say, the idealized benchmark case constitutes real structure. This structure will appear as salient or dominant at a particular characteristic scale. It must fit into a larger framework that tells a coherent story about why it is that a complex system behaves the way it does, and different properties and behaviors may appear as dominant at different scales. For instance, we can tell a coherent story about steel grain size and beam elasticity—it is fundamentally a story that has to make reference to different scale lengths. In an analogous sense, we can tell a coherent story about the economy that in part refers to (say) population distribution, excess sensitivity, or even the time scales at which agents do (or don't) appear to act in a forward-looking manner.[26,27]

12.4 Modest (Pragmatic and Perspectival) Structural Realism

The kind of realism I favor is modest; it doesn't even claim that a theory marches steadily toward a unique true theory even in the limit. All that it requires is that at least sometimes, we're justified in thinking that we are getting things right.[28] I propose that the perspectival realism we endorse corresponds to a pragmatic interpretation of Lyons's axiological realism, capturing what is intuitively attractive about the view in addition to seamlessly providing an epistemic counterpart to his position.

Lyons (2005) suggests that "science is interested, not in true statements, per se, but a certain type of true statements, those that are manifested as true" (174). These statements partake in theory complexes—groups of theories—and are such that they have bearing on our world of experience; they can *manifest* as true.

[25] That is, economic structures are structures relative to perspective (scale). Our account is also different from another attractive account of realism, *selective realism*. First, perspectival realism gives us something to make sense of theories synchronically—the attitude that we have at any given time for a theory under consideration. On the other hand, selective realism is a diachronic account—it explains something about a persistent element of a theory (or theories) over time.

[26] It is within this larger framework that we can ask further questions such as: Is wealth a factor that we should have included in our macroeconomic analysis? Is income inequality a structural feature that matters? And so on. See, for work along these lines, Carroll, Slacalek, and Tokuoka (2014).

[27] But articulating multi-scale relationships is a tricky matter in itself. Bursten (2018) argues in the case of nanoscience that articulating the relationship between scales may require conceptual maneuvers that resist traditional categorization (e.g. reductionist ones). I would like to leave space, for economics, to take advantage of these oddball conceptual maneuvers, though this may lead to further complications for thinking about realism down the line.

[28] Note here a sympathy with something like Mäki's (2009) realistic realism.

Lyons (2011) has called these statements, as a subclass of true statements, those that are "experientially concretized" truths (XT statements): "true statements whose truth is made to deductively impact, is deductively pushed to and enters into, documented reports of specific experiences" (329).[29] True statements are candidates for—but not all are—XT statements, whose truth is experientially concretized as true. And while actual XT statements are required to be true in order to qualify as being so, we do not necessarily think ourselves to have achieved truth per se. That is, despite the fact that that they inform the articulation of or imply some claim that corresponds to an experience:

> Crucially, the experiential concretization of XT statements is non-epistemic: While XT statements are such that their truth is deductively pushed down to and enters into such documented reports, no claim is being made here that the fact of this relation to documented reports *informs* us of the truth of XT statements. (329)

The emphasis here is on the pursuit of truth, rather than justification for believing we have achieved it. That is, we can be committed to seeking truth and yet be agnostic about whether we have achieved it or not.

Macroeconomics can be considered a complex of theories and models. Statements such as "the proportion of income accrued to forward-looking agents is λ" have implications for what we expect to see in the macroeconomic aggregate. The Euler equations are such that (given some additional context) they imply particular empirical consequences that may or may not hold.[30]

However, fixtures such as the Euler equations cannot be XT statements according to Lyons's account. And fundamental theoretical posits like the Euler equation on its own are not even really treated as genuine candidates for truth. Some may simply assume that they are outright false because they're often considered idealizations. After all, the Euler equation approach is probably not a good way of describing aggregate macroeconomic behavior. It is not even a good way to think about individual behavior (people often simply fail at intertemporal utility maximization, never mind that they are not infinitely lived agents anyway). So in order to make Lyons's insights relevant to economics, I propose that we step away from the foundational conception of theory complexes as lying on a bedrock of (manifest) truth from which we can infer other (manifest) truths and think of theoretical posits as doing something other than serving as such

---

[29] In his earlier work he tells us that a truth is manifest when "it has bearing on, makes a difference to, the world of experience and is transmitted to tested predictions derived from the complex" (2005, 174). Such truth is "not manifested as true *to us* but in the consequences derived from the theory and in the world of our experience" (177).

[30] See, for this position, Hoover (2009).

ground-level manifest truths. The Euler equation has a shaping or framing role, not a grounding one.

The Euler equation, in and of itself, is not literally true (nor, arguably, is it false by itself either), so it cannot be an XT statement.[31] But we use it to formulate claims that we do compare against empirical data to determine whether it explains or describes our data well. And if we take Lyons's account seriously, we'd actually have what he calls an "evident XT-deficiency" where "it is evident that non-XT statements are present in the complex" (2011, 330).

But it's consistent to claim that the Euler equation doesn't usually hold of anything by itself and yet also claim that this approach is part of a methodology that helps us find out what's true about the world, by allowing us to anchor ourselves in a particular economic context. Used wisely, I think such theoretical posits embody strategies that help us find manifest (or experientially concretized) truths— truths that make a difference to our world of experience. The way it is doing so, then, can't be hashed out in terms of deductive informativeness (however way we spell that out). Such non-XT statements figure as central to theory. For example, Campbell and Mankiw do not discard the Euler equation approach from their toolbox in lieu of a more realistic equation. Nor is it an idealization that simply needs to be filled out with more detail in order to account for empirical data. They use it, in fact, to construct a larger generalized framework around it. It remains central to their reasoning, and such statements are never quite eradicated from the science. But their primary role is not to deductively impute truth.[32]

Chao was right to emphasize the persistence of the Euler equation. I am suggesting that the role of such equations is methodological rather than representational; they serve to locate and articulate structure and serve as diagnostic tools. They are fixtures that are central to analysis. When it comes to complex systems, we had to identify scale-dominant behavior in order to diagnose the ways

---

[31] If anything, I would argue that the Euler equation has a role in shaping the complex of XT statements, so that its role is prior to the question of whether or not a particular model represents phenomena accurately. To put it in Lyons's terms, we could understand the Euler equation as an instrument that even helps "remedy evident XT deficiencies by increasing the number—and/or the extent, degree, or exactitude of the experiential concretization—of XT statements" (2011, 330).

[32] Lyons (2017) anticipates the kind of suggestion that I'm making.

> [P]atently false posits can serve toward "bringing the truth" of high level posits "down" to statements that describe experiences. Once novel predictions are derived, even from false hypotheses, and new data is gathered, then, on this view, those mediating blatantly false constituents previously deployed are eliminated in want of increasing the experientially concretized truth of the accepted system. (3213)

So false posits can have a central role, making the distinction between our positions a subtle matter. But for us the posit is never really eliminated. This is because (for instance) Mankiw and Campbell do not attain their results on the basis of false assumptions; rather, the posit fully serves as a foundation for the model they eventually articulate, so it is in effect built into the final product they offer. But I think ultimately that the difference between our positions isn't so much of a disagreement so much as we are simply pointing at the nuanced guises that posits take in scientific reasoning.

the actual behavior of a system deviates from the idealized model. We might start an enquiry by examining whether the Euler equation fits our population-wide data; when it fails to do so, we must figure out where else (if anywhere) it can have explanatory or descriptive power—where those causal relationships stipulated by the equation actually hold. The Euler equation is more like a navigational tool (that helps us answer questions such as, where does it hold, and where does it fail to hold? Why not?). And the more we are able to diagnose different kinds of deviations from the ideal and able to tell when that benchmark is or isn't the appropriate one to use, the more our science progresses. In a way, economists too are in the business of seeking and maximizing manifest truth.

Readers will note that this view, given its endorsement of Hoover's perspectival realism, also shares kinship with Chang's (2016) pragmatic realism, where the pragmatic refers to the practical aspect of *activity*:

> I define (pragmatist) coherence as a harmonious fitting-together of actions that leads to the successful achievement of one's aims. Such coherence may be exhibited in something as simple as the correct coordination of bodily movements needed in riding a bicycle . . . or something as complex as the successful integration of a range of material technologies and various abstract theories in the operation of the global positioning system (GPS). A coherent epistemic activity achieves its aim well, and avoids performative self-contradiction. (112–113)

With this notion of coherence in hand, Chang (2017) formulates his coherence theory of reality as follows: "a putative entity should be considered real if it is employed in a coherent epistemic activity that relies on its existence and its basic properties (by which we identify it)" (15).

So for the economist, a coherent activity might be implementing a bit of policy, which in turn requires conducting a bit of counterfactual analysis in a model. Policymaking would require an economist to take seriously how behavior at the macroscale might change in response to changes at other scales. I might want to know how overall consumption changes when the underlying distribution of agents changes. That is, exercises of counterfactual reasoning about complex systems behavior rely on the existence of a (multiscale) framework. While something like the Euler equation need not be considered real, it partakes in framing a coherent epistemic project—this coherent multiscale framework where goings-on at different scales must coordinate with one another. The scale-specific structure that we discover in our investigation, insofar as it partakes in such a framework, is what is real. Grasping the causal structure of the world is partly a matter of trying to figure out how fundamental equations like the Euler equation fit in and help articulate such a framework.

This modest realism is nothing new. What I am proposing here is that such a realism supplies what's missing in Lyons's axiological account—the epistemic *oomph* of what is experientially concretized as true. Experientially concretized truth can be legitimate, plain old truth that we can have at hand—there's nothing that is truth per se in the pragmatic account apart from what is experientially concretized. Positing the Euler equation is the beginning of an enterprise that is open to the possibility that multiple perspectives of a system have to be taken into account, and that the parcels of information we obtain from different perspectives must cooperate with one another. While how we place those perspectives (and they very well may be separate models) into a coherent framework is a question for another time, that complex is one—insofar as it allows us to successfully conduct something like counterfactual analyses—that is apt for a realist interpretation.

A final note. I have been deliberate in my noncommittal attitude to whether or not the realism I care about is epistemic or ontic. This is because for the most part, the distinction doesn't bear much on the suggestion I'm trying to make. Structural realism is, in both senses, a positive and a negative thesis. It is a positive thesis about what we can know, and what there is, and by parity about what we do not know, and what there isn't. My thesis is simply that there is structure that we can be realists about, namely causal structure, and that the Euler equation is one tool that economists use to uncover it. And given the pragmatic nature of our discussion, in that our knowledge of the economy involves how to do things, the simple answer is: I am a realist both about what we can know and about what there is at the same time, though I do not commit here to the negative thesis that that's all we can know or that that's all that there is. The realist attitude outlined here simply tries to get right how economists actually go about doing things, insofar as we conceive of economics as a policy science. This pragmatic view emphasizes the coherence of the multi-scale causal structure, where structure occupies a fairly fundamental place in economists' thinking. At least, given our investigation of economic methodology, we shouldn't take as granted the priority of (knowledge of) individuals over structure.

12.5   Against Reiss Against Realism

A few remarks before I conclude. I don't consider the Euler equation itself to be an approximation of the truth, nor something that is a pit stop for a model that's developing toward a future truer version of itself. The Euler equation aids us in finding the truth, but in itself is not representational—so it is not an approximation of anything. If the Euler equation fails to "fit the data" it may be for a number of different reasons; that the population is heterogeneous is a causally relevant feature about the structure of the population of interest. That is, the Euler equation's purpose is more diagnostic than representational.

Because this realism looks somewhat different from standard forms of realism, which are grounded on correspondence views of truth, the defenses that it can provide will look a bit different, too. In this section, we consider Reiss's (2012) criticism of realism; he argues that instrumentalism has at least three advantages over it. I argue that we can sidestep those criticisms.

The first criticism: "Once we have usefulness, truth is redundant" (370).

The redundancy shows up in our account in an automatic kind of way. If usefulness is what we mean by pragmatic coherence, then truth trivially follows because we conceived of truth in terms of being able to undertake coherent activity. So truth is built in to our notion of success rather than something vestigial.

The second criticism: "There is something disturbing about causal structure" (372).

Reiss worries that "if we seek predictive success, we don't necessarily want to build causal models" (373). I do not claim anywhere that predictive success is the mark of a successful model, nor does a model need be an accurate representation in order to be true. Reiss maintains that "a model is not predictively successful qua representing causal structure" (373–374). In fact, I agree! Reduced-form models can be perfectly suitable for prediction in some cases. And this is because a good reduced form model is, in fact, the collapsed form of a structural model. So the relevant structure is incorporated into the reduced-form construction.[33] And in order to use a model for *diagnostic* purposes, á la counterfactual reasoning, it ought to be causal.

Reiss's larger worry is that "even when we are able to establish causality, to make a causal model practically useful, additional facts about the represented relations have to be discovered. But once we've discovered the additional facts, it is irrelevant whether the relation at hand is causal or not" (374). That is, the fact that a relationship is causal is irrelevant as to whether it is successful or not. Given that diagnostic success for us is conceived in terms of possible (and possibly actual!) interventions, this seems implausible. If the structure of the economy is something that counts as a difference-maker to its overall behavior, then I consider that a causal contribution.

---

[33] Every structural model gives rise to different reduced forms, and there might be observational equivalence between some of them, but there remain legitimate questions about which one is a good one. These questions are going to ask which ones have got the structure right (where structure is understood in my sense!).

The third criticism: "It's better to do what one can than to chase rainbows" (375).

The final worry is that "building causal models has enormous informational requirements" (375), and so a different heuristic that one might use is what is known as Marschak's Maxim. Marschak (1953) observed that for many questions of policy analysis there is no need to identify fully specified models that are invariant to whole classes of policies. If the researcher instead focuses on particular interventions, all that may be needed are combinations of subsets of the structural parameters—those required to identify the effect of the policy intervention.

There's nothing stopping a structural realist of my stripe from abiding by Marschak's Maxim.[34] Think of the analogy with materials. We don't need all the information about all the atoms in the steel beam—we do, however, need to know about grain and how that granular structure affects macroscale behavior.[35] The engineer's version of Marschak's Maxim would tell us to focus our attentions at this scale. Something similar is the case in economics. It's not necessary to uncover all the individual preferences of individuals to get a sense of what consumption in the aggregate looks like or all the causes that might be at play in order to assess the effect of a particular disturbance to a system. We may need to know something about the *distribution* of preferences, and finding this information is a matter of finding the right *perspective*—that is, the appropriate *scale*. And insofar as it is specific to scale and partakes in a multi-scale framework that enables us to undertake counterfactual exercises, I call that structural and real.

12.6 Conclusion

I've argued thus far for a modest structural realism that accommodates at least some of the ways that economists actually deploy idealization in their search for the real causes underlying dynamic behavior. Making sense of these practices necessitates our moving away from correspondence notions of truth to coherence ones; aligning our realism with Lyons's axiological realism cannot be achieved otherwise. But a pragmatic spin on realism allows us to recognize the central methodological role that idealizations have to play in obtaining (in particular

---

[34] In Marschak's 1953 Cowles paper, he actually describes what can be recognized as the Lucas Critique. One way in which they differ is that Marschak is not concerned with expectations in the way Lucas is. Marschak's point, nonetheless, is that we need invariance—but the notion of being invariant is relative (as nothing is invariant to everything)—something that Lucas himself seems to accept! This theme reaches back to Haavelmo and Frisch (see Aldrich 1989).

[35] And the range of circumstances under which a particular granular configuration will persist; for instance, with extremely high temperatures we can induce *grain growth*, which in turn affects macroscale properties by making steel more elastic.

causal) knowledge. It is not the Euler equation that is real but the structure that the Euler equation helps us uncover and use to explain economic behavior.

Acknowledgments

I am grateful to audience members at Duke's Center for the History of Political Economy for their helpful discussion on an earlier draft of this paper. In addition, I am especially indebted to Hsiang-Ke Chao, Kevin Hoover, Timothy Lyons, and an anonymous referee for their extensive and invaluable comments.

References

Aldrich, J. (1989). Autonomy. *Oxford Economic Papers, 41*(1), 15–34.
Attanasio O. P. (1999). Consumption. In, Taylor, J. B., & Woodford, M. (eds.) *Handbook of Macroeconomics,* Vol. 1B. Amsterdam: Elsevier Sci, pp. 741–812.
Batterman, R.W. (2013). The Tyranny of Scales. In, *The Oxford Handbook of Philosophy of Physics.* Oxford: Oxford University Press, pp. 255–286.
Blinder, A. S., Deaton, A., Hall, R. E., & Hubbard, R. G. (1985). The time series consumption function revisited. Brookings Papers on Economic Activity, *1985*(2), 465–521.
Brady, D. S., & Friedman, R. D. (1947). Savings and the income distribution. In, *Studies in Income and Wealth,* Vol. 10. National Bureau of Economic Research, pp. 247–265.
Bronfenbrenner, M. (1948). The consumption function controversy. *Southern Economic Journal, 14*(3), 304–320.
Bursten, J. (2018). Conceptual strategies and inter-theory relations: The case of nanoscale cracks. *Studies in History and Philosophy of Science Part B: Studies in History and Philosophy of Modern Physics, 62,* 158–165.
Campbell, J.Y. and Mankiw, N.G. (1991). The response of consumption to income: A cross-country investigation. *European Economic Review,* 35(4), 723–756.
Campbell, J. Y., & Mankiw, N. G. (1990). Permanent income, current income, and consumption. *Journal of Business & Economic Statistics,* 8(3), 265–279
Campbell, J. Y., & Mankiw, N. G. (1989). International evidence on the persistence of economic fluctuations. *Journal of Monetary Economics, 23*(2), 319–333.
Canzoneri, M. B., Cumby, R. E., & Diba, B. T. (2007). Euler equations and money market interest rates: A challenge for monetary policy models. *Journal of Monetary Economics,* 54(7), 1863–1881.
Carroll, C. D. (2001). Death to the log-linearized consumption Euler equation! (And very poor health to the second-order approximation). *Advances in Macroeconomics, 1*(1), 1–38.
Carroll, C. D., Slacalek, J., & Tokuoka, K. (2014). The distribution of wealth and the MPC: Implications of new European data. *American Economic Review, 104*(5), 107–11.
Cate, T. (ed.). (2013). *An encyclopedia of Keynesian economics.* Edward Elgar Publishing.
Chang, H. (2017). Operational Coherence as the Source of Truth. *Proceedings of the Aristotelian Society, 117*(2), 103–122.

Chang, H. (2016). Pragmatic realism. *Journal of Humanities of Valparaiso, 0*(8), 107–122.

Chao, H. K. (2007). A structure of the consumption function. *Journal of Economic Methodology, 14*(2), 227–248.

Chao, H. K. (2003). Milton Friedman and the emergence of the permanent income hypothesis. *History of Political Economy, 35*(1), 77–104.

Deaton, A. (1992). *Understanding consumption.* Oxford University Press.

Dennett, D. (1991). Real patterns. *Journal of Philosophy, 88:* 27–51.

Duesenberry, J. S. (1949). *Income, saving and the theory of consumption behavior.* Cambridge, Mass: Harvard University Press.

Estrella, A., & Fuhrer, J. C. (1999). Are "Deep" Parameters Stable? The Lucas Critique as an Empirical Hypothesis, Working Papers 99-4, Federal Reserve Bank of Boston; paper available at: fmwww.bc.edu/cef99/papers/efpaper.pdf

Estrella, A., & Fuhrer, J. C. (2002). Dynamic inconsistencies: Counterfactual implications of a class of rational-expectations models. *American Economic Review, 92*(4), 1013–1028.

Ferber, R. (1953). *A study of aggregate consumption functions.* New York: National Bureau of Economic Research.

Flavin, M. (1985). Excess sensitivity of consumption to current income: Liquidity constraints of myopia?. *Canadian Journal of Economics, 18*(1), 117–136.

Friedman, M. (1957). The Permanent Income Hypothesis. In, Friedman, M. (ed.) *A Theory of the Consumption Function.* Princeton, NJ: Princeton University Press, pp. 20–37.

Friedman, M., and Kuznets, S. (1945). *Income from independent professional practice.* New York: National Bureau of Economic Research.

Haavelmo, T. (1947). Methods of measuring the marginal propensity to consume, *Journal of the American Statistical Association, 42*(237), 105–122.

Haavelmo, T. (1944). The probability approach in econometrics. *Econometrica: Journal of the Econometric Society,* Supplement (July 1944), iii–115.

Hall, R. (1978). Stochastic implications of the life cycle-permanent income hypothesis: theory and evidence. *Journal of Political Economy, 86*(6), 971–987.

Hall, R. E. (1990). *The rational consumer: theory and evidence.* Cambridge, MA: MIT Press. Chapter 7: 'Survey of research on the random walk of consumption', pp. 133-158.

Hall R., & Mishkin F. S. (1982). The sensitivity of consumption to transitory income: estimates from panel data on households. *Econometrica, 50,* 461–481.

Harker, D. (2012). How to split a theory: Defending selective realism and convergence without proximity. *The British Journal for the Philosophy of Science, 64*(1), 79–106.

Hart, A. G. (1946). National Budgets and National Policy: A Rejoinder. *The American Economic Review, 36*(4), 632–636.

Heckman, J. J., & Vytlacil, E. J. (2007). Econometric evaluation of social programs, part I: Causal models, structural models and econometric policy evaluation. *Handbook of econometrics, 6,* 4779–4874.

Hendry, D. F. & Mizon, G. E. (2005). Forecasting in the presence of structural breaks and policy regime shifts. In, Andrews, D. W., and Stock, J. H. (eds.) *Identification and Inference for Econometric Models: Essays in Honor of Thomas Rothenberg.* Cambridge, UK. Cambridge University Press, pp. 480–502.

Hoover, K. D. (2015) Reductionism in economics: Intentionality and eschatological justification in the microfoundations of macroeconomics. *Philosophy of Science, 82*(4), 689–711.

Hoover, K. (2011). Counterfactuals and Causal Structure. In, Illari, P. M., Russo, F., & Williamson, J. (eds.) *Causality in the Sciences*. Oxford: Oxford University Press, pp. 338–360.

Hoover, K. (2009). Microfoundations and the Ontology of Macroeconomics. In, Kincaid, H., & Ross, D. (eds.) *Oxford Handbook of the Philosophy of Economic Science*. Oxford: Oxford University Press, pp. 386–409.

Hoover, K. (2001). Is Macroeconomics for real? In, Maki, U. (ed.) *The Economic World View: Studies in the Ontology of Economics*. Cambridge: Cambridge University Press, pp. 225–45.

Havranek, T. (2015). Measuring intertemporal substitution: The importance of method choices and selective reporting. *Journal of the European Economic Association, 13*(6), 1180–1204.

Jhun, J. S. (forthcoming). Economics, equilibrium methods, and multi-scale modeling. *Erkenntnis*. https://doi.org/10.1007/s10670-019-00113-6.

Jhun, J. S. (2018). What's the point of ceteris paribus? or, how to understand supply and demand curves. *Philosophy of Science, 85*(2), 271–292.

Keynes, J.M. (1936) *The General Theory of Employment, Interest and Money.* Cambridge: Macmillan Cambridge University Press.

Kuznets, S. (1952). Proportion of capital formation to national product. *The American Economic Review, 42*(2), 507–526.

Kuznets, S. (1942). *Uses of national income in peace and war.* New York: National Bureau of Economic Research.

Ladyman, J. (1998). What is structural realism?. *Studies in History and Philosophy of Science, 29*(3), 409–424.

Lucas, R. E. (1976). Econometric policy evaluation: A critique. In, Brunner, K., and Meltzer, A. H. (eds.) *The Phillips Curve and Labour Markets.* Amsterdam: North Holland, pp. 19–46.

Lyons, T. D. (2005). Toward a purely axiological scientific realism. *Erkenntnis, 63*(2), 167–204.

Lyons T.D. (2011). The problem of deep competitors and the pursuit of epistemically utopian truths. *Journal of General Philosophy of Science 42*, 317–0338.

Lyons, T. D. (2017). Epistemic selectivity, historical threats, and the non-epistemic tenets of scientific realism. *Synthese, 194*(9), 3203–3219.

Lucas, R. E., & Sargent T. J. (1979). After Keynesian macroeconomics. *Federal Reserve Bank of Minneapolis Quarterly Review, 3*(2), 1–16.

Mäki, U. (2009). Realistic realism about unrealistic models. In, Kincaid, Harold, & Ross, Don (eds.) *The Oxford Handbook of Philosophy of Economics.* New York: Oxford University Press, pp. 68–98.

Mankiw, N. G., Rotemberg, J. J., & Summers, L. H. (1985). Intertemporal substitution in macroeconomics. *The Quarterly Journal of Economics, 100*(1), 225–251.

Marschak, J. (1953). Economic measurements for policy and prediction. In, Hood, W., & Koopmans, T. (eds.) *Studies in Econometric Method.* New York and London: John Wiley and Sons, 1–26.

Modigliani, G., & Brumberg, R. (1954). Utility analysis and the consumption function: An interpretation of cross-section data. In Kurihara, K. K. (ed.) *Post- Keynesian economics.* New Brunswick, NJ: Rutgers University Press, pp. 388–436.

Molana, H. (1993). The role of income in the consumption function: A review of on-going developments. *Scottish Journal of Political Economy, 40*(3), 335–352.

Morganti, M. (2011). Is there a compelling argument for Ontic Structural Realism? *Philosophy of Science, 78*(5), 1165–1176.

Mumtaz, H., & Surico, P. (2011). *Estimating the aggregate consumption Euler Equation with state-dependent parameters* (No. 8233). CEPR Discussion Papers.

Orcutt, G. H., & Roy, A. D. (1949). *A bibliography of the consumption function.* Cambridge University, Dept. of Applied Economics.

Parker, J. A. (1999). The reaction of household consumption to predictable changes in social security taxes. *American Economic Review, 89,* 959–73

Reiss, J. (2012). Idealization and the aims of economics: Three cheers for instrumentalism. *Economics & Philosophy, 28*(3), 363–383.

Ross, D. (2008). Ontic structural realism and economics. *Philosophy of Science, 75*(5), 732–743.

Runkle, D. E. (1991). Liquidity constraints and the permanent-income hypothesis: evidence from panel data. *Journal of Monetary Economics 27,* 73–98.

Shea, J. (1995). Union contracts and the life-cycle permanent income hypothesis. *American Economic Review, 85,* 186–200

Smithies, A. (1945). Forecasting postwar demand. *Econometrica, 13,* 1–14.

Wilson, M. (2017). *Physics avoidance: And other essays in conceptual strategy.* Oxford: Oxford University Press.

Worrall, J. (1989). Structural realism: The best of both worlds? *Dialectica, 43*(1–2), 99–124.

Woodward, J. (2005). *Making things happen: A theory of causal explanation.* New York: Oxford University Press.

Yogo, M. (2004). Estimating the elasticity of intertemporal substitution when instruments are weak. *Review of Economics and Statistics, 86*(3), 797–810.

Zeldes, S. P. (1989). Consumption and liquidity constraints: an empirical investigation. *Journal of political economy, 97*(2), 305–346.

13

We Think, They Thought

A Critique of the Pessimistic Meta-Meta-Induction

Ludwig Fahrbach

University of Duesseldorf
ludwig.fahrbach@gmail.com

13.1 Introduction

The aim of this paper is to analyze and assess a certain argument used by antirealists in the scientific realism debate, the pessimistic meta-meta-induction (two "meta"s). Let scientific realism be the view that our current successful theories are probably approximately true. Antirealists challenge realism by presenting the pessimistic meta-induction, PMI (one "meta"), according to which many theories in the past of science were also successful for a while, but were refuted later on. Typically the very first response by scientific realists to the PMI is that science has improved a lot since the times of the past refuted theories and that these improvements block the PMI and save realism.[1] Antirealists then often reply that *past realists could have said the same thing*, namely that science has improved a lot in the same manner as realists claim for the recent past, but those improvements didn't help past realists, as the subsequent theory refutations show; hence, the recent improvements likewise don't help current realists to block the PMI and to save realism, so the realists' defense against the PMI fails. It is this argument by antirealists which is the focus of this paper. I call it the pessimistic meta-meta-induction, PMMI.[2] I will analyze the PMMI and attempt to show that it does not succeed as a reply to the realist's defense.

---

[1] Stathis Psillos calls this response "the Privilege-for-current-theories strategy." (Psillos 2018, Sect. 2.8). Mario Alai calls it the "discontinuity strategy": "there are marked differences between past and present science," which imply that "pessimism about past theories can[not] be extended to current ones." (Alai 2017, p. 3271/67).

[2] The term "pessimistic meta-meta-induction" is also used by Devitt (1991, p. 163) and by Psillos (2018, Sect. 2.8). The PMMI is expounded by Wray (2013, 2018, Ch. 6) to whom this paper is a rejoinder, and Psillos (2018). A related version directed at the realist defense based on the improvement of scientific methods is criticized by Devitt (1991, p. 163/4). See also Doppelt (2014, p. 286), Müller (2015), Stanford (2015, p. 3), Alai (2017, p. 3282), Vickers (2018, p. 49), Rowbottom (2019, p. 476), and Frost-Arnold (2018, p. 3).

Ludwig Fahrbach, *We Think, They Thought* In: *Contemporary Scientific Realism*. Edited by: Timothy D. Lyons and Peter Vickers, Oxford University Press. © Oxford University Press 2021. DOI: 10.1093/oso/9780190946814.003.0013

The defense of realism just presented, which invokes recent improvements in science to block the PMI and to which the PMMI is a reply, is currently not the most prominent realist response to the PMI; rather that is selective realism. This paper is not about selective realism, however, so I will only briefly note some problems of selective realism, which are meant to show that it is not unreasonable to explore alternative responses to the PMI.

Selective realism typically consists of two claims: Theories are required to enjoy novel predictive success, and novel predictive success of a theory only allows us to infer some sort of partial truth of the theory which concerns specific types of parts of a theory, such as structure, entities, or success-conferring parts.[3] It has become quite clear that many past refuted theories also enjoyed novel predictive success, hence the defense of selective realism has to rely primarily on the respective notion of partial truth.[4] Such notions have likewise been extensively discussed, but one problem has not received much attention.

Selective realism implies that currently accepted scientific theories contain specific types of false parts which can be identified today and which can be expected to be replaced in future theory changes. However, selective realists have rarely made any effort to actually determine those false parts of current theories. They should do so. They should spread out into the scientific world, examine the numerous statements accepted as scientific fact today, and try to find the false ones. An important task would be the examination of scientific textbooks containing what scientists take to be established scientific knowledge in disciplines such as chemistry, biology, geology, and so on. Selective realists should check all statements in the textbooks, determine the parts that are idle, or about non-entities, or about non-structure (nature, substance, content, interpretation, . . .), inform the textbook authors of their findings, and urge the correction or deletion of the problematic statements. Of course, if they actually did so, they would meet with surprise and resistance from scientists who deem the established scientific knowledge to be confirmed by compelling empirical evidence and would not be prepared to rewrite their textbooks.[5]

What this shows is that selective realism is highly revisionary. It is in deep conflict with the judgments of practicing scientists. The conflict is not just theoretical, but may also concern practical matters such as decisions about the direction of future research and concrete applications of scientific knowledge in modern society. These short remarks have to be worked out in detail. Still, if one

---

[3] For a list of versions of (selective) realism see Lyons (2017, p. 3215). For an overview of structural realism see Frigg and Votsis (2011).

[4] See, for example, Lyons (2002, 2006, 2017) and Vickers (2013, 2016, footnotes 6 and 10). Novelty of predictions is here understood in the usual way, as use-novelty: The prediction is not used in the construction of the theory.

[5] In Section 13.11 I will hint at some reasons why I think scientists are right on this matter.

is bothered by the revisionary nature of selective realism, as I am, then one will find it worthwhile to explore alternative responses to the PMI.

The dialectic in which the PMMI arises has to be developed with some care (Sections 13.2 to 13.5, and 13.7). In Section 13.2 I define scientific realism and present the PMI. In Section 13.3 I analyze the PMI. Then I discuss the realist's defense against the PMI, which consists in claiming that science has improved a lot since the times of the refuted theories, and formulate three assumptions on which the defense is based (Section 13.4). Finally Section 13.5 introduces the PMMI. I analyze it, and discuss some objections (Section 13.6). Then comes the central part of the paper. I show how the assessment of the PMMI is related to the assessment of the PMI, and present my objections to the PMMI (Section 13.7 to 13.9). Section 13.10 deals with the charge of ad hocness. In Section 13.11 I briefly discuss how the PMI and the PMMI fare, if it is assumed that there has been a large increase in degrees of success in the recent past. A short conclusion summarizes the argument against the PMMI.

The dialectic can be developed in a number of different ways. Obviously I cannot discuss every possible branch of the dialectic, but have to make some choices and assumptions. Some assumptions constitute, or rely on, strong simplifications. For example, as will become apparent later on, I rely on a strongly idealized picture of the history of science. The assumptions and simplifications are always meant to serve the goal of understanding and assessing the PMMI. Thus I won't necessarily use the most prominent or plausible versions of the positions and arguments in the realism debate, but ones that are useful and convenient for the purposes of this goal.

13.2 Realism and the PMI

In this Section I set up the dialectic between the realist and the antirealist. Let scientific realism be defined as the position that our current successful theories are probably approximately true. (In the following I will omit "probably" and "approximately.") Examples of such theories are plate tectonics, the heliocentric system, and the theory of evolution. There are many versions of scientific realism, of course, but this definition is a fairly standard one. A theory enjoys empirical success iff it has made sufficiently many sufficiently significant true predictions and no important false predictions, in other words, iff it has passed sufficiently many sufficiently severe tests and not failed any important tests. This is not very precise, but will do for our purposes.

Scientific realists support their position with the No-Miracles Argument (NMA), according to which the success of our current successful theories would be a miracle if they were false (Putnam 1978, Smart 1963, p. 39). Although this argument is usually spelled out as an inference to the best explanation, I will only work with the basic intuition of the NMA. Antirealists usually do not accept the NMA, but considering challenges to the NMA other than the PMI would lead us too far afield.

Antirealists attack realism by means of the PMI. I begin with a version of the PMI that highlights the commonalities between the PMI and the PMMI, and therefore looks a bit different from the ways it is usually presented. This version starts with the observation that *many realists in the past said the same thing* as today's realists do: They also believed that their successful theories were true, supporting their realism with arguments resembling the NMA. For example, Clavius, Kepler, and Whewell defended realism about their theories by alluding to the empirical success of their theories (Musgrave 1988, p. 229/30). However, so the attack goes, many of their successful theories were later refuted. For example, the Ptolemaic system endorsed by Clavius was refuted, and Kepler relied on all sorts of later-abandoned assumptions about the sun and the planets (Lyons 2006). All theories just mentioned (heliocentric system, theory of evolution, plate tectonics) had precursors that were successful to some degree and held by many scientists to be true, but were later replaced.[6] Thus, past realists were proven wrong about their realism. It follows that today's realists will probably share the same fate as past realists, implying that realism about current successful theories is likewise wrong. So, this is the PMI. In the next section we will see that it is not essentially different from more familiar versions of the PMI. Wray captures the spirit of the PMI nicely:

> The Pessimistic Induction asks us to see ourselves as similar to the scientists of the past. We are not to be Whigs or deluded, assuming that we are not prone to make the same sorts of mistakes that they were prone to make. . . . The pessimistic induction is . . . designed to aid us in recognizing the similarities between our predicament and the predicament of our predecessors. (2018, p. 95/6)

There is another objection to realism. Realism as just defined refers only to present, but not to past successful theories. The reason is obviously that past successful theories were often refuted, hence cannot be included in the scope of realism. This invites the objection that this is not a good reason to restrict the scope of realism to current successful theories. The realist has to offer an independent reason for the restriction, or else the restriction is ad hoc, only serving

---

[6] Other examples famously include the phlogiston theory, the caloric theory of heat, the ether theory of light, and some more (see Laudan 1981 and the references in footnote 4). Note that the successful refuted theories do not necessarily refer to unobservables, rather some refer chiefly or exclusively to observables (Stanford 2006, Ch. 2, Fahrbach 2011, 2017, Sect. 3.3). For this reason the PMI is also a threat to realism about theories of the latter kind.

the purpose of making realism compatible with the existence of past refutations.[7] This objection differs from the PMI since it does not use past refuted theories to undermine realism directly, instead it only uses them to point out a weakness in the definition of realism, namely an unprincipled restriction of the scope of realism. I will discuss the charge of ad hocness in Section 13.10.

The PMI as just presented is an instance of a general argumentative strategy on the antirealist's side that has the following form: The target of each instance of the strategy is some claim or piece of reasoning endorsed by the realist which refers to current theories. The strategy consists in first observing that past realists made, or could have made, the same claim (employed, or could have employed, the same piece of reasoning) about past theories as the realist does about current theories, and, second, pointing out that past realists were, or would have been, proven wrong by later theory refutations or rejections. We are then invited to conclude that the claim (or piece of reasoning) offered by today's realist about current theories is wrong as well. I will call any instance of this general strategy an *argument from the past*. Different arguments from the past differ with respect to which claim or piece of reasoning is inserted into the general form. The PMI is obviously an argument from the past. It targets the realist claim that current successful theories are true. Later we will encounter further arguments from the past, in particular the PMMI.

13.3 Analyzing the PMI

I will now provide a detailed analysis of the PMI. This will pave the way for the later analysis of the PMMI. The analysis consists of two remarks and a procedure to simplify the PMI.

First remark. The PMI starts from the observation that many past realists said *the same thing* as current realists do: they endorsed realism and supported it with arguments like the NMA. The PMI goes on to state that past realists were proven wrong by subsequent theory refutations and that today's realists will share *the same fate* as past realists. Now, past realists obviously didn't say the *very same* thing as current realists; rather they referred to different theories, namely past successful theories. Furthermore, there is an inductive step at the end of the PMI which carries us from the statement that past realists were wrong about past realism to the conclusion that current realists are wrong about current realism. This inductive step is obscured when the PMI states that past realists said "the

---

[7] The charge of ad hocness has several variants, see Stanford (2006, p. 10), Devitt (2011, p. 292), Vickers (2018, p. 52), and Fahrbach (2017).

same thing" as present realists, and present realists "will share the same fate" as past realists. The inductive step will be examined more closely in Section 13.7.

Second remark. The claim of the PMI that past realists were shown to be wrong by subsequent theory refutations should not be taken to mean that past realists were unjustified or irrational when they endorsed realism for their theories. After all, the relevant refutations could not be known by past realists at their respective times, because they occurred at respective later times. Generally, the conditions which determine whether a position held by some people at some time is justified or not should be knowable by those people at that time, and should not depend on empirical information which only becomes available later on.[8] Hence, the PMI should not be understood to be talking about the *justifiedness* of the assertions of past realists, but about the *truth or falsity* of those assertions. We learn empirically, using information acquired later on, that their assertions were false. Whether or not their assertions where justified can be left open. The PMI then invites us to infer that current realism is also false, and this *is* intended to imply that endorsing current realism is unjustified.

Let us now simplify the PMI. First, referring to present and past realists and their respective assertions makes the argument more vivid by turning it into a story about the misfortunes of real people, but this is not essential to the argument. Rather we can focus on the contents of the assertions and just talk about theories and the properties of theories such as success, truth, and falsity. Then the PMI is the following inference:[9]

<u>Many past successful theories were refuted.</u>	(Premise)
<u>Many past realisms are false.</u>	(Intermediate conclusion)
Current realism is false.	(Final conclusion)

Second, the intermediate conclusion is likewise not essential to the argument and can be removed. The refuted theories are the "negative thing," the source of all trouble, hence occur in the premise. Take them away and there is no argument. In contrast, the intermediate conclusion is just that: an intermediary for transmitting the trouble to the final conclusion (the falsity of current realism). Hence, instead of first using past refutations to confute past realisms and then

---

[8] I use an internalistic notion of justification.

[9] Single underlines indicate a deductive inference, double underlines indicate an inductive inference. Unless stated otherwise I assume that the inductive inferences we discuss are strong enough for detachment, i.e., the premises justify believing the conclusion. In the intermediate conclusion, "past realisms" is plural, since past theories were successful and accepted at different times. I mostly ignore this complication.

inferring therefrom that present realism is false, we can use past refutations to undermine present realism directly. Then the PMI becomes:

<u>Many past successful theories were refuted.</u>
Current realism is false.

This is a more familiar version of the PMI.[10] I will use it from now on. Note that the two simplifications mean that it is not essential to the PMI that we "see ourselves as similar to the scientists of the past" or that we "recognize the similarities between our predicament and the predicament of our predecessors" (Wray 2013).

If both the PMI and the NMA are still operative at the end of the dialectic, i.e., neither is taken to be defeated by some argument or other, then they have to be balanced with each other to reach a final verdict on realism. To simplify the discussion, I will assume that the PMI always trumps the NMA.[11] Hence as long as the PMI is operative during the dialectic, the NMA can be ignored. If the PMI is not operative at the end of the dialectic, e.g., judged to be blocked by the improvements of science, then the realist may invoke the NMA to support realism, but doing so still faces the charge of ad hocness, which I discuss in Section 13.10.

Both simplifications may also be applied to other arguments from the past. Recall that arguments from the past are instances of the following general strategy: The aim is to undermine a certain target claim made by a current realist about current theories by noting that past realists made (or could have made) the same claim about past theories, but were (or would have been) shown wrong by later theory refutations. The conclusion we are meant to draw is that the target claim is wrong as well. The first simplification consists in abstracting from present and past realists and their respective assertions, and just working with the *contents* of the assertions, which refer to theories and their properties. The second simplification cuts out the intermediate conclusion that the claim about past theories was undermined by the ensuing refutations. Instead past refutations (possibly combined with further information) are used to undermine

---

[10] Current realism is the claim that current successful theories are probably true. Hence the negation of current realism, which is the conclusion of the PMI, can be taken to be the statement that many current successful theories are false. Then the PMI is the inference from "Many past successful theories were refuted" to "Many current successful theories are false."

[11] This assumption is actually quite strong. The NMA is arguably equivalent to, or reducible to, a first-order argument about the support of theories by observation (see e.g., Bird 1998), while the PMI is a second-order argument (on the meta-level), and one might hold that first-order arguments are generally stronger than second-order arguments, or at least not that easily trumped by the latter. The whole dialectic can be easily modified to accommodate a different judgement about the balancing of the two arguments, although things get a bit more complicated.

the present realist's claim about present theories directly. Later we will apply both simplifications to the PMMI.

13.4 The Defense of the Realist

Realists have developed a number of responses to the PMI. One response is selective realism, on which I commented earlier. In this paper I focus on another defense, the one that gives rise to the PMMI. This defense claims that science has improved a lot since the times of the refuted theories, and these improvements block the PMI. There are several versions of this defense, which differ with respect to the precise nature of the invoked improvements. One version claims that the methods of science have improved.[12] Other versions hold that current theories are better than past theories in this or that respect, e.g., have fewer rival theories, have more "explanatory success," are more mature, or have more empirical success.[13] I will use the last version, involving the notion of empirical success. Then the realist defense against the PMI can be taken to make two claims: First, present theories are significantly more successful than past refuted theories, and second, this difference in success between present and past theories blocks the PMI. Both claims will be explained over the course of the paper.

The realist's defense is based on three assumptions. I will adopt them without much discussion, because they are quite plausible, and defending them is not our topic here. The first assumption is that empirical success comes in degrees. Earlier I defined a theory to be successful if it makes sufficiently many sufficiently significant true predictions and only negligible false ones. Accordingly we can understand the degree of success of a theory to be determined by the number and quality of true predictions of the theory, in other words, by the number and severity of the empirical tests the theory has passed. I use the notion of degree of success because the realism debate is usually framed with the notion of success, but one could just as well use other, closely related notions from confirmation theory, such as the empirical support enjoyed by a theory from empirical evidence, or some notion of confirmation of a theory by the empirical evidence.[14] The notion of degree of success may be identified with some probabilistic notion

---

[12] "[N]ot only are scientists learning more and more about the world, but also . . . they are learning more and more about how to find out about the world; there is an improvement in methodology." (Devitt 1991 p. 163). See also Devitt (2007, p. 287, 2011, Sect. 3), Roush (2009), and Wray (2018, Ch. 6).

[13] For fewer rival theories ("attrition") see Ruhmkorff (2013, 412), for "explanatory success" see Doppelt (2007), for maturity see Vickers (2018, 49), for empirical success see, e.g., Bird (2007, p. 80), Devitt (2011, p. 292), and Fahrbach (2017). It is an interesting question how the different versions are related to each other, but I won't discuss that here.

[14] Compare Vickers (2013, Sect. 3) and Vickers (2019, Sect. 4).

such as the likelihood ratio $\Pr(O|T)/\Pr(O|\neg T)$ or the posterior $\Pr(T|O)$, but we need not discuss such proposals here.

Generally the degree of success of a theory at some time[15] depends on the computing power available at the time (needed to produce predictions from quantitative theories) and the amount and quality of the observations gathered by scientists until that time, where the quality of the observations consists in good-making features such as diversity, specificity, precision, and so on.[16] A theory that makes non-negligible false predictions still counts as refuted.

The second assumption is that the accepted scientific theories have by and large become continuously more successful over the history of science (except when they were refuted). This is very plausible, because the ability to produce predictions from theories has generally been growing due to more computing power, better mathematical techniques to solve equations, and so on, and the amount and quality of observations and experimental results has generally been growing continuously due to better scientific instruments, more scientists, better data-gathering techniques, and so on. Observations were generally only lost or forgotten when better observations became available. Given more and better predictions from theories and more and better observations, tests of theories can be more severe and diverse, leading to higher degrees of success for the theories that pass the tests.

In addition to the amount and quality of the empirical evidence, the degree of success of a theory could be taken to depend on further factors such as the intrinsic properties of the theory, like simplicity and scope, and the nature of the relation between the theory and the empirical evidence (entailment or probabilistic relations). However, I will ignore these other factors here, because past and present theories mostly don't differ very much in these regards, certainly not enough to be of help for scientific realism. If anything, later theories are often less simple, more general, more precise, and more probabilified than earlier theories. For example, quantum mechanics and general relativity are considerably more complicated than Newtonian mechanics and Newtonian gravitational theory (although the losses in simplicity and so on are usually far outweighed by the gains in empirical support from true predictions, or so the realist defense has to claim).

The third assumption is that different theories, for instance from different scientific areas (with disjoint domains), can be compared with respect to degree of success, at least roughly. Statements such as "The theory of evolution enjoys a

---

[15] Strictly speaking, degrees of success are not ascribed to theories simpliciter, but to temporal stages of theories.

[16] If one thinks that novel predictive success comes in degrees (Vickers 2013, p. 196, 2016, Alai 2014, p. 298, 312), then one may construct a version of the realist defense against the PMI that relies on degrees of novel predictive success. Such a defense also faces a PMMI-like argument.

higher degree of success than phlogiston theory" are meaningful and justifiable. The third assumption may be deemed especially plausible if the notion of degree of success is identified with some probabilistic notion such as the likelihood ratio. If one does not accept that theories from different specialties can be compared with respect to degree of success, or only accepts comparisons inside some discipline or subdiscipline of science, then one may pursue the whole discussion at a more local level.

The increase in success over the history of science has obviously been far from uniform. Different scientific areas started life at different times, and have been developing at different rates. Hence, the theories accepted by scientists today occupy a broad range of degrees of success. Some areas, especially in the natural sciences, have produced theories with enormous success, while many other areas are stuck with theories at medium or lower levels of success. The realist defense against the PMI as I am developing it here focuses on the theories with the highest degrees of success. These are "our current best theories." They constitute the easiest kind of case for defending realism. Once secured, such a realism may serve as a bridgehead for advancing to realist positions for harder cases, i.e., to develop appropriate (possibly weaker) realist positions for theories with medium and lower degrees of success.

The realist's defense against the PMI can now be taken to consist of two claims: Our current best theories are significantly more successful than practically all refuted theories, and this difference in success blocks the PMI. The first claim presupposes a way to delineate the set of "our current best theories," a notion of difference in success between theories, and an explanation of the term "significant." However these notions are understood, it is clear that the first claim is a substantial empirical claim requiring support from the history of science. I will return to all these issues later. The second claim (that the difference in success blocks the PMI) means that the difference in success prevents past refutations from undermining realism about our current best theories, i.e., the inductive step of the PMI does not go through. Given the realist's defense we can finally formulate the PMMI.

13.5 The PMMI

Here is the PMMI:

> Realists in the past could have reasoned in the same way as today's realist does in response to the PMI: "Our current best theories are significantly more successful than past refuted theories, and this difference blocks the PMI." But look what subsequently happened, many of their theories were refuted. Hence, the

reasoning of today's realist fails, and the PMI is not blocked by the increase in success.

Most people seem to find the PMMI to be intuitively quite compelling.[17] However, getting clear about its actual import is not so easy. To simplify the discussion let us fix on a specific date at which the imagined realists of the past do their reasoning. One plausible such time is the year 1900, at least for physics, because Newtonian mechanics and the Newtonian theory of gravitation had been quite successful and stable until around 1900, so that Max Planck famously got the advice, "in this field [physics], almost everything is already discovered, and all that remains is to fill a few unimportant holes" (Lightman 2005, p. 8). Then the imagined realists from 1900 don't argue against our PMI, or PMI_{today}, but against a version of the PMI for the year 1900, call it PMI_{1900}. This leads to a second rendering of the PMMI:

> Realists in 1900 could have reasoned in the same way as today's realist does in response to the PMI_{today}: "Our current best theories are significantly more successful than past refuted theories, and this difference blocks the PMI_{1900}." This reasoning by realists in 1900 would have been proven wrong by subsequent theory refutations. Hence the reasoning of today's realist in response to the PMI_{today} also fails and the PMI_{today} is not blocked by the increase in success.

The second rendering makes it clear that the realists of 1900 would not have reasoned in *exactly* the same way as the realist does today, only in an *analogous* way, namely at a lower level of success and targeting a different PMI. The PMMI is obviously an argument from the past, i.e., an instance of the general antirealist strategy mentioned earlier: A piece of reasoning about current theories is attacked by transferring it to the past and noting that the past version of the reasoning is undermined by subsequent theory changes.

The PMMI states that the reasoning of the imagined realists in 1900 "would have been proven wrong" by subsequent refutations. This should not be taken to mean that their reasoning would have been unjustified or irrational. The refutations which proved their reasoning wrong occurred subsequently, hence were not known in 1900. Therefore, the PMMI should not be understood as talking about the justifiedness of the PMI_{1900}, but about the reliability of the PMI_{1900}.[18] The PMMI then invites us to infer that the PMI_{today} is likewise a reliable inference.[19]

---

[17] Most references in footnote 3 seem to endorse the PMMI. Exceptions include Devitt and Psillos.

[18] For our purposes we can define an inductive inference to be reliable, if its conclusion is true given true premises. An inductive inference understood as a type is reliable, if the conclusions of its instances are mostly true when its premises are true.

[19] I assume that the statement that a given version of the PMI is not blocked by the increase in success is equivalent to the statement that it goes through (is reliable).

We may also describe the PMMI as follows: Using historical information acquired after 1900, we learn empirically which kind of meta-inductive inference, pessimistic or optimistic or neither, is reliable in 1900; we find that a pessimistic meta-induction is reliable and project the finding to the present. In other words, the PMMI uses historical information to *calibrate* the meta-inductive inference for 1900, and uses the result to *calibrate* the meta-inductive inference for today. Note that the conclusion of the PMMI *is* intended to mean that the realist's assertion today, that the PMI_{today} is blocked by the increase in success, is unjustified.

Like the PMI, the PMMI can be simplified in two steps. First, let the PMMI refer to theories and their properties rather than to imagined people and their assertions. Then the premise of the PMMI marshals the historical evidence and can be taken to state that the past of science exhibits a certain pattern, namely ⟨refutations before 1900, successes until 1900, refutations after 1900⟩ (Figure 13.1). This premise supports the intermediate conclusion that the PMI_{1900} is a reliable inference (is not blocked by the increase in success until 1900), where the premise of the PMI_{1900} refers to the first and second component of the pattern, and the conclusion of the PMI_{1900} refers to the third component of the pattern. The intermediate conclusion supports the final conclusion that the PMI_{today} is a reliable inference (is not blocked by the recent increase in success).

Second, we can remove the intermediate conclusion from the PMMI. Then the PMMI looks like this:

The past of science exhibits the pattern ⟨refutations before 1900, successes until 1900, refutations after 1900⟩.

The PMI_{today} is a reliable inference.

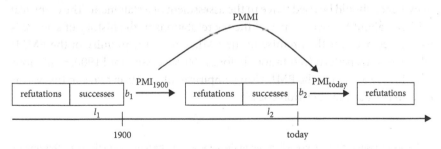

Figure 13.1 The pessimistic meta-meta-induction, PMMI. The depiction of the history of science is strongly idealized, serving the goal of understanding the PMMI. l_1, l_2, b_1, and b_2, the lengths and breadths of the two rectangles dubbed "successes," are explained in Section 13.9.

So, instead of using the pattern of the premise to first calibrate the meta-inductive inference for 1900 and then using the result to calibrate the meta-inductive inference for today, we use the pattern of the premise to calibrate directly the meta-inductive inference for today. If this contraction of the PMMI is rejected, the discussion that follows gets a bit more complicated, but yields essentially the same conclusions.

Figure 13.1 invites a quick recapitulation of the whole discussion so far. The realist is impressed by the successes of our current best theories (the second rectangle from the right) and wants to infer realism about these theories. The antirealist is impressed by the previous refutations and recommends the projection of theory failure to the future instead (the three rectangles to the right).[20] The realist replies that degrees of success have increased from past to present, blocking the projection of theory failure. The antirealist responds that the whole situation already occurred in the past (the three rectangles to the left): First, past successes also suggested realism about past theories; second, previous theory refutations suggested the projection of theory failure instead; third, past realists could have replied that the projection of theory failure is blocked by the increase in success; and fourth, they would have been proven wrong by subsequent refutations, showing that in 1900 the projection of theory failure would have been the right thing to do. It follows (big curved arrow) that the projection of theory failure goes through today. Unsurprisingly the big curved arrow will be a topic later.

13.6 The Double-Counting of Evidence

In this section I discuss an objection to the PMMI which is not related to my main critique, but is interesting and puzzling in its own right. Moreover, it highlights further similarities between the PMI and the PMMI. The objection is that endorsing both the PMI and the PMMI violates the principle that no piece of evidence should be used twice in the assessment of a statement. The empirical evidence about the occurrence of theory refutations in the history of science is used twice over, for the premise of the PMI, and for the premise of the PMMI which uses the pattern ⟨refutations before 1900, successes until 1900, refutations after 1900⟩ to support the PMI. Using empirical information twice in this way is fallacious, or so the objection goes.[21]

---

[20] We can assume that science will go on in the same way in the future as in the past. This implies that the conclusion of the PMI that realism is false is equivalent to the statement that many current best theories will be refuted in the future, and similarly for the PMI_{1900}.

[21] The principle against the double-counting of evidence may be seen as the complement of the principle of total evidence, according to which every piece of evidence relevant for the truth of a statement should contribute to our epistemic assessment of the statement. Together the two principles

One might respond to the objection by pointing out that using empirical evidence twice is legitimate in this case, because it is used for two different purposes. The existence of refuted theories is used, first, for the *premise* of the PMI, and second, to support via the PMMI that the *inference* of the PMI is reliable, and these are two different purposes. However, one may retort that using empirical information twice in this manner is still dubious. One should not use the same empirical information both as a premise for an inference and to help determine what to infer from this premise (to calibrate the inference). What is more, both purposes serve the same *final* purpose, namely to assess the conclusion of the inference, in our case the conclusion of the PMI that current realism is false, so it is true after all that one piece of evidence is used twice for the same purpose.

I want to offer two considerations indicating that this kind of double-counting may not be illegitimate in the case of the PMMI. First, the objection of double-counting of evidence can also be directed against endorsing both the PMI and antirealism. Take antirealism to be the inference from the success of current successful theories to their falsity.[22] Let the PMI be the inference from the existence of successful-but-refuted theories to the conclusion that antirealism understood as an inference is reliable. (To simplify things I use a somewhat stronger version of the PMI here than in the rest of the paper.) Then the observations that are used to show that current successful theories are indeed successful—which is the premise of antirealism understood as an inference—are also used both to show that the successful-but-refuted theories were successful and to refute those theories—which is the premise of the PMI. Hence, these observations are used both for the *premise* of antirealism and to determine via the PMI what to *infer* from the premise of antirealism. They are used twice for the same final purpose, namely to assess current successful theories (in the conclusion of antirealism understood as an inference). If we reject the PMMI for the reason that it uses the same empirical evidence for both the premise of an inference and to determine what to infer from the premise, then we should also reject the PMI for the same reason. Given the intuitive plausibility of the PMI, this is a high price to pay.

Second, a version of the PMMI can be constructed that avoids the double-counting of evidence. Divide the set of scientific areas randomly into two subsets

assert that when assessing a statement every relevant piece of evidence should count at least once and no more than once.

[22] This sounds like an unduly strong version of antirealism, but it is not, if one assumes ordinary Bayesian probabilities for example. The priors of typical scientific theories are very low, very near to zero, because scientific theories are typically quite general and have numerous non-negligible rivals. Favorable observations eliminate many rivals, increasing the probabilities of the theories, but the question is by how much. Antirealism can be taken to be the position that current successful theories should still be expected to have many non-negligible rivals not eliminated by the observations, hence their posteriors, although higher than the priors, are still near zero, hence our current successful theories should be judged to be false.

of equal size. Use the first subset to run the PMI, i.e., use the theory refutations in the histories of the scientific areas of this subset for the premise of the PMI and let the conclusion of the PMI refer to the current theories of this subset. Use the second subset for the premise of the PMMI, i.e., to calibrate the meta-inductive inference for the PMI. So, the premise of the PMMI refers to patterns such as ⟨refutations before 1900, success until 1900, refutations after 1900⟩ in the histories of the second subset, and the conclusion of the PMMI states that the PMI, which runs in the first subset, is reliable.

One may object that this solution comes at the price of weakening the premises of the PMI and the PMMI: the strength of both premises is halved, intuitively speaking. However, this is not the case. The relevant quantity of both premises is a frequency, the frequency of refutations in the two sets, i.e., the ratio of the number of refuted theories in the respective periods of time to the number of scientific areas. If the set of scientific areas is cut in half, the ratios will not change much, because both the numerators (the number of refuted theories in the respective periods of time) and the denominators (the number of scientific areas) are cut in half. Hence, this proposal might work, and I will not hold the double-counting of evidence against the PMMI. However, in order to keep the discussion simple let us put dividing all scientific areas into two subsets right out of our minds.

13.7 Back to the PMI

Before assessing the PMMI, we have to take a closer look at the assessment of the PMI. This is the aim of this section. The aim is not to reach an actual assessment of the PMI, e.g., to argue on the realist's behalf that the PMI is blocked; I only want to get clear about which issues need to be addressed when assessing the PMI, without actually addressing them. At the end of the section we will see how the PMMI arises in this context.

The PMI of 13.3 states that many past successful theories were later refuted, therefore current realism is false. It is an induction: its premise refers to once-successful theories and its conclusion refers to current successful theories. It uses a binary notion of success where each theory at a given time is either successful or is not.[23] Then the induction proceeds inside the set of successful theories and seems quite plausible. However, we now operate under the three assumptions of the realist's defense, in particular that empirical success is graded and that it

---

[23] Most of the literature on the PMI starting with Laudan (1981) relies on a binary notion of (novel predictive) success.

has increased over the history of science. Given these assumptions, assessing the PMI becomes much more difficult.

At this point I will make a fourth assumption. It states that there is a reasonable definition of a global difference in success between past refuted and present best theories such that this difference is non-zero. This assumption is quite vague, but will do for our purposes.[24] It is a substantial assumption about the history of science, for which the realist has to provide empirical support, but the realist defense is making a considerably stronger claim anyway, namely asserting a "significant" difference in success between the two sets of theories, on which more later. The antirealist may attack the fourth assumption, of course. She may argue, for example, that a substantial number of refuted theories enjoyed degrees of success comparable to the theories of the set of our current best theories, on any reasonable definition of this set. This would imply, she could argue, that, correctly understood, the global difference in success between past refuted and current best theories is zero, and the realist's defense against the PMI fails. However, this would amount to an objection to the realist's defense that is different from the PMMI, hence will not be pursued here.

The fourth assumption (that there is a non-zero global difference in success between past refuted and present best theories) has the important consequence that the inductive step of the PMI is an *extrapolation*. It extrapolates the occurrence of theory failure along degrees of success from past to present levels of success. It is then not obvious that the extrapolation goes through, that theory failure can be extrapolated in this way.

The realist may reject the extrapolation of theory failure along degrees of success altogether. He may argue that from the existence of refuted theories at a given level of success one can only infer something about the assessment of theories at that level of success, not about any higher levels of success. This would mean that the PMI is blocked by *any* non-zero global difference in success. The antirealist may counter this move by arguing that theory failure *can* be extrapolated along degrees of success, and this may then be the debate between the two sides. However, I don't want to pursue this branch of the dialectic; rather I will assume that the realist grants that extrapolations of theory failure are possible in principle. The question is only whether the extrapolation goes through in the particular case at hand.

---

[24] The global difference in success between past refuted and present best theories can be defined in a number of ways. For example, start from today's most successful theory overall, then descend degress of success until sufficiently many theories for a set of "current best theories" have been gathered. Determine the lowest degree of success of the theories in this set. Below that, determine the highest degree of success above which there are only a negligible number of later refuted theories. The difference between these two degrees of success may be taken to define the global difference in success between past refuted and present best theories.

Whether this extrapolation goes through will depend on a number of factors. For example, it may depend on the frequency of past refutations, on the severity of past refutations (whether past refuted theories were completely false or only partly false), and on further aspects of the distribution of theory stability and theory change over degrees of success. To simplify the discussion, I will focus on just one factor: the global difference in success between past refuted and current best theories. Then it is plausible that the bigger the global difference, the smaller the bearing of past refutations on the assessment of our current best theories. Let us call a global difference in success "small" if it allows the extrapolation of theory failure, and "big" if it blocks the extrapolation of theory failure.

It is not hard to see that big differences in success are possible. Imagine a future scenario in which science will keep growing and degrees of success keep increasing in similar ways in the next 100 years as they did in the past of science. Imagine that scientific data, instruments, techniques, computing power, and so on, will keep improving at the same pace as they have done until now. Then the extrapolation of theory failure from our past to the levels of success of the year 2100 is surely very implausible. If the increase in success expected for the next 100 years doesn't seem high enough to block the extrapolation, then surely there are increases in degrees of success, 1000 years or 1 million years from now, that would block the extrapolation.

It is also worth considering the maximal level of empirical success that is in principle possible for scientific theories at which they accord with *all possible* empirical evidence. If there are any false theories at that level of success in some scientific field, then we have underdetermination of theories by all possible evidence in that field. Hence, if past refutations supported the existence of false theories at such a maximal level of success, then they would support the occurrence of underdetermination of theories by all possible evidence. However, as far as I am aware, nobody in the literature on underdetermination has argued in this way to support underdetermination, using a PMI-like argument referring to past refutations. The arguments offered in the literature are of a quite different sort, either using artificially constructed empirically equivalent rivals to theories, or referring to concrete examples of incompatible empirical equivalent theories in fundamental physics.[25] This indicates that nobody thinks that past refutations can be extrapolated that far along degrees of success.

The future scenario case and the underdetermination case show that mere quantitative differences in success can suffice to block the PMI. Such differences presumably count as "fundamental" differences in the sense of Wray:

---

[25] See for example Norton (2008).

[I]f realists are to blunt the threat of the Pessimistic Induction, they must identify some significant difference between today's theories and past theories. Without an argument to the effect that there is a *fundamental* difference between the theories we currently accept and the once successful theories we have since rejected, we have little reason to believe that today's theories will not end up on the pile of ruins to which Poincaré drew attention. (2018, p. 93, emphasis in original)

We have established that there are differences in success for which the extrapolation of theory failure is highly implausible. However, we are interested in the difference between past and *present* levels of success and *today's* PMI. The question is whether this difference is small, so as to allow the extrapolation of theory failure to present levels of success, or big, so as to block the extrapolation. As the discussion so far already suggests, this is a hard question. To answer it we have to engage with difficult empirical and epistemological issues. We have to find some way to actually measure or estimate the degrees of success of past refuted and current successful theories, at least roughly. Then we have to find a suitable delineation of the set of present best theories. Then we have to settle on a suitable definition of global difference in success between past and present best theories, and have to determine this difference. Finally we have to formulate and defend a judgment concerning whether the global difference is small or big, whether it allows or blocks the extrapolation of theory failure. It seems that we can only assess the PMI, if we tackle these difficult issues.[26]

The realist defense against the PMI as presented here does not engage with the difficult issues. It just claims that the global difference in success between past and present theories is "significant" (meaning "big," i.e., capable of blocking the PMI), without telling us how the difference was determined, and why it is "significant." Hence, the realist has some work to do here. He has to engage with the difficult epistemological and empirical issues to defend his claim that the difference is "significant." What about the antirealist? She wants the PMI to succeed. Hence, she has to show that the difference in success between past and present theories is small, i.e., does not block the extrapolation of theory failure. To show this she, too, has to engage with the difficult epistemological and empirical issues. As long as she does not do that, she, too, has some unfinished business—or so it seems.

It is at this point that the antirealist may offer the PMMI. The PMMI supports that the PMI goes through, that is its conclusion, hence it promises to give the antirealist an advantage, possibly decisive, over the realist in the debate

---

[26] And I didn't even mention the other factors that may be relevant for deciding whether theory failure should be extrapolated to current levels of success, and the NMA which I assume for the sake of simplicity to be trumped by the PMI.

concerning the PMI. What is more, it avoids the difficult epistemological and empirical issues of the PMI, because it looks at the past of science to establish the reliability of the PMI. But I will now argue that despite first appearances, the PMMI does not succeed in giving the antirealist an advantage in the debate.

13.8 First Objection to the PMMI

I will offer two objections to the PMMI. The first concerns the inference, the second the premise of the PMMI. The first objection focuses on the inductive step at the end of the PMMI (the curved arrow in Figure 13.1). Given the fourth assumption, that past refuted and present best theories are separated by a non-zero global difference in degrees of success, the inductive step of the PMMI is an extrapolation from past to present levels of success. It extrapolates the pattern ⟨refutations before 1900, successes until 1900, refutations after 1900⟩ to the present to support the PMI. Then the PMMI faces problems very similar to those faced by the PMI. It is not obvious that the extrapolation goes through; it depends on the size of the difference between past and present levels of success (as well as on other factors, which I ignore here). Let us call the difference "small" if the extrapolation of the PMMI succeeds and "big" otherwise. Once again the very existence of big differences can be backed by considering future scenarios in which science will develop in a similar manner in the next 100 or 1000 years as it did up to now, producing an extremely big increase in success. For such an increase it is highly plausible that the extrapolation of the PMMI fails. The question, though, is whether the increase in success *so far* has been small or big. Once again this is a hard question. To find an answer, we have to engage with difficult empirical and epistemological issues: we have to determine the degrees of success of the involved theories, define and determine the global difference, and formulate and defend a judgment whether the global difference in success is small or big, i.e., whether the extrapolation of the PMMI is plausible or not. Only by engaging with these issues can we assess the PMMI, or so it seems.

The antirealist may evade the difficult issues of the PMMI in the same way as she evaded those of the PMI: She may present a PMMMI. What then looms is an infinite hierarchy PMI, PMMI, . . . PM^nI, . . ., in which the assessment of each PM^nI faces difficult epistemological and empirical issues, which the antirealist evades by offering another argument from the past, the $PM^{n+1}I$, which invokes the past of science to establish the reliability of the PM^nI. In this way the antirealist avoids the difficult epistemological and empirical problems at every level, but it is a bit unclear whether such an infinite hierarchy of arguments is legitimate in principle.[27] Anyway, it does not succeed for practical reasons: The

---

[27] A related infinite hierarchy of arguments and counterarguments arises as follows: The realist

premise of the PMI refers to the occurrence of past refutations. The premise of the PMMI is more complicated, invoking patterns such as ⟨refutations before 1900, successes until 1900, refutations after 1900⟩. The premise of the PMMMI refers to even more complicated patterns such as

⟨⟨refutations before 1850, successes until 1850, refutations between 1850 and 1900⟩,
⟨refutations between 1850 and 1900, successes until 1900, refutations after 1900⟩⟩.

(The conclusion of the PMMMI is the statement that the inference of the PMMI, the curved arrow in Figure 13.1, goes through.) As n increases, the premises of the PM^nI become more and more complex, presupposing more and more episodes of theory refutation. But the number of such episodes in the history of science is finite. Hence, for some level N, the antirealist will not be able to rise to the level $N+1$. It follows that she has to face the difficult empirical and epistemological issues at *some* level.[28] She may choose the level, say level n, with $n \le N$, but she cannot avoid the difficult empirical and epistemological issues altogether; at some point she has to engage with them. What is more, it does not seem advisable to go up the hierarchy of PM^nI's very far, because the complexity of the premises grows exponentially as n increases, making it ever harder for the antirealist to show that the premise of the chosen PM^nI is actually true (compare the next section). So let us return to the assessment of the PMMI.

My first objection to the PMMI is then as follows. The PMMI promised to help the antirealist to gain an advantage in the debate over the PMI. It seemed to show that the extrapolation of the PMI goes through by invoking patterns such as ⟨refutations before 1900, successes until 1900, refutations after 1900⟩, while avoiding the difficult epistemological and empirical issues arising in the assessment of the PMI. But we just saw that the inductive step of the PMMI is also an extrapolation along degrees of success, whose assessment faces very similar problems as the assessment of the extrapolation of the PMI. Thus, invoking the PMMI just trades the problems of the PMI for very similar problems that are at

objects to each PM^nI by arguing that it is blocked by the increase in success, and the antirealist responds with the $PM^{n+1}I$ according to which realists in the past could have objected to the past version of the PM^nI in the same way, namely that it is blocked by the increase in success, but would have been proven wrong by later theory refutations. This infinite hierarchy has the same problem as the one in the text, namely the exponentially increasing complexity of the premises of the PM^nIs.

[28] Note that the antirealist only has to show for one level n that the PM^nI succeeds, because the conclusion of the PM^nI is that the $PM^{n-1}I$ succeeds, which has the conclusion that the $PM^{n-2}I$ succeeds, and so on, until the PMMI shows that the PMI succeeds, which has the conclusion that realism is false. In contrast, the realist has to argue that all PM^nIs fail.

least as hard as those of the PMI. It follows that the PMMI does not succeed in helping the antirealist to gain an advantage in the debate over the PMI.

13.9 Second Objection to the PMMI

For the second objection, consider the PMMI in its unsimplified form. It starts by claiming that people in the past, e.g., in 1900, could have reasoned in the same way as today's realist does. They, too, could have said: "Our current best theories are significantly more successful than past refuted theories, and this difference blocks the PMI_{1900}." This beginning of the PMMI presupposes that the situation in 1900 was actually similar in relevant respects to the situation today. That is a substantial claim, which needs empirical support from the history of science. Likewise for the simplified PMMI, whose premise refers to the pattern ⟨refutations before 1900, successes until 1900, refutations after 1900⟩. It is a substantial empirical claim that the first and second component of the pattern are sufficiently similar in relevant respects to the corresponding situations in the recent past. More generally, the PMMI requires the premise that there were times in the history of science in which the situation was sufficiently similar in relevant respects to the situation today.[29]

Which respects are relevant for the similarity between the past and present situations? I will examine two respects, the *global difference* in success between the respective sets of best theories and the respective sets of refuted theories, and the *number* of the respective scientific fields. Then the PMMI requires that there were times in the history of science in which the global difference in success and the number of relevant fields were both comparable to the corresponding global difference and number in the recent past. Figure 13.1 makes clear what I mean: the two rectangles dubbed "successes" should have roughly similar lengths l_1 and l_2 (which represent the respective global differences in success) and roughly similar breadths b_1 and b_2 (which represent the respective numbers of scientific fields with best theories). If, on the other hand, the lengths and breadths of successes in the past were never similar to those today, for example, if l_1 was always much smaller than l_2, and b_1 was always much smaller than b_2, then the premise of the PMMI is never true.[30]

We have identified a substantial empirical assumption of the PMMI about the existence of periods of time relevantly similar to the recent past.

---

[29] I want to thank Ioannis Votsis for a number of challenging comments at this and other portions of my argument.

[30] If at some time l_1 is smaller than l_2, but b_1 is bigger than b_2, and some suitable function of l_1 and b_1 such as their product $l_1 \cdot b_1$ is roughly the same as that of l_2 and b_2, then the two situations may also justifiably count as similar.

Earlier, we took 1900 to be a suitable time, for the reason that central theories of physics had been quite successful and stable for a while up until around 1900. But the realist may argue that this is only then-fundamental physics; the simultaneous stability of theories in recent times has been much broader, encompassing the best theories of a far bigger number of scientific fields (mostly from the natural sciences) than just then-fundamental physics. The antirealist either has to show that theory stability occurred on a much broader scale in 1900 than just then-fundamental physics, or she has to present some other historical periods at which the increase in success and the breadth of stability was comparable to today. Once again we find that the antirealist, if she wants to hold on to the PMMI, has some unfinished business.

It is worth noting that both objections may also be directed against other arguments from the past. Recall that arguments from the past aim to undermine a target claim made by a realist about current theories by noting that past realists could have made the same claim about past theories, but would have been proven wrong by later theory refutations. One objection consists in the challenge to show that there actually were situations in the past of science that were sufficiently similar in relevant respects to the situation today. The antirealist cannot just claim that this is so, but has to provide adequate evidence that such situations really occurred in the past. The other objection is that there is one respect in which all past situations are *dissimilar* to the present situation, namely with respect to degrees of success. If the fourth assumption (about a non-zero global difference between past and current best theories) is correct, then the inductive step of any argument from the past will be an extrapolation from past lower to current higher degrees of success. Whether the extrapolation is plausible will depend on several factors, such as the global difference in success between past and present. The antirealist has to show that the factors are such as to allow the extrapolation of theory failure.

13.10 The Charge of Ad Hocness

In this section I discuss the charge of ad hocness. In my definition of realism the scope of realism is restricted to current successful theories, at the exclusion of past successful theories. The obvious reason is that past theories were often refuted. The charge of ad hocness states that defining realism in this way is ad hoc, merely serving the purpose of making realism compatible with the existence of past refutations. The realist has to offer independent reasons why the definition of realism excludes past successful theories. To simplify the discussion in this section, I will set the PMI aside, ignoring the question whether the existence of past refutations can be used to attack realism about current theories. Only the seemingly arbitrary restriction of realism to current theories is at issue.

The realist may offer a similar response to the charge of ad hocness as to the PMI: our current best theories are more successful than past refuted theories, and this difference in success provides an independent reason to treat past theories differently from our current best theories in the definition of realism. To this move Stanford responds: "More success gives *no* reason to believe that we have now crossed over some kind of threshold . . . such that these predictive powers [of scientific theories] are now finally substantial enough [to accept them as true]" (2009, 384 footnote, emphasis added). At this point we need to reintroduce the NMA intuition, as it is generally taken to be the source of support for realism. Then it is plausible that the NMA intuition gets stronger as degrees of success increase: the higher the success of a theory, the bigger the "miracle" if the theory were false, and the higher the probability that the theory is true. It follows that Stanford's claim that more success gives us *no* reason to believe that we have crossed over the threshold for truth is false. Rather, more success gives us *some* reason.[31] The question is, only, whether the reason is strong enough to support realism as defined above.

At other places Stanford's wording is a bit more cautious. In his (2015), he says "little reason":

> Of course there are always important differences between each successive generation of theories (including our own) and their historical predecessors in a given domain of scientific inquiry, but there seems *little reason* to think that such differences are sufficiently categorical to warrant the conviction that contemporary scientific theories have now finally managed to more-or-less sort things out at last, given that the same inference as applied to earlier theories, predicated on the salient advances and advantages of those theories over their predecessors, has turned out to be so repeatedly and reliably mistaken. (p. 3, emphasis added)

"Little reason" is more cautious than "no reason," but it is still clearly meant to imply "not enough reason." To support this claim, Stanford merely offers an argument from the past (starting with "given that" in the quote), which runs into the problems noted earlier.[32] He does not offer any first order reasons concerning

---

[31] Even if the NMA intuition is not granted, more empirical success of a theory means that the theory has passed more tests, which generally implies, on standard accounts of confirmation such as Bayesianism, that the theory is incrementally confirmed. Hence, more generally, such additional success offers *some* reason to think that we have crossed the threshold for truth.

[32] It is not clear whether one should interpret Stanford's argument from the past as the PMI, the PMMI, or a further argument from the past concerning only ad hocness. Maybe it is a combination of these. In any case, it runs into the problems noted earlier.

the quantity and quality of the empirical evidence, and the strength of the empirical support for our current best theories. Hence, concerning first order reasons, his claim is as unprincipled as the opposite claim of the realist that present levels of success suffice for the NMA to support realism. Whether either claim is justified, or no claim at all can be justified, surely depends on how big the increase in success from past to present has actually been. If it has been small, then Stanford's claim may be justified and the charge of ad hocness may be plausible, but if it has been big, then not. Consider once again the future scenario in which degrees of success keep growing for another 1000 years at the same rate as they have done so far. In such a scenario the NMA would surely give us more than "little reason" to believe that we have crossed over the threshold for truth.

But the question is, of course, whether *current* levels of success are high enough for the NMA to establish realism. The realist answers this question in the affirmative, but according to the charge of ad hocness the affirmative answer requires a justification, which the realist has not yet given. To find a justification, the realist must engage with difficult empirical and epistemological issues: he has to determine and compare the degrees of success of past and present best theories and formulate and defend judgments based on the NMA intuition about the plausibility of restricting realism to present levels of success at the expense of past levels of success. So, the realist finds himself in a situation not unlike with respect to the PMI as detailed earlier. In the next section I will hint at one possible way how the realist may proceed in this situation.[33]

13.11 The Big Increase

We saw at several points that the debate can only progress if we engage with difficult empirical and epistemological issues, namely compare the degrees of success of past and present theories and formulate and defend judgments about the plausibility of the PMI, the PMMI, and the charge of ad hocness.

Actually some such work has already been done.[34] It strongly suggests that our current best theories have enjoyed a very large increase in degree of success in the recent past and swim in a sea of evidence today. One reason is, very roughly, that the amount of scientific research in general has been growing at very high rates in the recent history of science, which has led to huge improvements in

---

[33] Alternatively, the realist may search for some deep epistemic difference between past refuted and current best theories. However, I don't think that there is such a magic bullet, which scientists are unaware of, which is deeply buried in scientific practice, requiring the analytic digging powers of philosophers for its unearthing. To my mind, there is just ordinary empirical evidence and ordinary empirical success, but with large differences between different theories and different times, as I indicate in the next section.

[34] See, for example, Doppelt (2007, 2014), Park (2011), Mizrahi (2013), and Fahrbach (2017).

computing power and the amount and quality of empirical evidence. This has translated into ever-increasing degrees of success for the best theories. While acquiring their tremendous success, practically all of them have been entirely stable. There are no convincing examples of theories that enjoyed comparable successes and were later refuted. At several points in the discussion we imagined scenarios in which science develops in the same manner in the next few centuries as it has done so far. The work just mentioned shows that the state of science today is not so far from what we imagined in those scenarios. (Thus, Figure 13.1 offers a totally distorted view of the history of science.) Much more work needs to be done to develop this approach further. Here I will just assume that it is generally correct, and confine myself to noting the implications for the assessment of the PMI, the PMMI, and the charge of ad hocness.

First, the large increase in success and simultaneous stability of our current best theories implies that the extrapolation of theory failure from past to present levels of success is highly implausible. Our current best theories are far more successful than any past refuted theories; hence the existence of the latter should have no bearing on our assessments of the former. So the difference in success indeed blocks the PMI.

Second, the large difference between past and present levels of success also blocks the PMMI. It prevents the extrapolation of patterns like ⟨refutations before 1900, successes until 1900, refutations after 1900⟩ to the present. What is more, it is highly doubtful that the premise of the PMMI can be made true. To make it true we have to find times in the history of science in which the situation was sufficiently similar to the recent past, in which the then-best theories were similar in number and made comparable gains in success as in the recent past. However, the recent past is quite unprecedented in this regard. No other period comes anywhere close. (Again the picture of the history of science as presented by Figure 13.1 is much distorted.) Thus, the PMMI does not succeed in supporting the PMI and undermining the realist's defense against the PMI.[35]

Third, the tremendous success of our current best theories makes the application of the full NMA intuition to these theories very plausible: their success would indeed be a miracle if they were false. By contrast, the success of past theories, whether refuted later on or not, does not suffice to trigger the full NMA intuition: their success was far more moderate, and it was not such a big miracle when they occasionally turned out to be false. So, the NMA fully supports the inference to truth for our current best theories, but does so only rather weakly for past

[35] Similar remarks apply to the PM^{n}Is with $n > 2$. Note that, the bigger and broader the recent increase in success, the more the PMI faces competition from an optimistic meta-induction, OMI, which uses the recent stability of our current best theories to project the stability of these theories into the future, and likewise for higher-level optimistic meta-inductions, OM^{n}Is.

theories.[36] It follows that the realist's restriction of realism to current and future highest levels of success is not ad hoc but supported by good reasons. What is more, the big increase in success in recent times means that the realist need not commit himself to a precise threshold beyond which success suffices for truth, but can be very vague in this matter. All he needs to make plausible is that the threshold is *somewhere* in the large interval of degrees of success between past and present, and the NMA is able to do that. Thus, some of the difficult empirical and epistemological issues turn out to be not so difficult after all, but are readily solvable, because the recent increase in success has been so enormous.

13.12 Conclusion

Let me summarize my critique of the PMMI. (I ignore the previous section.) According to the PMI, we should extrapolate theory failure from past to present levels of success, thereby undermining realism about present theories. The realist reacts to the PMI by claiming that our current best theories are significantly more successful than past refuted theories, blocking the PMI. The antirealist denies this. She claims that, despite the difference in success, past refutations are severe enough to allow the extrapolation of theory failure to go through. In this situation, either the realist is right and the PMI is blocked by the increase in success, or the antirealist is right and the PMI succeeds. It then seems that the matter can only be decided if the two sides engage with difficult empirical and epistemological issues: they have to determine and compare the degrees of success of past refuted and current best theories and formulate and defend judgments concerning the plausibility of the extrapolation of theory failure and the undermining of realism.

At this point the antirealist presents the PMMI in the hope of gaining an advantage in the debate. The PMMI seems to avoid the difficult empirical and epistemological issues while implying that the PMI succeeds. It states that past realists could have reasoned in the same way as today's realist just did, namely that their theories were significantly more successful than earlier refuted theories, blocking the PMI of their day, but this didn't prevent further refutations from happening; hence the defense of today's realist fails and the PMI is not blocked by the increase in success. We analyzed the PMMI and saw that it faces

---

[36] Most of past refuted theories were not completely false, but contained a substantial amount of truth (judged from today), as selective realists have shown in numerous case studies. Furthermore many theories have been completely stable for centuries now, such as the heliocentric system, oxygen chemistry, the reduction of temperature to mean kinetic energy, and so on (for more examples see Bird 2007, 73).

very similar problems as the PMI. It also involves an extrapolation from past to present levels of success, which gives rise to empirical and epistemological issues very similar to those of the PMI: determining the degrees of success of the involved theories and reaching and justifying judgments concerning the plausibility of the extrapolation. So, the PMMI trades one set of problems, those of the PMI, for another set of very similar problems (as well as some additional problems). Therefore, the PMMI does not succeed in helping the antirealist to gain an advantage in the debate over the PMI. The upshot is that both realists and antirealists can only make progress in the debate if they engage with the difficult empirical and epistemological issues.

Acknowledgments

For discussion and support I would like to thank Claus Beisbart, Christopher von Bülow, Matthias Egg, Christian Feldbacher, Berna Kilinc, Tim Lyons, Sam Ruhmkorff, Yafeng Shan, Corina Strößner, Paul Thorn, Peter Vickers, Ioannis Votsis, Brad Wray, two anonymous referees, and audiences in Lausanne and Durham.

References

Alai, M. (2014). "Novel Predictions and the No Miracle Argument." *Erkenntnis*, 79(2), 297–326.
Alai, M. (2017). "Resisting the Historical Objections to Realism: Is Doppelt's a Viable Solution?" *Synthese*, 194(9), 3267–3290.
Bird, A. (1998). *Philosophy of Science*. Montreal & Kingston: McGill-Queen's University Press.
Bird, A. (2007). "What is Scientific Progress?" *Noûs*, 41(1), 64–89.
Devitt, M. (1991). *Realism and Truth*. 2nd edn. Oxford: Basil Blackwell.
Devitt, M. (2007). "Scientific Realism," in: Frank Jackson and Michael Smith (eds) *The Oxford Handbook of Contemporary Philosophy*. Oxford: Oxford University Press, 767–791.
Devitt, M. (2011). "Are Unconceived Alternatives a Problem for Scientific Realism?" *Journal for the General Philosophy of Science*, 42(2), 285–293.
Doppelt, G. (2007). "Reconstructing Scientific Realism to Rebut the Pessimistic Meta-Induction." *Philosophy of Science*, 74, 96–118.
Doppelt, G. (2014). "Best Theory Scientific Realism." *European Journal for Philosophy of Science*, 4(2), 271–291.
Fahrbach, L. (2011). "How the Growth of Science Ended Theory Change." *Synthese*, 180(2), 139–155
Fahrbach, L. (2017). "Scientific Revolutions and the Explosion of Scientific Evidence." *Synthese*, 194(12), 5039–5072.
Frigg, R., & Votsis, I. (2011). "Everything You Always Wanted to Know about Structural Realism but Were Afraid to Ask." *European Journal for Philosophy of Science*, 1(2), 227–276.

Frost-Arnold, G. (2018). "How to be a Historically Motivated Antirealist: The problem of Misleading Evidence." (ms)

Laudan, L. (1981). "A Refutation of Convergent Realism." *Philosophy of Science*, 48(March), 19–49.

Lightman, Alan P. (2005). *The Discoveries: Great Breakthroughs in Twentieth-Century Science, Including the Original Papers*. Toronto: Alfred A. Knopf Canada.

Lyons, T. D. (2002). "Scientific Realism and the Pessimistic Meta-Modus Tollens," in: S. Clarke and T. D. Lyons (eds.) *Recent Themes in the Philosophy of Science: Scientific Realism and Commonsense*. Dordrecht: Kluwer, 63–90.

Lyons, T. D. (2006). "Scientific Realism and the Stratagema de Divide et Impera." *The British Journal for the Philosophy of Science*, 57(3), 537–560.

Lyons, T. D. (2017). "Epistemic Selectivity, Historical Threats, and the Non-Epistemic Tenets of Scientific Realism." *Synthese*, 194(9), 3203–3219.

Mizrahi, M. (2013). "The Pessimistic Induction: A Bad Argument Gone Too Far." *Synthese*, 1–18.

Musgrave, A. (1988). "The Ultimate Argument for Scientific Realism," in: Robert Nola (ed.) *Relativism and Realism in Science*. Dordrecht: Springer, 229–252.

Müller, F. (2015). "The Pessimistic Meta-induction: Obsolete Through Scientific Progress?" *International Studies in the Philosophy of Science*, 29(4), 393–412.

Norton, J. (2008). "Must Evidence Underdetermine Theory?" in: M. Carrier, D. Howard, and J. Kourany (eds.) *The Challenge of the Social and the Pressure of Practice: Science and Values Revisited*. Pittsburgh: University of Pittsburgh Press, 17–44.

Park, S. (2011). "A Confutation of the Pessimistic Induction." *Journal for the General Philosophy of Science*, 42(1), 75–84.

Psillos, S. (2018). "Realism and Theory Change in Science." The Stanford Encyclopedia of Philosophy (Summer 2018 Edition), Edward N. Zalta (ed.), URL = <https://plato.stanford.edu/archives/sum2018/entries/realism-theory-change/>.

Putnam, H. (1978). *Meaning and the Moral Sciences*. Boston: Routledge and Kegan Paul.

Roush S. (2009). "Optimism about the Pessimistic Induction," in: P. D. Magnus and M. Busch (eds.) *New Waves in Philosophy of Science*. Palgrave MacMillan, 29–58.

Ruhmkorff, S. (2013). "Global and Local Pessimistic Meta-Inductions." *International Studies in the Philosophy of Science*, 27(4), 409–428.

Rowbottom, D. P. (2019). "Scientific Realism: What It Is, the Contemporary Debate, and New Directions." *Synthese*, 196(2), 451–484.

Smart, J. J. C. (1963). *Philosophy and Scientific Realism*. London: Routledge and Kegan Paul.

Stanford, P. K. (2006). *Exceeding Our Grasp: Science, History, and the Problem of Unconceived Alternatives*. Oxford University Press.

Stanford, P. K. (2015). "'Atoms Exist' Is Probably True, and Other Facts That Should Not Comfort Scientific Realists." *Journal of Philosophy*, 112(8), 397–416.

Vickers, P. (2013). "A Confrontation of Convergent Realism." *Philosophy of Science*, 80(2), 189–211.

Vickers, P. (2018). "Historical Challenges to Realism," in: Juha Saatsi (ed.) *The Routledge Handbook of Scientific Realism*. London: Routledge, 48–60.

Vickers, P. (2019). "Towards a Realistic Success-to-Truth Inference for Scientific Realism." *Synthese*, 196(2), 571–585.

Wray, K. B. (2013). "The Pessimistic Induction and the Exponential Growth of Science Reassessed." *Synthese*, 190(18), 4321–4330.

Wray, K. B. (2018). *Resisting Scientific Realism*. Cambridge: Cambridge University Press.

14

The Paradox of Infinite Limits

A Realist Response

Patricia Palacios[1] and Giovanni Valente[2]

[1]Department of Philosophy
University of Salzburg
patricia.palacios@sbg.ac.at
[2]Department of Mathematics
Politecnico di Milano
giovanni.valente@polimi.it

14.1 Introduction

Scientific realism is a central topic in philosophy of science. Although there are many different formulations of this concept in the literature, most scientific realists are committed to the idea that we have good reason to believe that the content of our best scientific theories, regarding both observable and unobservable aspects of the world, is true or at least approximately true. According to most realists, scientific realism involves a semantic dimension, according to which one is committed to a literal interpretation of scientific claims about the world (Chakravartty 2017). Perhaps the strongest argument in favor of scientific realism is the "no miracles argument," (NMA) which asserts that the success of our well-established scientific theories would be a miracle if the content of such theories were not true or at least approximately true (Putnam 1975; Boyd 1983). The notion of "approximate truth" plays an important role in current approaches to scientific realism, since it is widely held, even by realists, that our best scientific theories are likely false, strictly speaking. Important technical work has been developed to make the notion of "approximate truth" precise and we will address part of this work in the present paper.

As plausible as it is, a scientific realist position faces various difficulties that have cast doubt on the no miracles argument and have motivated an anti-realist attitude toward our most successful theories (van Fraassen 1980; Rosen 1994). One of the most outstanding challenges for scientific realism is the use

Patricia Palacios and Giovanni Valente, *The Paradox of Infinite Limits* In: *Contemporary Scientific Realism*.
Edited by: Timothy D. Lyons and Peter Vickers, Oxford University Press. © Oxford University Press 2021.
DOI: 10.1093/oso/9780190946814.003.0014

of idealizations in scientific theories, which involves the assumption of fictional systems that are intended to resemble the real-world systems we are interested in (Godfrey-Smith 2009). An important example of idealizations in physics is the use of "infinite idealizations," which involves the introduction of infinite systems that can be constructed by means of mathematical limits that are invoked to explain the behavior of target systems, notwithstanding the fact that the latter are considered to be finite according to our most successful background theories.[1]

The problem with infinite idealizations, as it has been presented in the literature, is that it apparently leads one to what we refer to as the "Paradox of Infinite Limits," which poses a challenge to scientific realism. Informally, the intended paradox can be formulated as follows. On the one hand, a scientific realist must believe that real physical systems are finite, as it is indeed suggested by some of our most successful background theories such as the atomic theory of matter and general relativity. On the other hand, she must believe that the content of scientific theories invoking infinite idealizations is true or approximatively true insofar as they are indispensable to recover empirically correct results. Allegedly, that calls scientific realism into question.[2]

This paradox was first introduced by Callender (2001) in the context of phase transitions and then further discussed, for instance, by Butterfield (2011), Norton (2012), and Shech (2013).[3] Recently, interesting attempts to generalize the problem for scientific realism have been made by Baron (2019), Liu (2019), and Shech (2019). However, the extent to which such proposals provide a definite solution is still open. Indeed, it remains unclear under what conditions the assumption of infinite limits leads one to paradoxes of this form. The present paper aims to offer a general formulation of the Paradox of Infinite Limits. In the attempt to show how it can be resolved, we elaborate a taxonomy of the different uses of infinite limits in physics, which is partially based on distinctions made by Norton (2012) and Godfrey-Smith (2009). Specifically, we point out that when being understood as approximations and abstractions, infinite limits do not pose any substantial problems to scientific realism. Yet, when they give rise to

---

[1] It is important to point out that one can also "construct" infinite systems without taking any limits such as with the assumption of an infinitely long cylinder or two dimensional systems. Although such examples deserve attention and have been discussed in the philosophical literature (e.g. Earman 2017; Shech 2018), we will restrict our analysis to infinite idealizations resulting from the use of infinite limits.

[2] For completeness, let us clarify that in the literature one can find at least three different versions of scientific realism: Explanatorianism, Entity realism and Structural realism (Chakravartty 2017). In this paper we consider a general characterization of scientific realism, leaving the question of how these particular versions can deal with infinite idealizations for future work. However, we point out that the Paradox of Infinite Limits is especially problematic for Explanatorianism (Kitcher 1993; Psillos 1999), which recommends a realist commitment with respect to those parts of our best theories that are indispensable to explaining empirical success.

[3] See also (Fletcher, Palacios, Ruetsche, and Shech 2019) for a recent collection of papers on infinite idealizations in science.

infinite idealizations they can actually lead one to the Paradox of Infinite Limits, depending on whether the idealization is regarded as *essential* for the explanation of the physical phenomenon under investigation. We then argue that, even in the case of essential idealizations, there are ways of coping with the alleged incompatibility between infinite idealizations and scientific realism, which ultimately rely on empirical considerations.

We organize the paper as follows. In Section 14.2, we distinguish between idealizations, approximations, and abstractions broadly constructed. This distinction is partially based on Norton's (2012) distinction between idealizations and approximations and on Godfrey-Smith's (2009) distinction between idealizations and abstractions. In the following section, we formulate the general Paradox of Infinite Limits and we explain in greater detail the sense in which it raises a challenge for scientific realism. In particular, in sub-section 14.3.3 we state the condition of Empirical Correctness in precise terms, and then we proceed to analyze the various possible understandings of the use of infinite limits in physics. Specifically, in Section 14.4 we develop the concept of approximation and argue that it is not problematic from a scientific realist perspective by means of concrete examples: in fact, the paradox does not arise in the case of approximations without idealizations, whereas it can be readily disarmed in the case of idealizations yielding approximations. In Section 14.5, we address the use of infinite limits as essential idealizations, explaining why that would lead the scientific realist to a paradox. Yet, by focusing on the controversial example of first-order phase transitions, we show that there is available a procedure to dispense the infinite idealization "on the way to the limit," thereby avoiding a contradiction with the claim that real target systems undergoing the phenomenon to be explained are finite. Finally, in the last section, 14.6, we address the case of continuous phase transitions to illustrate the use of infinite limits as abstractions, and we also explain why these limits would not raise any conflict with scientific realism.

14.2 Idealizations, Approximations, and Abstractions

14.2.1 Preliminary Concepts

In the scientific practice it is ubiquitous to, so to speak, "modify" the systems encountered in the world with the goal of making our theories computationally tractable, pedagogically accessible, or explanatorily rich. The representation of real planes as frictionless planes in which objects can move uniformly and perpetually is a prototypical example of this practice, in which real systems are modified with the purpose of making the theory manageable. Looking at the specific ways in which real systems are represented in scientific theories, one can distinguish at least between three different strategies labeled by philosophers of science as *idealizations, approximations,* and *abstractions.*

Although there is no consensus about the nature of scientific idealizations, most philosophers agree that *idealizations* involve a misrepresentation of a real system (i.e. the target system) driven by pragmatic concerns, such as mathematical tractability. Recent accounts provide a more precise characterization of idealizations by arguing that they always refer to imaginary systems considered to be analogues of the real-world systems of interest. For example, Godfrey-Smith (2009, p. 47) says "I will treat [idealizations] as equivalent to imagining the existence of a fictional thing that is similar to the real-world object we are interested in." In a similar vein, Norton (2012, p. 209) defines an idealization as "a real or fictitious system, distinct from the target system, some of whose properties provide an inexact description of some aspects of the target system." According to these authors, there will be an idealization only when there is reference to a novel (fictional) system, which has properties that resemble the properties of real-world systems. In other words, when the misrepresentation of the properties of a target system coincide with the exact properties of a fictional system.

An example of idealizations is the frictionless plane just mentioned. This fictional system was firstly introduced by Galileo to derive the equations of motion of an object moving down an inclined plane. Although no such planes exist in reality, they have proven to be extremely useful to predict the behavior of real world systems. Why can these fictional systems explain the behavior of concrete systems observed in the world, and how can we justify their use from a scientific realist perspective? The standard justification for idealizations of this kind is that they provide approximately true descriptions of real world systems, where approximation to truth is simply understood as a relation of similarity between the properties of fictional systems and the properties of real world systems (Godfrey-Smith 2009). It is also believed that these idealizations are dispensable, in the sense that it is possible in principle to de-idealize the theory by systematically eliminating distortions and by adding back to the model details of concrete systems (McMullin 1985).[4]

This characterization of idealizations puts us in position to distinguish them from other strategies of theory construction such as approximations and abstractions. For instance, according to Norton (2012), *approximations* are inexact descriptions of certain properties of the target system, which are given in terms of propositions expressed in the language of a theory, that do not need to correspond to the relevant properties of some other fictional system. In fact, the

---

[4] This kind of idealization corresponds to what has been called "Galilean idealization" (e.g. McMullin 1985; Weisberg 2007)).

use of approximations does not require one to make reference to any new system different from the target systems. In this sense, one can say that approximations involve distortions or misrepresentations of the target system but, differently from the case of idealizations, these distortions are merely propositional. For Norton (2012), idealizations can be demoted to approximations by discarding the idealizing system and extracting the inexact description, but the inverse promotion will not always succeed. However, we will see that there seem to be cases of essential idealizations in which idealizations cannot be easily demoted to approximations.

In recent years, the topic of essential idealizations has generated a great deal of excitement among philosophers of science. In particular, it has been argued that infinite idealizations arising from mathematical limits that are ineliminable cannot provide approximations of realistic systems, because the latter exhibit behavior that is qualitatively different from the behavior exhibited by the idealized infinite systems (e.g. Batterman 2005). We will examine some possible cases of ineliminable idealizations in Section 14.5, but before doing so, let us mention some examples of approximations and abstractions.

Let us recall that an approximation is an inexact description of the target system that does not (necessarily) involve the introduction of a fictional system. This means, there is no appeal to fictional system in which the inexact properties of the target system are true. A good example of an approximation without idealization also mentioned by Norton (2014) is the case of a mass falling in a weakly resisting medium. As we know, the speed v of a falling mass at time t is given by:

$$dv \, / \, dt = g - kv,$$

where g is the gravitational acceleration and k the friction coefficient. The speed of the mass at time t, as it starts from rest, is given in terms of the Taylor expansion series by:

$$v(t) = (g \, / \, k)(1 - exp(-kt)) = gt - gkt^2 \, / \, 2 + gk^2t^3 \, / \, 6 - \dots$$

If we assume that the friction coefficient is small, we can approximate the previous expression by taking only the first term of the series expansion:

$$v(t) = gt$$

When we use this expression to describe the behavior of a real mass falling in a resisting medium, we do not intend to give a literal description of the situation, but rather to give a good approximation of it. This strategy allows us to simplify problems that may be otherwise intractable. It is important to note, however, that one can promote this approximation to an idealization by introducing a fictional system in which a body falls under the same gravity in a vacuum, so that the fall is described exactly by $v(t) = gt$ (Norton 2014).

So understood, idealizations can also be distinguished from abstractions. According to GodfreySmith (2009), an *abstraction* is the act of "leaving things out while still giving a literally true description of the target system" (p. 48). In contrast to idealizations, abstractions do not intend to state claims that are literally false and do not make reference to fictional systems.[5] Instead, they involve the omission of a truth by leaving out features considered to be irrelevant. Abstractions also differ from approximations in that the former do not involve propositional misrepresentations of the target system whereas the latter do.

14.2.2 Approximate Truth in the Context of Idealizations, Approximations, and Abstractions

We will argue next that all the forms of inaccurate representations mentioned in the previous section, i.e. idealizations, approximation, and abstractions, can be made compatible with the more relaxed realist framework that accepts the content of scientific theories to be at least approximately true. However, in order to arrive at that conclusion, we need to offer a more precise definition of the notion of "approximate truth." Chakravartty (2010) distinguishes between three kinds of approaches for approximate truth in the standard philosophical literature: the verisimilitude approach, the possible-world approach, and the type hierarchy approach. The verisimilitude approach, elaborated by Popper (1972), consists in comparing the true and false consequences of different theories. In the possible word approach, which is meant to be an improvement of the verisimilitude approach, the truth-likeness is calculated by means of a function that measures a mathematical "distance" between the actual world and the possible worlds in which the theory is strictly correct (Tichy 1976; Oddie 1986) so that one can generate an ordering of theories with respect to truth-likeness. Finally, in the type hierarchy approach, truth-likeness is calculated in terms of similarity relationships between nodes that represent concepts or things in the word (Aronson 1990). As Chakravartty (2010) points out, the problem that all these approaches have in common is that they do not pay attention to the different ways in which scientific representations give inaccurate account of the target systems. This is an important limitation of these approaches, because the notion of approximate truth

---

[5] A similar view is defended by Jones (2005)

is best understood differently in different circumstances, especially in cases of idealizations yielding approximations, essential idealizations, approximations and abstractions.

Let us discuss first the notion of approximate truth in the case of idealizations. As said, idealizations involve a misrepresentation of a real system (i.e. the target system) by means of introducing an imaginary system considered to be an analogue of the real-world systems of interest. Here we follow (Chakravartty 2010, p. 40) in considering that the adequate notion of approximate truth concerns the degree to which this fictional system that successfully captures aspects of the target system resembles a non-idealized representation of that system. In some cases, especially when the idealization is a limit case of the de-idealized system, the degree of resemblance can be specified mathematically so that one can even quantify the degree of misrepresentation. All cases of idealizations yielding approximations can be put in this category and will be discussed in greater detail in the next sections. These cases do not represent a challenge for scientific realism since the notion of approximate truth can be adequately quantified.

A similar notion of approximate truth can be given in the case of approximation that do not involve idealizations. The difference is that degree of resemblance should not be evaluated between the properties of a fictional system and those of a target system, but rather between the misrepresentation of the properties of the target system and the actual properties of the real target system. In cases like the example of a mass falling in a weakly resisting medium, we can quantify the degree of misrepresentation by considering the terms of the series expansion that have been taken into account. In fact, more accurate descriptions will imply incorporating more terms of the series expansion. The more terms we consider, the better the approximation of the real properties of the target system will be. In this sense, approximations are straightforwardly compatible with realism, in that it is possible to quantify the degree of misrepresentation.

The challenge for scientific realism comes instead from the possibility of essential idealizations, which appear to be ineliminable to the explanation of a certain phenomenon and cannot be de-idealized toward more faithful explanations. In other words, they cannot be demoted to approximations. In these cases, there do not seem to be a straightforward notion of approximate truth and the degree of resemblance between the fictional system and the target system do not seem to be easily quantifiable. We will discuss such cases in Section 14.5.

Finally, let us refer again to abstractions. As said earlier, abstractions omit details that are considered to be irrelevant but do not involve any misrepresentation of the target system. For example, in trying to describe the behavior of a cannonball that has been fired on some particular day, there are innumerable features that will not be taken into account such as the composition of the cannonball, its color, its temperature, or the mechanism by which the initial velocity is conferred to the ball. The scientific model that predicts where the cannonball will land may include a number of distortions with respect to other properties like the gravitational force, which is generally assumed to have the same magnitude and direction at all points of the trajectory. However, such a model does not involve misrepresentation with respect to the properties that do *not* make a difference for the occurrence of the phenomena. In fact, the model is simply silent about them, and hence it does not say anything false as regards these irrelevant properties (Jones 2005). Although this strategy does not involve misrepresentation of certain properties it does give an inexact description of the target system because it leaves out factors that are irrelevant for the behavior under consideration. One then needs an appropriate notion of approximate truth in this case too. The intended notion will differ from the one involved in cases of idealizations and approximations in that there is no misrepresentation of properties. Chakravartty (2010, p. 39) suggests an articulation of the notion of approximate truth *qua* abstractions connected with the notion of comprehensiveness: "The greater the number of factors built into the representation [i.e. the more comprehensive is the description], the greater its approximate truth." In so far as one can quantify these factors, there will be a precise notion of approximate truth applying also to the case of abstractions. It is important to note that sometimes abstractions may be even crucial or essential for the explanation of certain classes of behavior.[6] For instance, if we want to explain the behavior of cannonballs in general, we should not include details that are specific of a certain fire. This is possible just because abstractions generally aid in the explanation of certain phenomena by enabling us to account for some common behavior that is generated among disparate systems (see also Weisberg 2007).

14.3 The Paradox of Infinite Limits: A Challenge for Scientific Realism?

Mathematical limits are vastly used in physics. For instance, in the statistical mechanical theory of phase transitions, it is assumed that the number of particles as well as the volume of the system goes to infinity; similarly, in the ergodic theory of equilibrium and in the explanation of reversibility in thermodynamics it is assumed that time goes to infinity. Appealing to infinite limits has the technical advantage that they render the formal account of physical phenomena more tractable. Furthermore, in some cases, like in the examples we just mentioned, it even

---

[6] This account of abstraction is consistent with Cartwright's (1994, p. 187) view, according to which an abstraction is a mental operation, where we "strip away—in our imagination—all that is irrelevant to the concerns of the moment to focus on some single property or set of properties, as if they were separate."

appears as a necessary condition to offer a mathematical description of real target systems in agreement with the empirical results. Nevertheless, when the variable growing to infinity corresponds to a physical parameter, e.g. the number of microscopic constituents, taking the limit introduces an unrealistic assumption, at least as long as real systems are believed to be finite. This raises many conceptual questions. For instance: How can we explain the empirical success of models that introduce such an unrealistic assumption? To what extent are models that use infinite limits compatible with scientific realism? In this section and the reminder of this paper we will address these questions based on the previous distinction between idealizations, approximations and abstractions.

14.3.1 A Taxonomy for Infinite Limits

The use of infinite limits has been frequently equated with the use of an infinite idealization. However, we want to stress a distinction between different uses of infinite limits as approximations, idealizations, and abstractions. In order to draw such a distinction, we begin by setting up the formal framework within which our discussion is cast. Let S_n represent a system characterized by some physical parameter n, which may take on discrete or continuous values: in particular, it could denote the number N of molecules constituting a gas system, so that the parameter takes on values in the natural numbers $\mathbb{N}$; or, it could denote the time t during which a physical process unfolds, so that the parameter takes on values in the real numbers $\mathbb{R}$. As the variable n increases, one defines the following sequence of systems

$$\{S_1, S_2, ..., S_n\} \quad n \in \mathbb{N}, \mathbb{R}$$

The limit of such a sequence for $n \rightarrow \infty$, if it exists, corresponds to the infinite system S_∞. Arguably, the latter is just a mathematical entity, and as such it would represent a fictitious rather than a real system. If so, by recalling the content of the previous section, taking the limit where the variable n goes to infinity gives rise to an idealization. We can thereby characterize the limit system S_∞ as an infinite idealization.

Now, limits can also be used as approximations without idealizations. Consider again the above sequence of systems $\{S_1, S_2, ..., S_n\}$, we can also define the following sequence of functions representing a putative physical quantity f:

$$\{f_1, f_2, ..., f_n\} \quad n \in \mathbb{N}, \mathbb{R}$$

where the notation f_n indicates the relevant property possessed by each finite system S_n, so that $f_n := f_{S_n}$. Here, some care should be taken when evaluating the infinite limit $n \to \infty$. In fact, as Butterfield (2011) suggested, there is a crucial difference between "the limit of a sequence of functions" and "what is true at that limit": that is, respectively,

- (*i*) the limit f_∞ of the sequence $\{f_1, f_2, ..., f_n\}n$ of functions, and
- (*ii*) the function f_{S_∞} associated with the limit system S_∞.

More to the point, if the parameter n takes on values on the natural numbers, (*i*) obtains when one adjoints infinity to the set $\mathbb{N}$, and hence the limit $f_\infty := \lim_{n \to \infty} f_n$ is defined as the function being the last element of this sequence with $n \in \mathbb{N} \cup \{\infty\}$. In this case, there is no reference to the infinite system S_∞, so that using the limit in this sense does not lead to an idealization. To the contrary, (*ii*) depends exactly on how the limit system is constructed. Indeed, the limit function f_{S_∞} represents the relevant quantity possessed by the infinite system S_∞: as such, it tells us just what is true at the limit. It should be emphasized, though, that its existence is contingent upon the type of convergence one adopts: this point will be useful for our discussion in Section 14.5.

The proposed distinction becomes less abstract if one casts it in terms of values of quantities. In fact, supplying numerical values is right what enables us to directly check whether or not the expected results turn out to be empirically correct. So, given that the putative function f takes on values $v(f)$, one should consider yet another sequence, namely the sequence of values

$$\{v(f_1), v(f_2), ..., v(f_n)\} \quad n \in \mathbb{N}, \mathbb{R}$$

Of course, the actual value of each function f_n depends on the state s_n in which the system S_n is: in fact, there is also a sequence of states $\{s_1, s_2, ..., s_n\}$ implicitly understood along with the sequence of systems.

As above, one needs to distinguish between (*i*) the limit of the sequence of values of the function f, i.e. $\lim_{n \to \infty} v(f_n)$, and (*ii*) the value $v(f_{S_\infty})$ of the natural limit function computed when the limit system S_∞ is in the limit state s_∞.

Before addressing the issue whether, and how, it is possible to dispense from the infinity system in the explanation of some physical phenomenon, let us conclude this section by noting that the formal setting we have just presented enables us to sharply distinguish between different ways to characterize the use of infinite limits in physics. The proposed taxonomy identifies three main types, that is:

1. *Approximations without idealizations,* where (ii) is not well defined, or (i) is empirically correct but (ii) is not;
2. *Idealizations yielding approximations,* where (i) and (ii) are well defined and equal; and
3. *Essential idealizations,* where (i) and (ii) are well defined but are not equal, and (ii) rather than (i) is empirically correct.

Beside these cases, one can add to this taxonomy the use of *infinite limits as abstractions,* which does not directly follow from such a scheme: in fact, as we will see in greater detail in Section 14.6, these are cases in which the variable n does not represent any physical parameter of the target system. For example, the parameter can represent the number of times that one has to apply a transformation that successively coarse grains the system. Here the appeal to an infinite limit is merely instrumental in that it allows us to find fixed points in a topological space.

In the rest of the paper, we evaluate how each type of infinite limit fares against the so-called Paradox of Infinite Limits and its consequences for scientific realism, which we present next.

14.3.2 The Paradox of Infinite Limits

Suppose that in order to represent a physical phenomenon P we define a system of the form S_n, then the Paradox of Infinite Limits can be essentially formulated as a combination of the following statements:

- (I) *Finiteness of Real Systems*: If S_n represents a real system, then the variable n corresponding to some physical parameter cannot take on infinite values.
- (II) *Indispensability of the Limit System*: The explanation of the phenomenon P can *only* be given by means of claims about an infinite system S_∞ constructed in the limit $n \to \infty$.
- (III) *Enhanced Indispensability Argument (EIA)*: If a claim plays an indispensable role in the explanation of a phenomenon P we ought to believe in its existence.

The ostensive problem can be further articulated and made more precise when dealing with the description of particular phenomena, as we will see in greater detail for the Paradox of Reversible Processes and the Paradox of Phase Transitions in Sections 14.4 and 14.5, respectively. But the tension between these statements captures the core idea of the paradox. Intuitively, the fact that real systems are finite in the sense expressed by statement (I) can be understood as

a basic desideratum of scientific realism. Indeed, according to our most successful theories, such as the atomic theory of matter, real gases contain only a
finite, albeit extremely large, number N of molecules, just as real thermodynamical processes, even when being very slow, always take a finite amount of
time t to complete. However, if one commits to the reality of an infinite system
as demanded by statement (II) together with statement (III), then one infringes
on such a basic desideratum. In fact, a seeming contradiction with statement
(I) arises insofar as appealing to the infinite idealization S_∞ proves necessary. In
their strongest form, claims that taking the limit is indispensable are backed by
no-go theorems, established within the formalism of a given theory, to the effect
that certain features of P cannot be recovered unless n is infinite. A threat to scientific realism can then be mounted when these results are coupled with statement (III), which Baker (2009) called the Enhanced Indispensability Argument
(EIA). According to the EIA, we ought to rationally believe in the existence of
claims (e.g. mathematical entities or idealizations) that play an indispensable
role in our best scientific theories (cfr. Shech 2019) for a discussion of EIA in
relation to infinite idealizations). In fact, it follows that, if the best explanation
available for the physical phenomenon P requires one to take the limit $n \to \infty$,
owing to EIA one is bound to ontologically commit to the infinite system S_∞,
even though statement (I) entails that the latter cannot correspond to any real
system. A realist stance toward our best physical theories is thus endangered by
the Paradox of Infinite Limits.

   In order to resolve the problem, one should jettison one of the three statements
that jointly engender the ostensive paradox, at least in the form they have been
presented here. The first statement seems rather uncontroversial, and as such it
is hard to give it up. Indeed, the finiteness of a real system S_n involved in the
phenomenon P to be explained is grounded in empirical considerations, so long
as the variable n denotes a physical parameter. Furthermore, in some cases the
truth of statement (I) is granted by the content of successful background theories: for example, the number of molecules in a thermal system being finite
is presupposed by the atomic theory of matter itself and general relativity (see
Baron 2019). Hence, insofar as the variable n represents a physical parameter as
in statement (I) in the paradox, the culprit must be traced back to statement (II)
or statement (III). Our own strategy will consist in giving up statement (II), but
let us first review some attempts to block statement (III).[7]

   Baron (2016) casts doubts on statement (III) by offering a helpful characterization of the sense in which idealizations can be regarded as indispensable to
the best explanation of a phenomenon. According to him, although idealizations

---

[7] See (Colyvan 2001), (Baker 2005; Baker 2009) and (Baron 2016) for a careful discussion of the
indispensability arguments.

may be indeed indispensable to the purported explanation, they do not carry explanatory load (understood as counterfactual dependence between the idealization and the effect) and therefore should not be reified. However, this strategy is questionable in the case of infinite idealizations, since the latter can have explanatory load, at least in the sense that we will expose in Section 14.5. In Baron (2019), he adopts an alternative strategy by distinguishing between constructive indispensability and substantive indispensability: given a mathematical entity that is explanatorily indispensable to our current best scientific theories, according to the former notion there is reason to believe that such a claim can actually be dispensed, although we do not know how to do it yet; according to the latter notion, instead, there is no reason to suppose that the claim can ever be dispensed. In other words, in the first case, indispensability is just a contingent matter, whereas in the second case it is an unavoidable fact. So, under this understanding, if an infinite limit that appears as necessary to explain a physical phenomenon P is constructively rather than substantively indispensable, the EIA does not apply with sufficient cogency to commit one to the reality of the infinite system constructed in the limit, thereby disarming the implications of the Paradox of Infinite Limits for scientific realism. The crucial question then becomes how to determine whether one is dealing with an instance of substantive indispensability or an instance of constructive indispensability. Although the exact answer can only be given on a case-by-case basis, there are two major strategies that one may adopt, according to Baron. One approach, which Baron himself favors, is grounded on the notion of coherence. Specifically, one ought to test the alleged indispensability of an infinite limit against other background scientific theories: if the claim under test is not consistent with other accepted theories, then there is reason to suppose that it is only constructively indispensable. For instance, when a gas system is contained in a finite region the limit for the number N of molecules going to infinity is at odds with the atomic theory of matter; moreover, Baron argues, it fails to cohere with the general theory of relativity in that the total mass of the molecules would become infinite despite the volume remaining finite, and hence the density would become infinite thereby giving rise to a black hole in the region where the gas is confined, which is not really observed in real-life phenomena. Although plausible, this solution seems to beg the question, since the problem that we are trying to solve concerns precisely the inconsistency between well-established background theories and mathematical entities used to explain certain phenomena. For us, the motivation for accepting statement (III) comes from the acceptance of the inference-to-the-best-explanation. Indeed, the best argument for scientific realism is the no miracles argument, which maintains that we ought to believe that abstract and theoretical claims about existing entities postulated by our most successful theories are (at least approximately) true because otherwise their success would be a miracle. Underlying this argument is the inference-to-the-best-explanation (Boyd 1983). By using a similar reasoning, if the best explanation available of a certain phenomenon P involves assuming an infinite idealization and there is no known way to de-idealize the model without losing explanatory power, then we should commit to the existence of such an idealization for the same reasons that ground scientific realism. This a particularly important for Explanationanists (Kitcher 1993; Psillos 1999), since they recommend holding a realist attitude toward entities that are indispensable to the explanation of certain phenomena.[8]

Therefore, we claim, a more satisfactory solution to the Paradox of Infinite Limits comes from the rejection of statement (II), rather than statement (III). To make our point, we need a criterion of empirical correctness that we introduce next.

14.3.3 The Condition of Empirical Correctness

Before spelling out our intended criterion for correctness, it should be stressed that the concept of explanatory indispensability heavily depends on what one means by scientific explanation, which is a huge and much debated topic in philosophy of science.[9] For our purposes, we restrict ourselves to what we conceive as a minimal requirement for a good explanation of some physical phenomenon P, namely that one recovers empirically correct results. Arguably, this yields a necessary condition in that, if an account fails to agree, at least approximatively, with the observed data then it cannot be said to explain P (whether it yields also a sufficient condition is less straightforward to establish, but our argument here does not really need that much). The relevant data are given by the values of physical quantities of interest, like energy, position, momentum, spin, etc., corresponding to properties of the physical system involved in the phenomenon. The proposed condition for empirical correctness is as follows:

---

[8] Shech (2013, p. 1177) makes a similar point, when he says:

> Insofar as arguments like the "no miracles argument" and "inference to best explanation" are cogent and give us good reason to believe the assertions of our best scientific accounts, including those about fundamental laws and unobservable entities, then in the case of accounts appealing to [essential idealizations], these arguments can be used via an Indispensability Argument to reduce the realist position to absurdity.

[9] See Woodward (2014) for an excellent review of the different approaches to scientific explanation in the philosophical literature.

Empirical Correctness: Let D be the observed data relative to a physical quantity represented by the function f, which characterizes the physical phenomenon P: then, the system S_n recovers *empirically correct results* for f just in case $v(f_n) \approx D$, in the sense that there exists an arbitrarily chosen real number $\varepsilon > 0$ such that $|v(f_n) - D| < \varepsilon$.

Let us explain the content of this definition. Note that one cannot expect that empirical data can be sharply determined, in that observations always involve some margin of error. Hence, our condition is formulated in such a way to allow for degrees of inexactness: in fact, it only requires that the value of f be approximately equal, rather than exactly equal, to the data D, that is $v(f_n) \approx D$ for some given n. Deciding the degrees of inexactness that one may tolerate is ultimately a pragmatic matter, and that is why the number ε is left unfixed in the definition: yet, by keeping ε very small, one ensures that the results produced by system S_n are empirically correct to a pretty good approximation.

Infinite limits can be used in the explanation of the physical phenomenon P when empirical correctness is satisfied by the infinite system S_∞ for some physically relevant function f. Accordingly, one has $v(f_{S_\infty}) \approx D$, meaning that the value at the limit, namely (ii) the function f_{S_∞} associated with the limit system S_∞, is the same, or at least approximatively the same, as the observed data D. More to the point, infinite idealizations are regarded as indispensable to the explanation of P insomuch as no finite system S_n can yield empirically correct results for f. That is, even though the parameter n grows while still remaining less than ∞, the values taken on by f_n fail to provide an approximation of the observed data for some sufficiently small ε. To put it in technical terms, this alleged indispensability of the infinite idealization typically manifests itself in the cases in which the limit is singular, and hence (i) the limit $\lim_{n \to \infty} v(f_n)$ of the sequence of values does not coincide with (ii) the value $v(f_{S_\infty})$ yielded by the limit system. (Butterfield (2011) actually makes a similar point when discussing the emergence of certain properties at the limit).

Armed with this stated condition of empirical correctness, we now proceed to show how infinite idealizations can be actually dispensed to the explanation of physical phenomena that seem to require one to take the limit $n \to \infty$, thereby allowing one to give up statement (II) in the Paradox of Infinite Limits. In fact, such a condition grounds the use of limits as approximations of the properties of real target systems. As we argue later, that is the basis to demonstrate that cases of idealizations yielding approximations as well as the more controversial cases of essential idealizations do not pose a threat to scientific realism.

14.4 Approximations with and without Idealizations

In Section 14.2 we gave a general characterization of the concept of approximations as inexact descriptions of target systems. Here, we can make this idea more precise thanks to the framework presented in previous section, according to which the relevant properties of a target system corresponding to physical quantities are represented by mathematical functions yielding numerical values: in fact, the very condition of empirical correctness rests on the possibility that such values match the observed data with a certain margin of accuracy. An approximation can therefore be defined as a formal description of some specific property of the target system that, despite being inexact, puts one in a position to recover empirically correct results. The choice of what properties are relevant to the purported description as well as the degrees of inexactness being allowed are dictated by pragmatic considerations regarding the physical phenomenon to be explained. Such an understanding of approximations is particularly suitable for our discussion of the use of mathematical limits, the more so because it is probably the most common attitude physicists tend to take on in their scientific practice. For, recall that in order to assuage the worries concerning scientific realism posed by the Paradox of Infinite Limits one ought to find a way to dispense from the infinite idealization introduced in statement (II) of the paradox. Interpreting infinite limits in terms of approximations, as contemplated by the first two classes listed in our proposed taxonomy, enables us to do so. Let us discuss both scenarios in general terms and then apply our reasoning to particular examples.

The scenario in which the limit $n \to \infty$ gives rise to an *idealization yielding approximations* may appear complicated in that it features a fictitious infinite system S_∞ that gives an inexact description of the real target system, and as such it effectively satisfies the condition of empirical correctness. However, given that in this case the limits (i) and (ii) in Butterfield's earlier distinction coincide, there is a strategy to dispense the infinite idealization that is readily available. Specifically, one can show that, as the variable n grows, for some finite value $n_0 < \infty$ the behavior of the corresponding system S_{n_0} is approximatively the same as the behavior of the limit system S_∞, thereby recovering empirically correct results for the relevant properties "on the way to the limit" rather than at the limit. More precisely, n_0 is supposed to be the actual value characterizing the real target system that undergoes the physical phenomenon P to be explained: so, if the value of the function $f_{S_{n_0}}$ is sufficiently close to the limit value of f_{S_∞}, one does not need to ontologically commit to S_∞ in order to fulfill empirical correctness. One can thus, in the same way as we presented earlier, dispense the infinite idealization to the explanation of P, in full compliance with statement (I) of the paradox asserting the finiteness of real systems. If idealizations yield approximations, though, one may still wonder why one should appeal to a mathematical limit in the first place. In other words, one may ask, how can one justify the use of an infinite limit to describe the target system? In order to answer this question, Butterfield (2011) puts forward what he calls a Straightforward Justification, which is based on two features that limits enjoy, namely mathematical convenience and empirical adequacy. Regarding the first feature, taking the limit often enables one to ignore some degrees of freedom that complicate the calculations, and so infinite systems, when they exist, turn out to be more tractable than finite systems for which the parameter n is very large.[10] As for empirical adequacy, arguably that assures that the values obtained at the limit are close enough to the real values. Hence, while the empirically correct results are given by the values of the function computed for the actual n_0, namely the values of the real target system, taking the limit for n growing to infinity still proves adequate within some acceptable margin of approximation. Thus, based upon these two desirable features proposed by Butterfield, one has both pragmatic and empirical reasons to justify the use of the infinite limit. That reinforces our claim that one does not need to commit to the reality of the infinite idealization S_∞, contrary to what is entailed by statements (II) and (III) in the Paradox of Infinite Limits.

The other scenario, namely the case in which taking the infinite limit $n \to \infty$ yields an *approximation without idealization*, is more straightforward to deal with. On the basis of our taxonomy, it occurs when the limit (i) rather than (ii) yields empirically correct results or (ii) is not well defined at all for some physically significant function f: as a consequence, the infinite system S_∞ does not satisfy empirical correctness. It follows that it cannot be even used to explain the relevant phenomenon. Indeed, an approximation without idealization should be understood as a misrepresentation of the target system whose properties are given an inexact description in terms of the limit f_∞ of the sequence $\{f_1, f_2, ..., f_n\}$ of functions representing the relevant physical quantities, rather than in terms of the properties of a fictional limit system. Accordingly, it would not make any sense to reify the infinite idealization, which means that statement (II) in the Paradox of Infinite Limits does not hold, and hence there cannot arise any ensuing problem for scientific realism. Very recently, Norton (2012) recognized that sometimes infinite limits are used precisely as approximations without idealizations. For him, there are two sufficient conditions under which an approximation cannot be promoted to the status of idealization, which are closely related to the formal conditions we stated earlier: that is, (1) the limit system

---

[10] In addition, Palacios (2018) points out that it does not suffice to show that the behavior that arises in the limit arises already for a large value of the parameter n, but one also needs to show that it arises for realistic values of n_0. This latter requirement ought to be empirically grounded and is meant to assure that the lower level theory that results from a limiting operation is empirically correct and therefore capable of describing realistic behavior.

does not exist in the sense that it is paradoxical, and (2) the limit system has properties that are inadequate for the idealization in that they do not match with the properties of the finite target system. The following quotation illustrates his proposed criterion for the failure of idealizations:

> Another type of limit used in thermodynamics cannot be used to create idealizations. Its limiting processes are beset with pathologies so that it either yields no limit system or yields one with properties unsuited for an idealization. (Norton 2012, p. 13)

Norton's goal is to show that, on the basis of these two conditions, some infinite limits that are regarded as idealizations in the literature do not deserve to be elevated to such a status and should in fact be demoted to mere approximations, as it happens in many examples in which mathematical limits are employed in thermodynamics and statistical mechanics.[11] In his view, an example in which an infinite limit does not give rise to a suitable idealization due to condition (2) is the account of phase transitions in the thermodynamical limit within classical statistical mechanics that we will address in the next section. Another controversial case that Norton claims to be an instance of failure of idealization due to condition (1) is the paradox of thermodynamically reversible processes arising in the infinite-time limit, which has recently drawn attention in the philosophical literature (Norton 2014; Norton 2016; Valente 2019). Let us discuss this case next.

Reversible processes are conceived as sequences of equilibrium states through which a thermodynamical system progressively passes in the course of time. They are typically constructed by means of the infinite-time limit, and as such they are interpreted as processes taking place infinitely slowly. They owe their name to the fact that, ideally, they could be traversed in both directions of time. However, reversible processes are fictitious processes: indeed, thermodynamical processes just take a finite amount of time t to complete, even if they proceed quasi-statically, and they can only unfold in one temporal direction but not in the reversed one. So, at best, taking the infinite-time limit $t \to \infty$ can yield approximations of real processes. The question that interests us is whether or not the limit processes can also be intended as infinite idealizations. According to Norton, reversible processes do not deserve to be elevated to such a status since they display contradictory properties. More to the point, he submits that they are

---

[11] To be sure, Norton concedes that, under certain circumstances, approximations can actually be promoted to idealizations. For instance, in his (2014) paper Norton presents the case of the law of ideal gases in thermodynamics as an example of an approximation which gives rise to an ideal system that can as well serve as an idealization. In fact, an ideal gas system is defined as constituted by a large number of non-interacting, spatially localized particles, and as such it can describe, at least approximately, the behavior of very rarefied real gases under appropriate circumstances.

plagued by a paradox: for, on the one hand, (I) thermodynamical processes are driven by a non-equilibrium imbalance of driving forces, which is necessary in order to enact the transition of the system from one state to another; on the other hand, though, (II) a reversible process amounts to a sequence of equilibrium states, for which there must be no imbalance of driving forces. By connecting the driving forces enacting the process with its duration in time, Norton's Paradox of Reversible Processes becomes somewhat similar to the general Paradox of Infinite Limits. For instance, when a fixed quantity Q of heat is exchanged between two bodies in thermal contact, the driving forces are given by the temperature difference ΔT between the bodies. So, if the latter is equal to zero, the process of heat transfer cannot take place. Now, by assuming Fourier law $Q = -k\Delta T\Delta t$ (with k being the heat transfer coefficient), the amount of heat that passes from the hotter to the colder body is also proportional to the time Δt that the process would take to complete. As a consequence, if one tries to construct a reversible process by letting time go to infinity, the temperature difference must vanish, as prescribed by statement (II), but then no heat would be exchanged between the two bodies, thereby raising a contradiction with statement (I). Based on such a paradox, Norton argues that reversible processes fail to be idealizations.

However, Valente (2019) objected to this conclusion, by offering a way to circumvent the alleged contradiction. As he pointed out, Norton's paradox arises due to the misconception that reversible processes should be treated as actual thermodynamical processes. Instead, they ought to be regarded as mere mathematical constructions that are introduced to apply infinitesimal calculus to thermodynamics. In fact, formally, they simply correspond to continuous curves in the space of equilibrium states of a thermal system: along these curves one can calculate the exact integrals of state-functions, such as entropy, and then compute the values of other physical quantities of interest, like heat and work. For this reason, as it was observed by the mathematical physicist Tatjana Ehrenfest-Afanassjewa (1956) in her book on the foundations of thermodynamics, reversible processes should better be called "quasi-processes," so as to avoid the misunderstanding that they would correspond to actual processes, which could even be reversed. Accordingly, while statement (II) holds by definition, it would be a mistake to ascribe to reversible processes the properties required by statement (I): that is, contrary to what happens during real thermodynamical processes, there cannot be any non-equilibrium imbalance of forces moving a thermal system from one state of equilibrium to another along the continuous curve representing a quasi-process. As a result, the ostensive paradox disappears, and hence one may as well regard reversible processes as idealizations, in accordance with Norton's own criterion. Notice that here, differently from the Paradox of Infinite Limits, in order to disarm the apparent contradiction we need to drop statement (I) rather than statement (II): the reason is that what is to be explained in this case is just the idealized object constructed in the limit. The connection with real thermodynamical processes can then be given thanks to the notion of approximations. To illustrate this idea, let us again refer to Ehrenfest-Afanassjewa's own work. After warning that, strictly speaking, infinitely slow processes cannot exist (cfr. p. 11), she went on to argue as follows:

> In order to make a quasi-process [i.e. a reversible process] open to experimental investigation or just to connect it with experiments in thought, one has to conceive it as approximated by quasi-static processes [i.e. extremely slow processes]. These are in turn real processes, if also idealized. (Ehrenfest-Afanassjewa 1956, p. 56)

Thus, even though an infinite-time limit process has no counterpart in reality, in practice it can serve to describe, albeit inexactly, the relevant properties of thermodynamical processes unfolding very slowly. To put it in terms of our characterization of idealizations yielding approximations: once a certain margin of accuracy is stipulated, the fictitious quasi-process constructed in the limit $t \to \infty$ gives approximatively true descriptions of real physical quantities, such as heat and work exchanged during real processes that take a large yet finite time t_0 to complete.

So, although Norton's distinction between approximations and idealizations may be fruitful, the extent to which the examples he addresses are genuine cases of approximations without idealizations is still subject to debate. In our opinion, a less controversial case where one can have approximations without idealization is the reduction of the classical equations of motion to relativistic equations in the Newtonian limit. As is well known, in the theory of special relativity physical quantities such as the energy and momentum of a moving body with an invariant mass m_0 can be expressed in terms of the so-called Lorentz factor γ:

$$E = \gamma m_0 c^2$$

$$p = \gamma m_0 v$$

Arguably, when the Lorentz factor goes to 1, one can recover the values of the classical counterparts of these quantities defined, respectively, as $E = m_0 c^2$ and $p = m_0 v$. Since the Lorentz factor takes on the form

$$\gamma(v) = \frac{1}{\sqrt{1-(v/c)^2}},$$

one can make this expression go to 1 in the limit $\left(\dfrac{v}{c}\right)^2 \to 0$, where c is the speed of light and v the velocity of a moving body in a given inertial frame. There are different possible ways of interpreting this limit. First, one can interpret it as taking the limit of a sequence of systems in which the velocity v goes to zero. The limit will then give rise to an idealization, namely to a fictional system in which the value of the velocity for all bodies is null in the given inertial frame. Nonetheless, understanding the limit in this sense has the problem that, e.g. the momentum p, which depends exactly on the velocity of the bodies, will be always zero. This naturally means that the quantities defined in the limit system will not provide a good approximation for the behavior of restless objects, in which momentum is different from zero. Therefore, the purported idealization would not serve to describe the behavior of moving objects. Likewise, a similar problem arises if one interprets the Newtonian limit as taking the limit of a sequence of systems in which c goes to infinity. In this case, the limit system will be an imaginary system in which the speed of light is infinite. In this system, quantities such as the kinetic energy that are defined as a function of the speed of light c will also go to infinity, thereby failing to give an approximation of the actual kinetic energy of real moving bodies, which is instead finite.

On the contrary, a much more suitable interpretation of the Newtonian limit $\left(\dfrac{v}{c}\right)^2 \to 0$ should not be given in terms of fictional systems, but rather as a mere approximation of the velocity of bodies that are moving slowly compared to the speed of light. This can be seen more clearly by noticing that the Lorentz factor γ can be expanded into a Taylor series:

$$\gamma(v) = \frac{1}{\sqrt{1-(v/c)^2}} = \sum_{n=0}^{\infty}\left(\frac{v}{c}\right)^{2n}\prod_{k=1}^{n}\left(\frac{2k-1}{2k}\right) = 1 + \frac{1}{2}\left(\frac{v}{c}\right)^2 + \frac{3}{8}\left(\frac{v}{c}\right)^4 + \frac{5}{16}\left(\frac{v}{c}\right)^6 + \dots$$

Accordingly, if one considers only the first term of this Taylor expansion, one will recover the exact values of the classical quantities of interest, and this will constitute just an approximation of the velocity v of the objects for which $v \ll c$. Taking more terms into account will give results that are more accurate from the relativistic point of view, but that depart from the classical values. It is important to stress that the thus-described process of approximation is analogous to the one employed to account for the behavior of a mass falling in a weakly resisting medium presented in Section 14.2, and therefore it lends itself to the same interpretation with respect to the notion of approximate truth. In fact, limits of this kind should be understood as a mere misrepresentation of the values of certain properties of the target system, which will gradually disappear if one takes more terms of the series expansion into account. Note that here, like in all other examples of approximations without idealization, one does not posit any fictional system that would yield empirically adequate results, and hence the Paradox of Infinite Limits does not arise. Since there are no infinite systems involved, the procedure we have just outlined offers a straightforward way to make the process of taking the Newtonian limit compatible with a form of scientific realism that allows for theories to be approximately true.

In the last analysis, using mathematical limits to provide approximations of properties of real finite systems enables one to dispense limit systems from the explanation of the relevant physical phenomena. Indeed, as we have argued throughout the present section, despite the fact that letting some physical parameter grow to infinity is tantamount to introducing a prima facie unrealistic assumption, cases of approximations without idealizations and even cases of idealizations yielding approximations should not pose any worry to a scientific realist. A more outstanding challenge is instead raised by cases in which an infinite limit is used as an essential idealization, to which we now turn.

14.5 Essential Idealizations

The claim that there are essential idealizations has been put forward by some authors, especially Batterman (e.g. 2001, 2005, 2011), in reference to singular limits, whereby empirically adequate results are obtained just for the infinite system and not for finite systems. In terms of Butterfield's distinction, this apparently mysterious case occurs when, given a function f representing some physical quantity, (i) the limit of the sequence of values and (ii) what is true at the limit for $n \rightarrow \infty$ differ from each other, but only (ii) is empirically adequate (these cases correspond to *essential idealizations* according to the taxonomy presented in Section 14.3.1). So, if the variable n growing to infinity represents some physical parameter, it means that the limit system would not yield an approximation of the relevant property of the target systems, which are instead finite. An infinite idealization is then deemed as essential in that, arguably, it is only the limit system S_∞ that allows one to explain the physical phenomenon for which the function f is relevant. So, if one further assumes EIA, then one falls into the Paradox of Infinite Limits, hence threatening scientific realism. In this section we present a possible solution to this paradox that casts doubt on the "essential character" of the idealization, which is partially based on Butterfield's (2011) results.

14.5.1 Resolving the Paradox for First-Order Phase Transitions

Phase transitions are sudden transformations of a thermal system from one state into another, occurring for instance when some material changes from solid to liquid state due to an increase of temperature. That is a much debated case study where, according to indispensabilists like Batterman, there arises an essential idealization when taking the so-called thermodynamic limit. Informally, the argument goes as follows. According to thermodynamics, which deals with the behavior of thermal systems from a macroscopic point of view, phase transitions occur when the function representing the derivatives of the free energy is discontinuous. Allegedly, such a discontinuity matches with the observed data, since sudden transitions from one phase to another appear to take place abruptly. Yet, when attempting to recover the same phenomena within statistical mechanics, which describes thermal systems at the microscopic level as being composed by a very large number N of molecules, one faces a technical impossibility: that is, if N is finite, the function representing the derivative of the free energy remains continuous no matter how large N is. Instead, one can recover the sought-after discontinuity by taking the thermodynamical limit, which prescribes that both the number of molecules N and the volume V of the system go to infinity while keeping its density fixed (Goldenfeld 1992). This motivates the persistent attitude among physics, such as Kadanoff (2009), to emphasize the importance of infinite systems to explain the phenomenon of phase transitions:

> Phase transitions cannot occur in finite systems, phase transitions are solely a property of infinite systems. (p. 7)

Accordingly, it seems that one is bound to assume the infinite system S_∞ constructed in the thermodynamical limit as being indispensable in order to account for thermodynamical phase transitions within statistical mechanics. Hence, the thus-defined infinite idealization is supposed to be essential. If so, though, the Paradox of Phase Transitions would present itself.

In the philosophical literature, the paradox was originally proposed by Callender (2001). In his formulation, a contradiction arises due to the joint combination of four conditions: (1) Phase transitions are governed by classical statistical mechanics, (2) real systems have finite N, (3) phase transitions occur when the partition function has a discontinuity, and (4) real systems display phase transitions. Let us stress that this is just a special case of the general Paradox of Infinite Limits. In fact, Callender's conditions can be explicitly connected with the three statements presented in Section 14.3.1. For, condition (2) is tantamount to statement (i) expressing the basic desideratum for scientific realism that real systems are finite, whereas condition (4) assures that the phenomenon to be explained, namely phase transitions, occurs for such systems, which is captured by statement (ii). On the other hand, condition (3) requires empirical correctness to be satisfied for the partition function being discontinuous: yet, as discussed earlier, if one tries to explain the phenomenon within the framework of statistical mechanics, in accordance with condition (1), then one is bound to reify the thermodynamical limit, which gives rise to an infinite idealization proving indispensable, exactly as our statement (iii) dictates. Other formulations of the paradox have been put forward in the literature, which differ from the one just presented only regarding how the allegedly conflicting conditions are stated, but they all basically agree on the content of the problem. There is also a variety of proposed solutions. Callender himself suggests that one should not take "too seriously" the theory that prompts one to introduce a discontinuity in the description of phase transitions, namely thermodynamics: accordingly, condition (3) can be dropped, which means that one does not need to appeal to the thermodynamical limit and therefore denies statement (ii) of our formulation of the paradox. Shech (2013), on the other hand, suggests that one can resolve the paradox by noticing that the terms related to the existence of infinite systems do not refer to concrete physical systems but just to mathematical objects, which do not carry any ontological significance. However, as he correctly recognizes, this cannot be the attitude of scientific realists, who are interested in our abstract scientific accounts getting something right about the real world. Alternatively, Liu (2019) suggests that the alleged contradiction disappears if one adopts a form of contextual realism, whereby realist claims should be evaluated relative to anchoring assumptions formulated within the background theory: in particular, the claim that the number of molecules grows to infinity can be regarded as true in the context of a microscopic theory holding that condensed matter is continuous. However, the jury is still out as to whether these proposals effectively dissolve the paradox of phase transitions.

Instead, our preferred strategy to elude the paradox and thus salvage scientific realism goes along the lines of Butterfield's (2011) own dissolution of the mystery of singular limits, which is endorsed, at least in connection with classical phase transitions, by Menon and Callender (2013) as well as by Norton (2014) and Palacios (2019). It develops into two steps: first of all, one ought to make a careful choice of the physical quantities to work with, so as to avoid those quantities for which the infinite idealization appears essential in that the limit is singular; then, one employs the notion of approximation to show that for the selected quantities empirically correct results obtain on the way to the limit, without having to commit to the reality of the infinite system. Thus, the underlying idea of the proposed strategy is that of focusing on just the properties that are relevant for the behavior that we intend to explain, instead of requiring that all properties of the finite target system extend smoothly to the limit system. More to the point, Butterfield observes that the alleged mystery of singular limits is simply a consequence of looking at functions that do not give information regarding the behavior of real systems as N increases toward infinity. For instance, if one restricts one's attention only to the fact that some function f be discontinuous, like in the case of the derivative of the free energy phase, one would lose insight of what happens for very large but finite N, since the expected discontinuity obtains only at the limit " $N = \infty$." To the contrary, one ought to turn one's attention to physical quantities represented by different functions, so that the corresponding properties of the actual target system S_{N_0} are approximated by the properties of S_∞. In other words, even though it appears to be an essential idealization for some quantities, the limit system is an idealization yielding approximations for other properties that one regards as physically salient for the phenomenon to be explained. In this way the desired behavior is recovered on the way to the limit, and hence one can dispense from the infinite idealization, by demonstrating that this apparent "essential idealization" is a case of idealizations yielding approximations.

In order to make this proposal more concrete, let us look at a specific example of first-order phase transitions, that is a ferromagnet at sub-critical temperature. The Ising model portrays a ferromagnet as a chain of N spins, wherein a physical quantity called magnetization is represented as a function of the applied magnetic field. At very low temperatures, there are two possible phases available: a state where all spins are oriented in the up-direction, for which the magnetization takes on the value +1; and a state where all spins are oriented in the down-direction, for which the magnetization takes on the value −1. When the sub-critical temperature is reached, even if the applied magnetic field is null, one observes an abrupt flip from one phase to the other. That is formally captured by the magnetization function being discontinuous. Arguably, one can describe this phenomenon just in case one takes the thermodynamical limit whereby the number N of spins grows to infinity. Butterfield's proposed strategy can then be illustrated by means of a toy model. He defines a sequence of real-valued functions $\{g_1, g_2, ..., g_N\}$ with argument x belonging to the real numbers $\mathbb{R}$, which take on the following form:

$$g_N(x):=\begin{cases} -1 & \text{iff } x \leq -\dfrac{1}{N} \\[2mm] Nx & \text{iff } -\dfrac{1}{N} \leq x \leq \dfrac{1}{N} \\[2mm] +1 & \text{iff } x \geq \dfrac{1}{N} \end{cases}$$

All such functions are continuous, in that they remain equal to –1 until $x = -1/N$ and then they start to grow linearly with gradient N up to the value +1 for $x = 1/N$, after which they constant again. However, when one takes the limit for $N \to \infty$, the resulting function is no more continuous: in fact, the limit is given by

$$g_\infty(x)=\begin{cases} -1 & \text{iff } x < 0 \\ 0 & \text{iff } x = 0 \\ +1 & \text{iff } x > 0 \end{cases}$$

and hence it exhibits a discontinuity at point $x = 0$. Concretely, in the example of ferromagnetism, the behavior of the functions g_N's mimics the magnetization function for a chain of N spins, with the argument x representing the applied magnetic field. A singular limit arises if one introduces a binary function $f_N : \mathbb{N} \cup \{\infty\} \to \{0,1\}$ whose numerical values are determined by whether g_N is continuous or not, i.e.,

$$f_N := \begin{cases} 1 & \text{iff } g_N \text{ continuous} \\ 0 & \text{iff } g_N \text{ discontinuous} \end{cases}$$

Here, in accordance with the condition for essential idealizations stated in Section 14.3, when N goes to infinity one distinguishes between two possible values that are quite different: that is, (i) the value 1 taken on by all the functions in the sequence $\{f_1, f_2, ..., f_N\}$ for finite N, which results from the fact that all f_N are continuous; and (ii) the value 0 taken on by the function f_{S_∞} evaluated on the limit system S_∞, which results from the fact that g_∞ is discontinuous. So, if one focuses on the function f_N, the values for finite N's will always be continuous from the value of the quantity evaluated on the limit system, no matter how large N is. In particular, no real system with a finite number N_0 of molecules would have values of f_{N_0} that are approximated by the function f_{S_∞}. For Butterfield, that is just what makes the limit $N \to \infty$ seem mysterious.

However, the mystery can be explained away if one looks at the behavior of the functions g_Ns instead of the functions f_N's. In fact, as N grows, the functions g_N become more and more similar to the step function g_∞, even though, contrary to the latter, they remain continuous. Specifically, when N_0 is extremely large, for most points x one has $g_{N_0}(x) = g_\infty(x)$, which is true in particular when the argument is $x = 0$; furthermore, whenever the values of these functions are not strictly equal, namely around the singular point $x = 0$, they will still be close to each other, so that $g_{N_0}(x) \approx g_\infty(x)$. Therefore, the empirically correct values of the magnetization function can be recovered by the properties of the real system S_{N_0} without having to resort to the infinite system S_∞. In this way, the values obtained when taking the limit $N \to \infty$ just yield approximations of the relevant property of the real target system. More importantly, as a theoretical analysis also shows for realistic values of N, the gradient in the derivatives of the free energy is sufficiently steep that the difference in the limit values of the thermodynamic quantities as $N \to \infty$ and realistic systems with finite N_0 becomes negligibly small (Schmelzer and Ulbricht 1987; Fisher and Berker 1982).

A straightforward justification similar to the one obtained in cases of idealizations yielding approximations can now be given also for infinite limits that appear to give rise to essential idealizations. In fact, the use of the latter is justified based on the two desirable properties of mathematical convenience and empirical adequacy. Here, though, we would like to emphasize a point that Butterfield does not develop, namely the fact that the choice of a certain topology over the other determines different degrees of empirical adequacy. For this purpose, let us consider the function f_n indexed by the parameter n that maps the independent variable x onto the real numbers: accordingly, the notation $f_n(x)$ indicates the value that the function takes on for each x in the domain, where in concrete physical cases the variable x would represent the possible states of the system under investigation. A natural topology is induced by the following type of convergence:

Pointwise Convergence: The sequence $\{f_n(x)\}_n$ converges pointwise to $f_\infty(x)$ if, given any x in $\mathbb{R}$ and given any $\varepsilon > 0$, there exists a natural number $n_0(\varepsilon, x)$ such that $|f_n(x) - f_\infty(x)| < \varepsilon$ for every $n > n_0(\varepsilon, x)$.

Translated into our framework, the fact that the sequence of functions $\{f_n(x)\}_n$ converges pointwise to the function $f_\infty(x)$ means that there is a real finite system S_{n_0} for which the value of the relevant function is approximated by the value of the limit function for a given state x, that is $f_{n_0}(x) \approx f_\infty(x)$. But one may as well adopt a different type of convergence, that is:

Uniform Convergence: The sequence $\{f_n(x)\}$ converges uniformly to $f_\infty(x)$ if, given any $\varepsilon > 0$, there exists a natural number $n_0(\varepsilon)$ such that, for any x in $\mathbb{R}$, $|f_n(x) - f_\infty(x)| < \varepsilon$ for every $n > n_0(\varepsilon)$.

Since in this case n_0 depends only on ε, and not even on the variable x like in the case of pointwise convergence, uniform convergence proves stronger than the latter (in fact, it is strictly stronger in that one can show by means of counterexamples that pointwise convergence does not imply uniform convergence). Indeed, here one can choose a natural number n_0 for which the sought-after approximation $f_{n_0}(x) \approx f_\infty(x)$ holds for all the possible states x. To the contrary, pointwise convergence allows for exceptions, in the sense that once n_0 is fixed together with the margin of approximation determined by $\varepsilon > 0$ there may be some state x for which $|f_n(x) - f_\infty(x)| < \varepsilon$ does not hold for any $n > n_0$. The upshot of this analysis is that the standard hierarchy of convergence conditions for functions representing physical quantities entails different degrees of accuracy up to which empirical accuracy is satisfied. Hence, just as the precise notion of approximation rests on what one means by the expression "sufficiently close," Butterfield's straightforward justification of infinite limits ultimately depends on the topology under which one takes the limits. In fact, in the explanation of first order phase transitions the criterion of empirical adequacy is satisfied only in its weak degrees of accuracy: for, it is just when one chooses the topology induced by pointwise convergence, and not by the stronger uniform convergence, that a sequence of continuous functions approaches a discontinuous function in the limit.

This analysis shows that the property of empirical adequacy is sensitive to the choice of topology. Thus, the extent to which one satisfies empirical correctness, namely the condition that the value of a function representing a given physical quantity is approximately equal to the empirical data, depends on how the limit is taken. On this point it is worth making an important clarification: the issue whether the use of the limit is justified and the issue whether the use of the limit raises a threat to scientific realism, even though they have a common root, they should be kept separate. The common root is that the problem that, if the variable n represents a physical parameter in that, then when n becomes infinite the limit system would not be real. The former issue asks the question: idealizations arising in the limit in physical applications? The second issue, instead, arises because it appears as if one violates statement (I) of the Paradox of Infinite Limits, which is a basic desideratum for scientific realism. Butterfield's two suggested properties, i.e. empirical adequacy and mathematical tractability, helps one answer the first question. Arguably, the use of an infinite limit is justified only if it is empirically adequate, that is if it yields empirically correct results, at least approximately. Furthermore, if one can recover empirically correct results also without taking the infinite limit but the latter is mathematically more tractable, then one has a pragmatic reason to favor the use of the unrealistic limit. That seems a prima facie reasonable justification for one to use an infinite idealization for practical calculations even though it does not, strictly speaking, represent the finite target system.[12] But, the fact that for pragmatic purposes one is justified to use the limit in physical applications is independent from the issue whether one is a scientific realist or not. In fact, contrary to the issue of justification, this second issue bears on whether or not one ontologically commits to the existence of the limit system. When one wishes to give an explanation for some physical phenomenon P, if the only way to recover empirical correct results is by assuming the limit system, the resulting infinite idealization appears explanatorily indispensable, as prescribed by statement (II) in the Paradox of Infinite Limits. Scientific realism is then called into question if one accepts the EIA, namely statement (III) in the paradox, whereby the limit system is supposed to exist. Thus, it would seem that a scientific realist is not just justified to use the infinite limit, but also that she ought to be ontologically committed to it, in conflict with statement (I). Our strategy to resolve the problem, at least for the purported form of scientific realism that allows for approximate truth, is to reject statement (II) in the paradox, in that empirical correct results can be approximately obtained already on the way to the limit for the relevant physical quantities of interest.

14.5.2 Quantum Phase Transitions and Other Cases of Essential Idealizations

Let us move on to address the question whether the purported strategy to cope with the indispensability of infinite limits can be generalized to other cases where there appear essential idealizations. Butterfield (2011) conjectures that the solution of the Paradox of Phase Transitions proposed in the classical context holds generally. However, matters are less straightforward in other cases such as in the

---

[12] For a complete investigation of this issue one should actually discuss in much greater details than we can here the sense in which the limit system can be said to *represent* the target system. To this extent, Shech (2014) put forward a distinction between epistemologically and ontologically faithful representations. However, it goes beyond the scope of the present paper to survey the notion of representation.

description of phase transitions in quantum statistical mechanics. In fact, in this case the indispensability of the thermodynamical limit seems to present itself in a stronger form than in classical statistical mechanics. A rigorous description of quantum phase transitions can be given within the algebraic approach to physical theories. In this framework, one represents a physical system by means of an algebra of observables, that is the set of physical quantities representing its observable properties. The possible states that the system can occupy are then defined on such an algebra. In some cases of interest, in order to describe the relevant physical quantities, instead of working directly with the algebra one needs to refer to its representation into a concrete Hilbert space (the so-called GNS representation), which is induced by a chosen state. For instance, in the Ising model, both the phase in which all spins in the chain are oriented in the up-direction and the phase in which all spins in the chain are oriented in the down-direction give rise to distinct concrete representations. Now, the familiar issue concerning the indispensability of infinite idealizations arises since a quantum observable representing magnetization can be rigorously defined only if one takes the thermodynamical limit. However, such an observable is state-dependent, in the sense that it is constructed only within the representation of one phase or, alternatively, within the representation of the other phase. This fact marks a difference with respect to the description of phase transitions in classical statistical mechanics. Indeed, in the classical context the values taken on by the magnetization function depend on the particular state of the system, which is determined by the applied magnetic field; instead, in the quantum context it is the magnetization function itself, and not just its values, that depends on a particular state.

Furthermore, in the case of quantum phase transitions, there is an additional problem arising since the representation induced by the state in which all spins are induced by the state in which all spins are oriented in the down-direction are not unitarily equivalent. Without entering into too many technicalities, this means that, if one defines the magnetization observable with respect to one phase by taking the infinite limit $N \rightarrow \infty$, one cannot recover the values of magnetization relative to the other phase (and vice versa), contrary to what happens for finite N's when the up and down representations always remain unitarily equivalent. Arguably, the issue of unitary inequivalence of the phase representations thus complicates the mystery of quantum phase transitions in the example of the Ising model. It goes beyond the scope of the present paper to settle this entire problem. For our purposes here, it is sufficient to point out that two contrasting positions can be identified in the literature: on the one hand, there are authors such as Liu and Emch (2005) and Ruetsche (2011), who claim that taking the thermodynamical limit to account for quantum phase transitions is indispensable; on the other hand, there are authors, most notably Landsman (2013) and Fraser (2016), who take a deflationary view toward the limit. In this respect, the jury is still out as to whether or not phase transitions in quantum statistical mechanics constitute a case of essential idealizations where one can provide an explanation of the phenomenon without committing to the reality of the infinite system constructed in the thermodynamical limit.[13]

Be that as it may, there is a further point that is worth emphasizing regarding the general strategy to cope with other cases of essential idealizations, besides classical phase transitions. As pointed out by Palacios (2018), it does not suffice to demonstrate that approximately the same behavior that occurs in the limit, also occurs "on the way to the limit," but in addition we need to demonstrate that it arises for realistic values of the parameter n that goes to infinity. This latter condition is important because there are cases, such as the ergodic approach to equilibrium, in which the expected behavior arises for finite but unrealistic values of the parameter that goes to infinity. In more detail, in the ergodic theory of equilibrium, one takes the infinite-time limit $t \rightarrow \infty$ in order to assure that phase-averages and time-averages coincide. Even in simple examples like a small sample of diluted hydrogen, though, one can estimate that the desired behavior can occur for times t_0 that are finite but unimaginably longer than the age of universe. In such cases, even if we can actually demonstrate that the behavior arises for finite values of the parameter, we would not have succeeded in demonstrating the empirical adequacy of the theory.

To conclude, the analysis we have developed in the present section indicates that the challenge posed by essential idealizations to scientific realism ought to be resolved on a case-by-case basis, and that ultimately calls for empirical considerations. In general, the strategy to dispense with the infinite-limit system in the explanation of a certain phenomenon involves a suitable selection of the physical quantities to focus on, so as to yield approximations of the relevant properties of the target system already "on the way to the limit," where the degrees of inexactness that one may accept depends on the particular situation at stake and the choice of an adequate topology. Accordingly, as the example of classical phase transitions shows, one can assuage worries concerning scientific realism. In fact, if the idealization is not genuinely essential, one does not need to assume statement (II) of the Paradox of Infinite Limits. Moreover, the limit values of the relevant physical quantities can still give good, albeit inexact, descriptions of the properties of realistic systems, which are supposed to be finite in accordance with statement (I), and so in order to evade the paradox we do not even need to cast doubts upon the validity of the enhanced indispensability argument contained in statement (III). The Paradox of Infinite Limits can thus be avoided in spite of

---

[13] Another example of apparent essential idealizations are continuous phase transitions. We will address this case in Section 14.6.

apparent essential idealizations emerging in the infinite limit, while endorsing a form of scientific realism that allows for our best theories to be sufficiently close to the truth. There now remains to discuss the last type of use of infinite limits, namely as abstractions. As we will see, differently from the case studies we have investigated so far, sometimes in such cases the variable n growing to infinity does not represent a physical parameter, and hence in principle one may not satisfy statement (I) in the paradox.

14.6 Infinite Limits as Abstractions

For the sake of completeness, in this last section we address a different role played by mathematical limits that has been much less discussed in the philosophical literature than approximations and idealizations, namely the use of limits as *abstractions*. We mentioned earlier that the thermodynamical limit, in which the number of particles N as well as the volume V go to infinity, can be used to recover the quantities that successfully describe phase transitions in thermodynamics. Another important role of the thermodynamic limit is to enable the removal of irrelevant contributions such as "surface" and "edge" effects. More to the point, any finite lattice system will include contributions to the partition function coming from the edges and surfaces, which may be considerably different from those coming from the center of the sample. Taking the thermodynamical limit allows one to treat the system as a bulk, leaving out surface effects that are mostly irrelevant for the behavior. In this sense, the thermodynamical limit enables one to abstract away details that do not make a difference for the phenomenon under investigation (see also Mainwood 2006; Butterfield 2011; Jones 2006).

   A more interesting case of infinite limits used as abstractions is given in the context of continuous phase transitions. In contrast to first-order phase transitions that involve discontinuities in the derivatives of the free energy, in the case of continuous phase transitions there are no discontinuities but rather there are divergences in the response functions (e.g. specific heat, susceptibility for a magnet, compressibility for a fluid). An example of a continuous phase transition is the transition in magnetic materials from the phase featuring spontaneous magnetization—the ferromagnetic phase—to the phase where the spontaneous magnetization vanishes—the paramagnetic phase. Continuous phase transitions are also characterized by the divergence of a quantity called the correlation length ξ, which measures the distance over which the particles are correlated. A typical way of dealing with these long correlations is by introducing renormalization group methods, which are mathematical and conceptual tools that consist in defining a transformation that successively coarse-grains the effective degrees of freedom while keeping the partition function and the free energy (approximately) invariant. Palacios (2019) pointed out that such a process can be interpreted as assuming an infinite limit, in which the number of iterations n of the renormalization group transformation goes to infinity. An important aspect of the infinite limit for $n \to \infty$ is that, at each step of the sequence, irrelevant coupling constants are abstracted away, so as to retain only factors that are relevant for the behavior under description. That follows because, in this process, the partition function and the free energy that determine the behavior at a phase transition remain (approximately) invariant. After an infinite number of iterations, the sequence of systems may converge toward non-trivial fixed points, where the values of the coupling constants no longer change by applying the transformation. Note that the limit for the number n of iterations going to infinite does not represent any physical parameter of the system under description, but rather the number of transformations that one needs to apply by means of the renormalization group procedure.

The beauty of renormalization group methods is that linearizing around fixed points allows one to calculate the critical exponents of the power laws that determine the behavior close to the transition. Furthermore, it allows one to give an account of universality, namely the remarkable fact that physical systems as heterogeneous as fluids and magnets exhibit the same behavior near a phase transition (details in Goldenfeld 1992). There is good reason to consider the infinite iteration limit $n \to \infty$ as an abstraction. In fact, in Section 14.2, we defined abstraction as a process that consists in leaving some factors out without misrepresenting the properties of the original system. In fact, when applying a renormalization group transformation to the description of a given system, one does exactly this: that is, at each stage n, one leaves out irrelevant details while retaining those factors that make a difference for the phenomenon to be explained. To be sure, in doing that, one might have to use approximations; yet, one does not introduce any idealization. Moreover, it is also important to point out that the renormalization group process has the same epistemic role attributed to abstractions, namely that it enables us to give an explanation of the common behavior generated by systems that are diverse from each other at a fine-grained level, thereby accounting for universality.

In recent philosophical literature, the use of renormalization group methods has often been considered as an example of essential idealizations, e.g. Batterman (2011), Batterman (2017) and Morrison (2012). Indeed, Batterman says in a footnote:

It seems to me that if one is going to hold that the use of the infinite limits is a convenience, then one should be able to say how (even if inconveniently) one might go about finding a fixed point of the RG transformation without infinite iterations. I have not seen any sketch of how this is to be done. The point is that the fixed point, as just noted, determines the behavior of the flow in its neighborhood. If we want to explain the universal behavior of finite but large systems using the RG, then we need to find a fixed point and, to my knowledge, this requires an infinite system. (2017, p. 571)

And, later on in the paper, he argues

Kadanoff's understanding of the new RG theory of critical phenomena reflects a different conception of the role of asymptotics and infinities. The kind of upscaling that leads to an understanding of the universal macroscopic behavior of micro-diverse systems is different than the upscaling provided in the ensemble averaging of mean field theory. (p. 573)

However, we resist the conclusion that the appeal to renormalization group theory gives rise to an essential idealization. For the purpose of our argument, it is useful to distinguish between the roles played by the different limits involved in this approach, namely the thermodynamical limit entailing $N \to \infty$ for the number of molecules and the limit $n \to \infty$ for the number of iterations employed in the use of renormalization group theory. As is well known, in order to define a system with an infinite correlation length, one needs to take the thermodynamical limit. Granted, the latter may perhaps be interpreted as an idealization, in the sense that it gives rise to a fictional system with an infinite number of particles. Nonetheless, as Palacios (2019) pointed out, this limit, which appears essential to find the non-trivial fixed points that explain critical phenomena, can be also demoted to an approximation. In order to understand why, one should note that the two limits involved in the explanation of critical phenomena, i.e. $N \to \infty$ and $n \to \infty$, do not commute with each other: indeed, taking the limit $N \to \infty$ followed by the limit $n \to \infty$ yields empirically correct results, whereas taking the limit $n \to \infty$ before $N \to \infty$ yields results that are not empirically correct. What Palacios emphasized is that for finite systems, i.e. when $N < \infty$, one should *not* take the second infinite limit if one wants to obtain empirically correct results. In fact, she showed—based on the work done by Wilson and Kogut (1974)—that in finite systems one can find effective fixed point solutions that approximate the desired behavior for a large but finite number n of iterations, which means that one obtains empirically correct results "on the way" to the second limit $n \to \infty$ as well as "on the way" to the thermodynamic limit $N \to \infty$.

These results are relevant for our answer to the aforementioned question because they indicate that the Paradox of Infinite Limits can again be resolved by resorting to the notion of approximations, like in the examples discussed in the previous sections. Note, first of all, that the variable n corresponding to the number of iterations is not a physical parameter, hence in principle one may not violate statement (I) even if one takes the limit $n \to \infty$. However, since this limit does not commute with the thermodynamical limit, taking the former implies taking the latter limit too, and thus if n grows to infinity so does the number N of particles in the gas, thereby violating statement (I). Instead, the resolution to the paradox comes once again by rejecting statement (II), which says that the explanation of the phenomenon P can *only* be given by means of claims about an infinite system constructed in the limit $n \to \infty$. In fact, the procedure outlined earlier shows how the two limits $N \to \infty$ and $n \to \infty$ can be dispensed by means of the notion of approximation: accordingly, one can satisfy the condition of empirical correctness without committing to the limit system S_∞. In light of this analysis, we conclude that one has good reason to believe that renormalization group methods, at least in the example we considered, do not pose by themselves any challenge for a form of scientific realism allowing for approximate truth.

14.7 Conclusion

In this paper, we addressed the issue of whether or not the use of infinite limits in physics raises a problem of compatibility with scientific realism. For this purpose, we surveyed various physical examples where infinite limits are invoked in order to describe real target systems, so as to offer a taxonomy of the different uses of infinite limits (Section 14.3). In particular, we distinguished between approximations that do not constitute idealizations, idealizations yielding approximations, essential idealizations, and abstractions, which we then discussed in greater details in the subsequent sections. We argued in Section 14.5 that a challenge for scientific realism arises just when infinite limits are intended as essential idealizations, due to the fact that it appears as if empirically correct results can be recovered only in the limit. However, if one commits to the limit system in that it seems indispensable, one runs against the Paradox of Infinite Limits since the limit system is infinite whereas real systems are necessarily finite. We then went on to suggest how the ensuing worries for scientific realism can be assuaged in concrete examples, e.g. in the much debated case of classical phase transitions. The strategy to do so is to show, without referring to the infinite system, that the limit values of some physical quantities of interest yield approximations of the values obtained for realistic systems. That is of course compatible with a form of scientific realism that allows for theories to be just sufficiently close to the truth. In the same vein, as we explained in Section 14.4 and Section 14.6 by means of physical examples such as that of thermodynamically reversible processes and that of continuous phase transitions, understanding infinite limits as approximations that do not constitute idealizations or even as abstractions, respectively, does not raise any further issue for scientific realism.

Acknowledgments

The authors thank two anonymous referees for helpful and constructive comments. Giovanni Valente acknowledges financial support from the Italian Ministry of Education, Universities and Research (MIUR) through the grant n. 201743F9YE (PRIN 2017 project "From models to decisions").

References

Aronson, J. (1990). Verisimilitude and types of hierarchies. *Philosophical Topics 18*, 5–28.
Baker, A. (2005). Are there genuine mathematical explanations of physical phenomena? *Mind 114*(454), 223–238.
Baker, A. (2009). Mathematical explanation in science. *British Journal for the Philosophy of Science 60*(3), 611–633.
Baron, S. (2016). The explanatory dispensability of idealizations. *Synthese 193*(2), 365–386.
Baron, S. (2019). Infinite lies and explanatory ties: Idealization in phase transitions. *Synthese 196*, 1939–1961.
Batterman, R. W. (2001). *The devil in the details: Asymptotic reasoning in explanation, reduction, and emergence*. New York: Oxford University Press.
Batterman, R. W. (2005). Critical phenomena and breaking drops: Infinite Idealizations in physics. *Studies in History and Philosophy of Science Part B: Studies in History and Modern Physics 36*(2), 225–244.
Batterman, R. W. (2011). Emergence, singularities and symmetry breaking. *Foundations of Physics 41*(6), 1031–1050.
Batterman, R. W. (2017). Philosophical implications of Kadanoff's work on the renormalization group. *Journal of Statistical Physics 167*(3–4), 559–574.
Boyd, R. (1983). On the current status of the issue of scientific realism. *Erkenntnis 19*(1–3), 45–90.
Butterfield, J. (2011). Less is different: Emergence and reduction reconciled. *Foundations of Physics 41*(6), 1065–1135.
Callender, C. (2001). Taking thermodynamics too seriously. *Studies in History and Philosophy of Science Part B: Studies in History and Philosophy of Modern Physics 32*(4), 539–553.
Cartwright, N. (1994). *Nature's capacities and their measurement*. New York: Oxford University Press.
Chakravartty, A. (2010). Truth and Representation in Science: Two Inspirations from Art. In R. Frigg and M. Hunter (Eds.), *Beyond Mimesis and Convention: Representation in Art and Science*, pp. 33–50, Boston Studies in the Philosophy of Science. Dordrecht: Springer.

Chakravartty, A. (2017). Scientific Realism. In Edward N. Zalta (Ed.), *Stanford Encyclopedia of Philosophy* (Summer 2017 Edition): available in: https://plato.stanford.edu/archives/sum2017/entries/scientific-realism/.

Colyvan, M. (2001). *The indispensability of mathematics*. New York: Oxford University Press.

Earman, J. (2019). The role of idealizations in the Aharonov-Bohm effect. *Synthese 196*(5), 1991–2019.

Ehrenfest-Afanassjewa, T. (1956). *Die Grundlagen der Thermodynamik*. Leiden: E.J.Brill.

Fisher, M. E., and A. N. Berker (1982). Scaling for first-order phase transitions in thermo-dynamic and finite systems. *Physical Review B 26*(5), 2507.

Fletcher, S., P. Palacios, L. Ruetsche, and E. Shech (2019). Special issue: Infinite idealizations in science. *Synthese 196*(5), 1657–2019.

Fraser, J. (2016). Spontaneous symmetry breaking in finite systems. *Philosophy of Science 83*(4).

Godfrey-Smith, P. (2009). Abstractions, idealizations, and evolutionary biology. In *Mapping the future of biology*, pp. 47–56. Springer, Dordrecht.

Goldenfeld, N. (1992). *Lectures on phase transitions and the renormalization group*. Westview Press.

Jones, M. (2005). Idealization and abstraction: A framework. *Poznan Studies in the Philosophy of the Sciences and the Humanities 86*(1), 173–218.

Jones, N. (2006). *Ineliminable idealizations, phase transitions, and irreversibility*. Ph.D. thesis, Ohio State University.

Kadanoff, L. P. (2009). More is the same: Phase transitions and mean field theories. *Journal of Statistical Physics 137*(5–6), 777.

Kitcher, P. (1993). *The advancement of science: Science without legend, objectivity without illusions*. Oxford: Oxford University Press.

Landsman, N. P. (2013). Spontaneous symmetry breaking in quantum systems: Emergence or reduction? *Studies in History and Philosophy of Science Part B: Studies in History and Modern Physics 44*(4), 379–394.

Liu, C. (2019). Infinite idealization and contextual realism. *Synthese 196*, 1885–1918.

Liu, C. and G. Emch (2005). Explaining quantum spontaneous symmetry breaking. *Studies in History and Philosophy of Science Part B: Studies in History and Modern Physics 36*(1), 137–163.

Mainwood, P. (2006). *Phase transitions in finite systems*. Ph.D. thesis, University of Oxford.

McMullin, E. (1985). Galilean idealization. *Studies in History and Philosophy of Science Part A 16*(3), 247–273.

Menon, T. and C. Callender (2013). Turn and face the strange . . . Ch-ch-changes: Philosophical questions raised by phase transitions. In R. Batterman (Ed.), *The Oxford Handbook of Philosophy of Physics*, pp. 189–223. New York: Oxford University Press.

Morrison, M. (2012). Emergent physics and micro-ontology. *Philosophy of Science 79*(1), 141–166.

Norton, J. (2016). The impossible process: Thermodynamic reversibility. *Studies in History and Philosophy of Science Part B: Studies in History and Modern Physics 55*(1), 43–61.

Norton, J. D. (2012). Approximation and idealizations: Why the difference matters. *Philosophy of Science 79*(2), 207–232.

Norton, J. D. (2014). Infinite idealizations. In M. C. Galavotti, E. Nemeth, and F. Stadler (Eds.), *European Philosophy of Science—Philosophy of Science in Europe and the Viennese Heritage*, pp. 197–210. Dordrecht: Springer.

Oddie, G. (1986). *Likeness to truth*. Dordrecht: Reidel.

Palacios, P. (2018). Had we but world enough, and time . . . but we don't!: Justifying the thermodynamic and infinite-time limits in statistical mechanics. *Foundations of Physics 5*(48), 526–541.

Palacios, P. (2019). Phase transitions: A challenge for intertheoretic reduction? *Philosophy of Science 86*(4), 612–640.

Popper, K. (1972). *Conjectures and refutations: The Growth of Knowledge*. 4th ed. London: Routledge & Kegan Paul.

Psillos, S. (1999). *Scientific realism: How science tracks truth*. London: Routledge.

Putnam, H. (1975). *Mathematics, matter and method*. Cambridge: Cambridge University Press.

Rosen, G. (1994). What is constructive empiricism? *Philosophical Studies 74*(2), 146–178.

Ruetsche, L. (2011). *Interpreting quantum theories*. Oxford: Oxford University Press.

Schmelzer, J., and H. Ulbricht (1987). Thermodynamics of finite systems and the kinetics of first-order phase transitions. *Journal of Colloid and Interface Science 117*(2), 325–338.

Shech, E. (2013). What is the paradox of phase transitions. *Philosophy of Science 80*(5), 1170–1180.

Shech, E. (2014). Scientific misrepresentation and guides to ontology: The need for representational code and contents. *Synthese 192*, 3463–3485.

Shech, E. (2019). Philosophical issues concerning phase transitions and anyons: Emergence, reduction, and explanatory fictions. *Erkenntnis 84*, 585–615.

Shech, E. (2019). Infinitesimal idealization, easy road nominalism, and fractional quantum statistics. *Synthese 196*, 1963–1990 forthcoming.

Tichy, P. (1976). Verisimilitude redefined. *British Journal for the Philosophy of Science 27*(1), 25–42.

Valente, G. (2019). On the paradox of reversible processes in thermodynamics. *Synthese 196*(5), 1761–1781.

van Fraassen, B. C. (1980). *The scientific image*. Oxford: Oxford University Press.

Weisberg, M. (2007). Three kinds of idealization. *The Journal of Philosophy 104*(12), 639–659.

Wilson, K., and J. Kogut (1974). The renormalization group and the ε expansion. *Physics Reports 12*(2), 75–199.

Woodward, J. (2014). Scientific explanation. *Stanford Encyclopedia of Philosophy*, available in: https://plato.stanford.edu/entries/scientific-explanation/.

15

Realist Representations of Particles

The Standard Model, Top Down and Bottom Up

Anjan Chakravartty

Department of Philosophy
University of Miami
chakravartty@miami.edu

> It is the sense in which Tycho and Kepler do not observe the same thing which
> must be grasped if one is to understand disagreements within microphysics.
> Fundamental physics is primarily a search for intelligibility—it is a philosophy
> of matter. Only secondarily is it a search for objects and facts (though the two
> endeavors are as hand and glove). Microphysicists seek new modes of concep-
> tual organization. If that can be done the finding of new entities will follow.
>   Norwood Russell Hanson (1965/1958, pp. 18–19)

15.1 Fixing the Content of Realism: Reference and Description

Scientific realism is commonly understood as the idea that our best scientific
theories, read literally as descriptions of a mind-independent world, afford
knowledge of their subject matters independently of the question of whether
they are detectable with the unaided senses or, in some cases, detectable at all.
It is a staple of the field of history and philosophy of science to wonder whether
any such prescription for interpreting theories (and models and other scientific
representations; I will take this as read henceforth) is plausible given the history
of theory change in specific domains of the sciences. A lot of ink has been spilled
on the question of whether, or under what circumstances, a realist interpretation of theories is reasonable. Antirealists of various kinds have argued that given
the lessons of changing descriptions of targets of scientific interest over time,
adhering to realism is something of a fool's errand. Conversely, realists of various kinds—often referred to as *selective* realists—have sought to identify some

Anjan Chakravartty, *Realist Representations of Particles* In: *Contemporary Scientific Realism*.
Edited by: Timothy D. Lyons and Peter Vickers, Oxford University Press. © Oxford University Press 2021.
DOI: 10.1093/oso/9780190946814.003.0015

principled part of theories regarding which there has been continuity across theory change in the past, thus fostering the reasonableness of expectations of continuity in the future.

My focus in this essay is not the historical framing of these particular debates about realism per se, but rather a key feature of them that amounts to a more general problematic for realism. The shared strategy among selective realists for dealing with descriptive discontinuity across historical theory change has been, unsurprisingly, *selectivity* in what they take to be correct about a theory, proposed in discussions of how certain claims have had or do have greater epistemic warrant than others, such that continuity regarding these claims may then serve as a bulwark for realism even while discontinuity rules more generally. Hence the now familiar maneuver of associating realism with only certain parts of theories, such as those involved in making successful novel predictions, or concerning experimental entities or mathematical structures. In each case we find concomitant arguments about the typical preservation of the relevant parts of theories across theory change, both as a reading of history and as a promise for the future.

Here emerges the key feature of debates surrounding the shared strategy of selective realism on which I will focus. The hope of selective realism is that less is more. By associating realist commitments with less, the hope is that it will become easier to defend—indeed, that it will amount to a plausible epistemology of science. It is by no means easy, however, to know how much is enough. Consider, in connection with any given theory, an imagined spectrum of epistemic commitments one might make regarding its content. At one end of the spectrum one believes almost nothing; at the other end, one believes everything the theory states or suggests. Arguably, if realism is purchased at the cost of believing almost nothing, it is largely empty; if instead realism is made more substantial by licensing ever greater quantities of substantive belief, it runs an ever greater risk of (for example) falling prey to concerns arising from theory change. The realist, then, in any given case, must perform a kind of balancing act appropriate to that case. Let me label this challenge *the realist tightrope*. On one side, there is the temptation to affirm less and less, and on the other, the temptation to affirm more and more. Giving in to either of these temptations may spell disaster, but it is no easy feat to get the balance *just right*.

The potential benefit to realism of walking the tightrope is wide-ranging, in that it is relevant to both historical and ahistorical defenses of the position. As noted, if the realist were able to get the balance just right in some particular domain of science, she might then be in a position to furnish a narrative of continuity of warranted belief across theory change in that domain, past, present, and future. But the tightrope is something that must be walked not only in connection with historical lineages of theories, but also in connection with any given theory, for it is often a challenge to work out how any one theory should be interpreted in a realist way. As Jones (1991) notes, it can be rather unclear how best to articulate the subject matters of theories in physics (as per his examples), where different interpretations can amount to different explanatory frameworks, each suggesting a different ontology. This he presents as a challenge to realism, which "envisions mature science as populating the world with a clearly defined and described set of objects, properties, and processes, and progressing by steady refinement of the descriptions and consequent clarification of the referential taxonomy to a full blown correspondence with the natural order" (p. 186). In this characterization of realism we catch a glimpse of how the tightrope must be walked both synchronically and diachronically.

How shall we understand the spectrum of commitment, from thinner to more substantial? I take it that an understanding of this is implicit in most discussions of realism; indeed, it is implicit in Jones's characterization. The thinnest possible realist commitment is to the mere existence of something, which we capture by speaking of successful reference. We say that the term "ribonucleic acid" or "black hole" refers, which is (typically) shorthand for saying that it refers determinately to something in the world. From here, commitment becomes increasingly substantial as realists assert descriptions of the properties and relations of these things. A very substantial commitment may involve asserting all of the descriptions comprising or entailed by a theory, but it is often expressed otherwise, not by believing everything—often not in the cards in any case, since theories often contain known idealizations and approximations—but by asserting increasingly detailed descriptions of whatever the realist does, in fact, endorse. For example, an entity realist might describe the natures of the entities she endorses in terms of certain causally efficacious properties. A structural realist might describe the natures of these same entities in terms of certain structural relations. One can imagine yet further, finer-grained descriptions of the natures of these properties and structures. On the thin end of the spectrum we have bare reference, and on the other end, ever more comprehensive or detailed descriptions of the natures of the referents.

With this understanding in hand, a clearer picture of the realist tightrope emerges. Believing as little as possible and thereby asserting successful reference alone, which might seem a more defensible position than believing significantly more, might also seem to run the risk of rendering realism empty of much content.[1] The more one believes, however, the wider one opens the door to both the epistemic peril of believing things that may be weeded out as theories develop and improve, and the metaphysical peril of defending finer- and finer-grained descriptions of the subject matters of realist commitment, in virtue of

---

[1] Stanford 2015 argues that it would make realism into something that is, if not empty, so weak that antirealists need not dispute it. I will return to this contention in section 15.5.

the inevitably and increasingly abstruse concepts and objections to which metaphysical theorizing is prone. In some cases the epistemic and metaphysical perils come together: in these cases, succumbing to the temptation to articulate the natures of the referents of a theory in some fine-grained detail yields descriptions that may become outmoded by subsequent developments in the relevant science. An examination of precisely this sort of case, to which I will turn now, forms the backbone of what follows.

The Standard Model of particle physics, one of the landmark achievements of twentieth-century science, itemizes a taxonomy of subatomic particles along with their properties and interactions. Beyond mere reference to these entities, however, their nature has been subject to realist wonderment and debate throughout the history of theorizing in this domain. In ways that are well known and which I will consider momentarily, the particles of the Standard Model are radically unlike what could be imagined in classical physics—thus providing an example of how descriptions of the physical and metaphysical natures of something conceived in connection with earlier theorizing would have to be relinquished in light of subsequent theorizing. Just what the natures of these things enumerated by the Standard Model are, however, is still far from clear. Indeed, if realist attempts to characterize them are any guide, there is no consensus at all. Advocates of different forms of selective realism, for example, have characterized them in very different ways. And all the while, realists and antirealists alike have suggested that in the absence of some unique characterization, realism is untenable. After considering various possibilities for thinking about the identity and individuality of particles, for instance, van Fraassen (1991, p. 480) concludes that we should say "good-bye to metaphysics." Ladyman (1998, p. 420) holds that tolerating metaphysical ambiguity regarding these issues would amount to a merely "ersatz form of realism."

In earlier work I have offered judgments that might be construed as echoing these kinds of sentiments. "One cannot fully appreciate what it might mean to be a realist until one has a clear picture of what one is being invited to be a realist about" (Chakravartty 2007, p. 26). But having a clear picture is compatible, I submit, with a number of different and defensible understandings of how best to walk the realist tightrope in any given case. In the remainder of this essay I endeavor to explain why this is so, taking particles as a case study. In the next section I briefly substantiate the contention that many questions regarding the natures of particles are still very much up for grabs in contemporary physics and philosophy of physics with a synopsis of a handful of the conceptual conundrums surrounding them. In sections 15.3 and 15.4, I examine, respectively, what I describe as the two main approaches to thinking about the natures of particles and their properties, which have in turn shaped varieties of selective realism—"top-down" approaches, emphasizing formal, mathematical descriptions furnished by theory, and "bottom-up" approaches, emphasizing causal interactions and manipulations at the heart of experiment. In the final section, I argue that realism is a commitment that can be shared, defensibly, by those who subscribe to different and even conflicting conceptions of the natures of particles.

15.2 Searching for a Realist Interpretation of "Particles"

The electron is the veritable poster child of scientific realist commitment. Most proponents of both realism and antirealism (rightly or wrongly) take various facts about observable phenomena as uncontroversial, but seriously contest the status of the unobservable. Though theories at every "level" of description—from social and psychological phenomena to biological and chemical phenomena through to the subject matters of physics—all theorize about putatively unobservable objects, events, processes, and properties, realists often cite subatomic particles as a shining example of entities conducive to realism. On the one hand, given the mind-boggling success of the uses to which we have put twentieth-century theories concerning them, not least in a host of startlingly effective technologies from computing to telecommunications to medical imaging, this may not seem surprising. On the other hand, perhaps it should be a cause for concern after all, because even a cursory foray into our best attempts to grasp the natures of particles and particle behavior are fraught with conceptual difficulties, and our best scientific theories are far from transparent on this particular score.

Trouble rears its head at the start with the term "particle." What is a particle in this domain? There is, of course, a classical conception of what a particle is, which is easily graspable and conceptually undemanding, relatively speaking, but this conception is simply inapplicable at atomic and subatomic scales in light of twentieth-century developments in theory and experiment. Classical particles are solid entities that can be envisioned colliding with and recoiling from one another in the ways that billiard balls appear to behave, phenomenologically. The natures of particles described by the Standard Model are not in this way intelligible. They appear to behave in the manner of discrete entities in some contexts but like continuous, wave-like entities in others. It is unclear whether all of their properties are well defined at all times, though we can ostensibly detect them and measure their values under certain conditions. As intimated earlier, there is controversy as to whether they can be regarded in any compelling way as individuals; if they can, it is certainly not in the way we commonly think about identity conditions and individuation in classical contexts, in terms of differential property ascription and allowing for their re-identification over time.[2] The natures

---

[2] Arguably, one may overstate this last point. Cf. Saunders 2006, p. 61: "there are many classical objects (shadows, droplets of water, patches of colour) that likewise may not be identifiable over time."

of particles conceived today may become enmeshed over arbitrarily large spatial distances—as per quantum entanglement—in ways not previously conceived.

All of this suggests the strangeness of particles from a classical point of view, but this should not be inimical to realism all by itself—if one adopts a thinner conception of realism about particles framed in terms of reference. It is in the nature of theoretical development that sometimes the features of target systems of scientific interest that come to be viewed in a new light will appear strange from the point of view of what came before. This by itself suggests only a common and understandable propensity to regard the unfamiliar as strange. The challenge to realism here is no mere strangeness *en passant*, but the fact that in this case in particular, our attempts to interpret our theories in more substantial ways, so as to make their content intelligible to ourselves, have resulted in a great deal of unsettled debate and lasting conceptual puzzlement. Taking this as a basis for realism regarding a more substantial conception of particles, an antirealist might be forgiven for wondering whether an *argument* here against realism is in fact required, since it would appear that collectively, realists cannot themselves decide what they should believe.

Consider, for example, the question of the basic ontological category to which particles belong. There is a long history here of being flummoxed, even upon careful consideration of the relevant physics, regarding what this might be. The term particle is commonly associated with the notion of objecthood, but we have already noted that particles cannot be objects if "objecthood" is allowed to carry classical connotations. In quantum field theory, particles are often described as modes of excitation of a quantum field. This does not sound very object-like in any traditional sense, which leads many to claim that particles are not objects after all. Yet even physicists who observe that with the advent of quantum field theory the ontology of the quantum realm might be thought of in terms of fields rather than particles are happy to talk about particles at will. This suggests that they either regard particle-talk as merely elliptical for states of fields, or that they do in fact regard particles as non-classical objects of some sort, perhaps standing in some sort of dependence relation to fields. Generally, there is nothing like a precise specification of a basic ontology to be found, thus leaving the answer to the question of an appropriate assignment of category ambiguous. Here one might think that philosophers would lend a hand, but a sampling of the views of philosophers merely reveals the trading of ambiguity for transparent disagreements.[3]

---

[3] The sample to follow is by no means comprehensive, given the long history of differing interpretations, but representative of some of the most recent literature. Not everyone cited is a realist, necessarily, but all are attempting to clarify the relevant ontology.

For example, Jantzen (2011) considers the permutation invariance of particles: representations of a physical state of particles of the same type that interchange the particles are taken to represent the same state. This is fatal, he thinks, to a particle ontology, where particles of a type are discrete objects that share state-independent properties and have state-dependent monadic properties, conceived as "approximately independent" (p. 42) of the properties of other things. Is it obvious, however, that objects need have precisely these features? Necessary conditions for objecthood are themselves up for grabs. Thus, while Bain (2011) acknowledges the prevalent view that relativistic quantum field theories are not amenable to particle interpretations if this requires that particles be localizable and countable (as expressed in terms of local and unique total number operators, which these theories do not support; see Fraser 2008), he argues that since the theorems on which this conclusion is based do not hold for *non*-relativistic quantum field theories, the characterization of localizability and countability at issue here must depend on classical features of these latter theories (specifically, regarding the structure of absolute spacetimes) that do not apply to the former theories. This suggests that the characterization of objects assumed here is inapplicable to non-relativistic quantum field theories, which then leaves open the possibility of a different conception of particles (or localizability and countability) in this different theoretical context.

No doubt the possibility of retooling our concept of objecthood in such a way as to admit particles will not appeal to everyone—perhaps it is too much of a promissory note on which to rest a substantial realism. In that case, perhaps a field interpretation of "particle" ontology is the way to go. Like many others, however, Baker (2009) and Bigaj (2018) hold that while particle interpretations of quantum field theory are problematic, the same is true of field interpretations. Perhaps we could simply stop worrying about what ontological category particles inhabit and instead satisfy ourselves with a clear understanding of the nature of their properties. Alas, further difficulties await, for the precise natures of these properties are themselves elusive. Consider the property of spin, which is one of a handful whose values are viewed as necessary and jointly sufficient for classifying particles into their respective kinds. What is spin? It is very difficult to say. Spin is usually described as a "sort of" internal or intrinsic angular momentum; the scare quotes are essential to the description, because there is no analogue of this property in everyday experience or otherwise familiar terms that allows for a more "visualizable," physical, dynamical (despite the term "spin" connoting some sort of rotation) understanding of it. Spin is causally exploited in many technologies including microscopy (more on which in section 15.3), and the Standard Model gives us a mathematical framework for discussing it (more on which in section 15.4), but it is difficult to say *what it is*.

At the very least, in the absence of an intuitive grasp of the natures of properties of the sort just suggested, perhaps we could say something about them in terms that philosophers find perspicuous. Are these properties monadic, dyadic, or polyadic? Are they intrinsic or extrinsic or essentially relational? Mass is commonly cited as an exemplar of an intrinsic property, but Bauer (2011) thinks that it is extrinsic, since it is "grounded" in and thus ontologically dependent on the Higgs field. French and McKenzie (2012) contend that there are no fundamental intrinsic properties by means of an argument appealing to gauge theory,[4] which is integral to the Standard Model, but Livanios (2012) does not find this argument compelling. Lyre (2012, p. 170) maintains that properties such as mass, charge, and spin are "structurally derived intrinsic properties," which suggests something of a hybrid, intrinsic-extrinsic nature. And in some cases an examination of the natures of these properties brings us full circle, back to a consideration of the ontological category of things that best corresponds to particle-talk, as when Berghofer (2018) holds that the relevant properties are, in fact, intrinsic and non-relational, but features of fields, not particles, and when Muller (2015, p. 201) contends that we would be better off with a new conception of objects: particles, he argues, are "relationals"; "objects that can be discerned by means of relations only and not by properties."

The purpose of the preceding whirlwind tour has not been to suggest that progress cannot be made on questions surrounding the ontological natures of particles. No doubt some and perhaps many of the issues disputed in the preceding discussion may ultimately be resolved in ways that produce a measure of consensus. The point here is a different one. If, in order that realism be a tenable epistemic attitude to adopt in connection with the Standard Model, we were to require a degree of communally sanctioned, fine-grained clarity regarding description that could only follow from having resolved all such debates, this would suggest a prima facie challenge to the very possibility of realism here and now. As the brief glimpse into a number of contemporary debates just presented makes plain, any attempt to clarify the ontology of the Standard Model quickly and inevitably draws one into contentious metaphysical discussions. In the following two sections I will examine the two overarching approaches to prosecuting these debates that have formed the basis of selective realist pronouncements regarding the natures of particles and their properties, with the eventual goal of arguing for a rapprochement between them *qua* realism—one that clarifies how realism can be a tenable, shared epistemic commitment even in cases where realists disagree about details of description.

---

[4] Offering a different argument, McKenzie 2016 takes this position with respect to various properties including mass, but excluding spin and parity.

15.3 The Nature of Particles I: Top Down

As it happens, the two approaches to thinking about how best to interpret the Standard Model I have in mind reflect a longstanding division of labor within the community of physicists. On the one hand there is theoretical physics, which views particles through the lens of formal, mathematical descriptions furnished by theory, and on the other hand there is experimental physics, which views particles through the lens of the sorts of detections and manipulations of them that are part and parcel of laboratory practice. These communities of scientists are not, of course, strictly isolated from one another; they must often work together. Nevertheless, their approaches to the subject matter are of necessity shaped by the kinds of work they do. Corresponding to this rough division, in the philosophy of science, there are what I will refer to as "top-down" approaches to interpretation, which place primary emphasis on mathematical descriptions of the properties and interactions of particles found in high theory as a source of insight into the natures of particles in the world; on the flipside there are what I will call "bottom-up" approaches to interpreting the natures of particles, which place primary emphasis on their behaviors in the trenches of concrete interventions characteristic of experimental investigation. As we will see, both approaches offer insight into the natures of particles, and both leave important questions open.

Let us begin with the top-down approach to thinking about particles. The Standard Model provides a remarkably elegant description of fundamental particles, their properties, and their (electromagnetic, weak, and strong) interactions, neatly systematizing them by means of symmetry principles. A symmetry is a transformation (or a group of transformations) of an entity or a theory in which certain features of these things are unchanged. That is, the relevant features are preserved or remain invariant under the transformation. The kinds of transformations relevant to physical descriptions include translations in space or time or spacetime, reflections, rotations, boosts of certain quantities such as velocity, and gauge transformations. When the states of systems are related by symmetries, they have the same values of certain quantities or properties—including those used to classify particles, such as mass, charge, and spin. To take an everyday example, if one rotates a square by 90 degrees, one gets back a square. "Squareness" is invariant under this transformation, which maps the square onto itself. With the notion of a symmetry in hand we may define a symmetry *group* as a mathematical structure comprising the set of all transformations that leave an entity unchanged together with the operation of composition of transformations on this set (satisfying the conditions: associativity; having an identity element; every element having an inverse).

There are a variety of accounts of realism about particles that one might characterize as top down. What they have in common is the (explicit or implicit) operating principle that insight regarding the natures of particles should be intimately and exclusively connected to interpreting the mathematical formalism I have just described. This is all we need to understand the natures of particles, nothing more. There is nothing in my description of the top-down approach that suggests that it should provide *exclusive* insight into the natures of particles, but in practice, this is how realists who take this approach proceed. To take this extra step from adopting a top-down approach to thinking, furthermore, that this is our best or only legitimate source of insight into the natures of particles requires some further motivation or argument. Let me now briefly consider a couple of arguments of this sort, and for each suggest one of two things: either the top-down description of particles provided does not preclude supplementation with bottom-up description; or if it does, it is unclear why the top-down characterization should be judged superior *qua* realism. Obviously, this will not amount to a comprehensive survey of all possible arguments for an exclusive commitment to the top-down approach. Nonetheless, I take it to be suggestive of a plausible general moral, that the necessity or irresistible appeal of this commitment is unproven.

Motivating at least some realists who are exclusively committed to top-down characterizations of particles are desiderata such as descriptive or ontological intelligibility or simplicity. In section 15.4 we will see in some detail how the bottom-up approach is typified by an ontologically robust understanding of the causal or modal natures of the properties of particles, but for the time being it will suffice to note that some realists who focus their attention on symmetries hold that this focus alone is sufficient for understanding the natures of these properties, thus precluding any "inflation" of our ontological commitments in ways recommended by bottom-up realists about particles. If the Standard Model is simply interpreted as describing properties such as mass, charge, and spin as invariants of certain symmetry groups, we might rest content with this purely mathematical, theoretical apparatus for describing the natures of properties. On this view it is unnecessary to appeal to the causal roles of things in order to identify or understand them. Armed with symmetries, we might then understand the natures of the relevant properties without appealing to the notion of causal features or roles at all, thus articulating realism in terms of a simpler ontological picture.

Let us then consider whether the content of a realism about particles can be provided solely through an examination of symmetries and invariants. It is difficult to see how it could, given that questions about what realists justifiably believe cannot be separated from matters of how evidence furnishes justification. Detection and measurement are intimately connected to determining what things there are, in fact, in the world. Are properties thus conceived, as things one might identify as existing or being exemplified in the physical world—things about which one might be a scientific realist—identified independently of their causal roles? Perhaps there are cases in which this happens, but the present case does not seem like one. While there is no doubt that symmetry groups comprise a beautiful framework for codifying particles and their properties, all that examining them can achieve in isolation is to generate descriptions of *candidate* entities that may then be put to the test of experimental detection. It is one thing to describe the natures of some target of realist commitment in terms of the formal or mathematical aspects of a theory, but generally, in order for descriptions to have the sort of content required to support realism, they must be taken to *refer* to some thing or things in the world, and establishing successful reference requires more than the examination of a formalism. Some supplementation seems necessary.

A nice illustration of this is furnished by permutation symmetry, which arose earlier in section 15.2. Recall that in quantum theory, state representations of particles of the same type in which the particles are interchanged do not count as representing different states of affairs. An examination of the permutation group yields certain "irreducible representations" corresponding to all of the particles, the fermions and bosons, populating the Standard Model. However, in addition to these fermionic and bosonic representations, there are also so-called "paraparticle representations," and unlike fermions and bosons, paraparticles do not appear to exist—at least, not in the actual world subject to scientific realism.[5] Thus, merely examining the mathematical formalism of the theory is insufficient for the identification of entities to which realists should commit. To avoid being misled about the ontology of the world, there would seem to be no substitute for getting one's hands dirty with the causal roles of properties in the context of experimental work using detectors, and this suggests that there may be something to the thought that properties of particles have some sort of causal efficacy after all, in virtue of which they are amenable to detection, measurement, manipulation, and so on. But now we have entered the territory of the bottom-up approach to understanding the natures of particles, to which we will return in the following section.

Let us consider a second possible motivation for an exclusive reliance on the top-down approach for the purpose of illuminating particles. Perhaps the boldest motivation yet proposed stems from a version of selective realism that was designed specifically (in the first instance) to serve as an account befitting

---

[5] The closest we have come to generating empirical evidence in this sphere is the detection of paraparticle-like states, though not paraparticles themselves, under very special conditions. For a brief discussion, see Chakravartty 2019, p. 14.

fundamental physics: ontic structural realism. There are many variants of the view, but generically, the common thread is what one might call a reversal of the ontological priority traditionally associated with objects and properties relative to their relations. In much traditional metaphysics, objects and/or properties are conceived as having forms of existence whereby their relations are in some way derivative (and not vice versa). Some variants of ontic structural realism simply boost the ontological "weight" of the relevant relations relative to their relata such that they are all on a par, ontologically speaking. Others take the relations to have greater ontological priority, and the most revisionary formulations do away with objects and properties altogether, eliminating them in favor of structural relations which are then viewed as ontologically subsistent in their own right, thus constituting the concrete furniture of the world.[6] If particles are conceived as being entirely dependent on relations described by symmetries—or stronger yet, as epiphenomena of these relations—it may well seem that a top-down approach to describing their natures should be sufficient.

As is true regarding any proposal for realism there are several aspects of this view that one might seek to clarify, but perhaps the most fundamental concern that has been raised is whether ontic structural realism can render intelligible the idea that things described in purely mathematical terms—such as symmetries and invariants, which are standardly regarded as (at best) abstract entities—can be understood to constitute the world of the concrete. Merely stipulating that some mathematical structures are subsistent appears to achieve no more than to substitute the term "concrete" with "subsistent." Something more is needed, and no doubt with this in mind, advocates of the position sometimes explicate the sense of concreteness or subsistence at issue by saying that the relevant structures are *causal* or *modal*.[7] Esfeld (2009, p. 180), for example, is explicit that on his variant of ontic structural realism, "fundamental physical structures are causal structures." French (2014, p. 231) is clear that on his, "we should take laws and symmetries—and hence the structure of which these are features—as inherently, or primitively, modal," and take this *de re* modality as serving the explanatory functions commonly associated with attributions of causality, such as helping us to explain what it means for something to be concrete.

If one goes this route, however, the first of our potential rationales for favoring a top-down approach to interpreting the natures of particles based on the promise of a comparatively simple or streamlined ontology is ruined, because it is difficult to see how a reification of symmetries and other mathematical structures endowed with causal or modal efficacy should count as less

---

[6] For a detailed and comprehensive exploration of the many variants, see Ladyman 2014/2007.

[7] Cf. Ben-Menahem 2018, p. 14: "causal relations and constraints go beyond purely mathematical constraints; they are (at least part of) what we add to mathematics to get physics."

ontologically inflationary than an understanding of particles based on an onto-logically robust conception of causal or modal properties, as suggested on the bottom-up approach, to which we will turn next. There is no obvious reason to think that Occam's razor should point us toward the former and away from the latter. As a guide to a description of the natures of particles that might satisfy the aspiration to walk the realist tightrope by adding something substantial to an otherwise spartan commitment to successful reference, the top-down approach furnishes a great deal, but not in so compelling a manner as to make it an irre-sistible choice of interpretation, or an exclusive choice, for realists. One reason for this, as I will now suggest, is that the natures of particles look significantly different from the bottom up.

15.4 The Nature of Particles II: Bottom Up

From the point of view of detection, which is intimately linked to many of the strongest cases that can be made for realism in specific instances, more abstract descriptions of the properties of things are somewhat removed from the work of physics. Where the focus of experimental work is the physical discernment of interactions between particles and between particles and detectors, often requiring extraordinarily precise adjustments and manipulations of both the ex-perimental apparatus and the target entities under investigation, more precise descriptions of concrete natures are necessary. It is here that the determinate properties of particles, whose values are detected and manipulated in such work, take center stage. This is not to say that group theoretic structures are irrelevant to describing these properties, but simply that the descriptions afforded by symme-tries and invariants are at a remove from the specificities of experimental work. As Morganti (2013, p. 101) puts it, "[w]hen one focuses on invariants . . . one moves at a high level of abstractness." In contexts of experimentation and detec-tion, it is necessary to move in the direction of more determinate description; the specific values of mass, charge, etc. (pertaining to different particles) at issue in these contexts are not given by descriptions of symmetries (cf. Wolff 2012, p. 617).

In the realm of experiment it is what we can *do* that is our best guide to what there is and what these things are like—that is, to the ontology of our targets of investigation. What we can do in the arena of particle physics is entirely de-pendent on the precise values of the properties of particles. Since all of this doing involves designing and engineering instruments to interact with those parts of the world we aim to explore, and generating certain kinds of effects, it is natural to describe it in terms of causal interactions, relations, and processes. Thus it is no surprise that selective realists who take a bottom-up approach to understanding the natures of particles typically emphasize the roles of determinate property values in causal interactions, relations, and processes. Perhaps the most obvious (but not the only) example of a position taking this approach is entity realism, wherein certain descriptions of the causal roles of entities are interpreted as the basis of experimental work that ostensibly generates an argument for this form of realism. Some who take these descriptions seriously as filling in our understandings of the natures of particles go further, giving more detailed characterizations of the natures of their properties as being inherently causal or modal, invoking conceptions of properties such as dispositions, propensities, and capacities—types of properties whose natures comprise abilities to do certain things.

Having seen just a moment ago how the top-down approach lends itself to increasingly detailed descriptions of the natures of particles, it may now be clear (on the basis of the preceding paragraph) that the same is true here. It is all too easy to drift from an initial question regarding the nature of some target of one's realism to further, deeper questions, and in attempting to answer these deeper questions, giving ever more detailed descriptions—all the while with little concern for the realist tightrope. Just as one might, from the top down, begin by thinking that the descriptive content of one's realism should be informed by a specification of the relevant symmetry groups, but then end up some way down the road, after twists and turns of elaboration, advocating for reified mathematical structures imbued with primitive causality or modality, one may perform analogous feats from the bottom up. One might begin by thinking that the causal roles of certain properties associated with particles are central to the descriptive content of one's realism, and then through earnest inquiry find oneself somewhere down the road defending one or another specific conception of causation, in just the way that some come to understand the natures of these properties as inherently modal or dispositional. Mirroring the moral of section 15.3, let me now suggest that bottom-up characterizations do not preclude supplementation with top-down description. And in some cases, it is unclear why either should be judged superior *qua* realism.

With a long history of empiricist concerns about the intelligibility of dispositional properties in the background, fueled by concerns about the metaphysical excesses of scholastic and neo-Aristotelian philosophy more generally, the notion that properties of particles should be understood dispositionally is unsurprisingly controversial. A dispositional *essentialist*, for example, takes the natures of the relevant properties to be exhausted by dispositions for certain kinds of behavior: the identities of these properties—their essences—are dispositional. Could we not simply deflate this scholastic-sounding reference to essences? Consider Livanios's (2010, p. 301) query: "if the identity of the fundamental physical properties . . . can be provided via symmetry considerations, why can't we claim that being invariant under the action of fundamental symmetries is an *essential* feature of the fundamental physical properties?" From the bottom up there is an immediate reply: what is the "action" of the symmetries? Presumably the intention here is not to claim that a mathematical *description*—a linguistic entity—is part of the essence of something in the world. This would be to conflate descriptions with that which they describe. Thus, the point must be that the symmetries are themselves things in the world. They are part of the ontology of the world and thus conceived, they are part of the essences of fundamental properties. But now the view is sounding a lot like dispositional essentialism, and certainly no less weighty as metaphysics!

Two points can be extrapolated from this brief illustration of how opposing metaphysical sensibilities sometimes play out in realist interpretations of scientific theories and models. In replying to the attempt to deflate the substantial metaphysical claim inherent in their proposal for how to understand the properties of particles, dispositional essentialists need not reject the top-down approach *simpliciter*, or broadly conceived, for as we have already acknowledged, there is nothing in group theoretic descriptions of symmetries and invariants that is incompatible with their view—on the contrary. Rather, it is simply the case that their entirely reasonable, bottom-up preoccupations regarding realism have not been well appreciated by their critics, which we noted earlier in terms of the relatively general or abstract knowledge afforded by symmetry groups and, in contrast, the utter centrality of the determinate values of properties in setting up, generating, detecting, and recording the effects of causal interactions and processes. A second point worth noting, though I will not detail it here in connection with this particular example, is how elaborate metaphysical proposals may become on *any* approach to explicating realism. Under the guise of empiricist or neo-Humean rejections of metaphysical excess, some pots call kettles black. Both pots and kettles, however, threaten to topple the realist off of her tightrope.

What, then, of the determinate property values of particles and their associated causal profiles? Perhaps the most serious concern about emphasizing these aspects of particles for the purpose of describing their natures is the worry that no such account can adequately explain certain constraints we find exhibited in their behaviors. For example, in any closed system, the values of properties such as mass-energy, momentum, charge, and spin are conserved—their totals remain constant. Emmy Noether proved in 1915 that for every continuous global symmetry of the Lagrangian there is a conserved quantity (and vice versa). But how might the causal efficacy of a particle, understood by means of an account of the causal profiles associated with its properties, explain the conservation of properties in an *ensemble* of particles (cf. Bird 2007, p. 213)? It is difficult to see how the properties of a particle can be parlayed into a constraint on a *collection* of particles, given that the relevant constraint must pertain to the collection, not to any given particle. Similarly, consider principles of so-called least action: for any given system and a specification of some initial and final conditions, the evolution of the state of the system will minimize a quantity referred to as "action." It is difficult to see how the causal profiles associated with a particle and its properties could somehow generate the minimization of action in systems more generally.

In order to explain constraints on behavior such as those expressed in principles of conservation and least action in terms of the causal natures or profiles of properties, it would seem we must think of these properties as belonging to the systems to which these principles apply, and this is inevitably controversial. Taking the dispositional variant of the bottom-up approach as an illustration once again, Harré (1986, p. 295) maintains that some dispositions may be grounded in "properties of the universe itself"—a phenomenon he labels "ultragrounding"—attributing the idea to Mach's discussion of inertia in the context of Newtonian thought experiments. Imagine two globes connected by a spring balance and rotating, alone in the universe. Newtonians held that there would be a force tending to separate the globes, registered in the spring balance, but on a Machian reading there is no reason to believe that the globes would have inertia in an otherwise empty universe; it is better to think of the disposition to resist acceleration as grounded in the universe itself. Bigelow, Ellis, and Lierse (1992, pp. 384–385; cf. Ellis 2005) go so far as to contend that the actual world is a member of a natural kind whose essence includes various symmetry principles, conservation laws, and so on. It is a short step from this to thinking that the system-level behaviors associated with these principles are properties of the world—a very large system indeed.

The standard objection to this family of speculations is that it is ad hoc.[8] Granted, explaining constraints on the behaviors of systems of particles in terms of properties of the entire world may seem, prima facie, rather convenient, especially in the absence of any independent motivation for the *explanans*. If one takes a bottom-up approach to understanding the natures of target systems of scientific interest, however, it is simply a mistake to suggest that there is no independent motivation. Just as particles are investigated empirically in carefully designed and executed experiments, systems of particles are likewise investigated. It is an empirical fact, not a convenient fact, that certain kinds of systems exhibit behaviors that conform to various principles of conservation and least action. Having adopted a methodology of associating causal profiles with certain types of particles and their characteristic properties, it is hardly an unmotivated extension of this methodology to do likewise in connection with certain types of systems, such as closed systems.

---

[8] For recent discussion on both sides of this fence, see Smart & Thébault 2015 and Livanios 2018.

Now, as it happens, the world itself is a system of this type. From the perspective of the bottom up, the attribution of a causal profile to it on the basis of a consideration of empirical investigations into members of the type cannot be said to be based on merely wishful speculation. And neither should it seem peculiar in the era of quantum theory, in which systems are routinely viewed as having properties, such as entanglement, that cannot be reduced to the properties of their parts. What may appear superficially as ad hoc speculation inevitably sounds more credible when the metaphysical terminology in terms of which it is sometimes expressed ("natural kinds," "essential properties") is given a plausible interpretation in the language of scientific description. Consider, for example, the possibility entertained earlier that particles are in fact best understood as field quanta. In that case the properties of systems suggested above would be properties of fields, and fields are global in the sense that they permeate the whole of the world. Anyone moved by *this* description would then be in a position to ask yet further questions about the natures of particles and their properties, depending on whether one is a substantivalist about fields, or interprets the values of field quantities as properties of spacetime points, or . . . . But let us stop here.

Having refined and extended a bottom-up description of particles in such a way as to answer a preeminent concern, let us once again inquire into how this approach fares in comparison to its counterpart, top down. Here, once again, it is very difficult to make a case one way or the other on the basis of some imagined criteria of descriptive or ontological intelligibility or simplicity. On a permissive enough conception of causation one may see it as an appropriate descriptor of many different things. Ben-Menahem (2018), for instance, applies the label "causal" to any general constraint on change, where constraints determine what may happen, or is likely to happen, or what cannot happen. On such a conception, symmetries, conservations laws, and variational principles (such as the principle of least action) all qualify as causal. But are they causal in the sense advocated by someone looking top down, as a primitive feature of certain mathematical structures, or are they causal in the sense of, say, a dispositionalist looking bottom up, where the (potential for) behaviors associated with the properties of various kinds of entities and systems determine their identities? And does anything hang on this choice, from the point of view of defending realism? In closing, let me attempt to shed some light on the latter question.

15.5 The Content of Realism
Redux: Anchoring Interpretation

I began this essay by citing a celebrated challenge to scientific realism. Given that no one thinks that most scientific theories and models (including many of our very best ones) are entirely correct, not merely in connection with the idealizations and approximations we know of, but also in other ways we have yet to discover, everyone appreciates that they will evolve over time as scientific inquiry proceeds. Hence the various strategies found among realists, especially selective realists, for identifying those aspects of theories and models that have sufficient warrant to command realist commitment, both as a guide to interpretation in the present and to reasonable expectations about what will survive into the future. From this I distilled a more specific challenge to anyone hoping to be successfully selective, which I called the realist tightrope: believing too little or too much of a theory that is not entirely correct may well appear to spell trouble. Believing too little—in the limit, the bare reference of central terms, or claims regarding the existence of their referents—may seem tenuous, but the more comprehensive or detailed the descriptions of such things one endorses, the more one runs the risk of falling foul of future developments in the relevant science, and/or metaphysical objections to the increasingly fine-grained natures proposed. Where does the proper balance lie?

In the case of the Standard Model, walking the tightrope seems especially fraught. Understanding the natures of the particles described by the theory has always been difficult, and serious proposals for illuminating these natures have inevitably required increasingly speculative and technical theorizing. Given this state of affairs, it is hardly surprising that there is so much disagreement among realists about how best to describe what particles are, exactly. On the surface this may appear a victory for antirealism, for if there is nothing determinate here to be found under the heading of "realism" to which all realists subscribe, but instead a thousand splintered commitments to conflicting and (what will appear to some as) increasingly esoteric interpretations, the camp of realism may look more like a ball of confusion than anything endorsing a shared epistemic commitment. This dismal portrait of the cognitive landscape of realism in connection with the Standard Model is, however, though perhaps understandable on the basis of what we have seen, entirely misleading. I submit that there is in fact something substantial to which all or most realists subscribe in the context of the Standard Model, even as they debate how this commitment is best elaborated. Here in conclusion I will attempt to explain how this can be so.

In order to understand how different and conflicting descriptive commitments among realists regarding particles are compatible with a shared commitment *qua* realism, we could do worse than to begin by looking at how scientists approach this area of physics. Here, just as Jones (1991, p. 191) notes in connection with the analogous case of multiple candidate interpretations of quantum mechanics, one may reasonably worry about "the failure of any interpretation to provide an 'explanatorily satisfactory' link between the mathematical formalism and the world of laboratory experience." He continues:

The general approach of one interpretation may suit a physicist more than the general approach of others, and he or she may spend some time adapting it to issues that he or she thinks particularly important and developing arguments as to why its lacunae are not devastating for its coherence. But every physicist will admit that such allegiance is to some degree a matter of taste. No physicist is unaware of competing interpretations, and none expects decisive evidence or arguments for one against the others.

Analogously, in the case of the Standard Model, the challenges of connecting the domain of abstract theorizing, conceived in terms of interpreting a mathematical formalism, and that of concrete experimentation, conceived in terms of interpreting laboratory experience, have a basis in the work of physics, all of which is mirrored in philosophers' attempts to elaborate the natures of the phenomena revealed by these practices. And as I will now contend, just as physicists across these domains can be realists despite differences in how they characterize their shared subject matter, philosophers of science can too.

To begin, note that terms like "physics" and even "fundamental physics" are rather broad designators. This is true not only in the sense that there are a variety of subareas of physics to which these terms are applied, but also in the sense that even within subareas, different approaches to one and the same subject matter can and often do take the form of highly disparate forms of scientific practice. Galison's (1997) detailed study of what he describes as the partly autonomous subcultures of physics in the twentieth century—experimenting; theorizing; and instrument making—furnishes a helpful and meticulous illustration. These subcultures, he contends, are "intercalated" in that they constrain, guide, and inspire one another, but they also develop and function significantly independently of one another and are thus identifiable as separate subareas of research and practice, with separate conferences, journals, and so on.

Most importantly for present purposes, the significant autonomy associated with these different subareas generates significantly different understandings of the subject matter. This is the source of Galison's provocative adaptation of the anthropological notion of a "trading zone": "an intermediate domain in which procedures could be coordinated locally even where broader meanings clashed" (p. 46). Subcultures of physics do not associate precisely the same meanings with the technical terms used in communication with one another: "Theorists and experimenters, for example, can hammer out an agreement that a particular track configuration found on a nuclear emulsion should be identified with an electron and yet hold irreconcilable views about the properties of the electron, or about philosophical interpretations of quantum field theory" (p. 46); when working together, they set aside "the 'deep' and global ontological problems of what an electron 'really' is" (p. 48). The upshot of a careful consideration of different approaches to the physics of particles is thus clearly and immediately consequential for philosophers interested in questions of scientific knowledge: "the significance of these partially separate lives is that—once one abandons 'observation' or 'theory' as the basis for a univocal account—no single narrative line can capture the physics of the twentieth century, even within a single specialty" (p. 9).[9]

What's good for the goose of particle physics, however, is good for the gander of philosophy of particle physics. Indeed, the various projects of interpretation of the natures of particles and their properties displayed in previous sections have demonstrated just this. Reflecting the different conceptions of particles adopted by physicists who approach them from the different vantages of mathematical theorizing and experimental detection, philosophers often view the natures and properties of particles in different ways, typically depending on the scientific practices on which they are most focused or with which they are most concerned. None of this all by itself is an argument for or against realism, but it does shed crucial light on the question of what it means to be a realist in this domain, if one is that way inclined. Just as scientists in different subareas of physics may believe in electrons—sharing an ontological commitment, but *under different descriptions* (more precisely: partially different and overlapping descriptions)— so too may philosophers of science. Realism, in the limit, is a commitment to the existence of something, to the idea that through theoretical descriptions and/or experimental detections, ideally both, we have picked out what Einstein, Podolsky, and Rosen (1935) described so evocatively as an "element of reality." Triangulating, using our best tools of mathematical and causal investigation, we have managed to pick something out in the world.

Thus we see how the challenge posed by the realist tightrope, with which we began, is misleading. Realism about x does not face mortal danger on either side by believing too much (believing increasingly refined descriptions of x) and believing too little (simply believing in the existence of x). A supplemental metaphor is needed. From a realist perspective, successful reference is, in fact, all that is required to *anchor* realism, and it is a shared judgment that such anchoring has been achieved that unifies different sorts of realists about any given x. This is compatible, of course, with further description rendering realist commitment more substantial, with all the risk and reward this entails. To the extent that further descriptions of the precise natures of things like particles are believed, the anchor of reference is compatible with there being different species of realist commitment (e.g., selective realisms), unified *qua* realism more broadly (as a

---

[9] Galison 1997, pp. 833–835, tells the story of how Sidney Drell and James Bjorken aspired to write a book on quantum field theory in the early 1960s, but ended up producing two separate volumes, one geared to experimentalists (concerned more with measurable quantities) and the other to theorists (concerned more with formal properties of theories, such as symmetries and invariances). Some of the differences between the volumes amounted to a "radical difference in the ontology" (p. 835).

genus) by a commitment to shared reference. It is also compatible with combining a high degree of confidence in our having picked something out in the world with lesser degrees of confidence in some or all of the descriptions of the nature of this thing elaborated in finer-grained ways by different versions of realism and in the metaphysics of science. As intimated earlier, talk of "particles" is loose—objects of some kind?; events?; some sort of hybrid?—and likewise, groups of cohering causal properties?; emergent or derivative features of an ontologically subsistent structure? Reference is the anchor.

Admittedly, the notion of anchoring is not by itself so comprehensive as to yield determinate answers to further questions that realists are often pressed to confront. Is a causal theory of reference best for anchoring? If so, the commitments shared by different sorts of realists may sometimes prove maximally thin, though depending on the strength of the evidence they may prove epistemically significant nonetheless. In many cases, as in the present case, a causal-descriptive or minimal descriptive theory (appealing to a shared subset of descriptions) may be appropriate, since physicists and philosophers alike generally agree on a number of features of particles, their properties, and interactions, with differences of interpretation emerging only in their finer-grained proposals for how best to understand the natures of these things. Should different species of realists, imbued in different ways by top-down and bottom-up approaches to particles, hold lower degrees of belief in their finer-grained interpretations than in the coarser descriptions they jointly affirm with others? If degrees of belief in finer-grained proposals are sufficiently low, this may suggest the wisdom of a pragmatic pluralism of accounts; if they are sufficiently high, this may suggest an agreement to disagree between different camps. Clearly, there is plenty of work here left to do in grappling with these issues.

All of this said, it is nevertheless the case that while the intuitive pull of the realist tightrope can be strong, it is properly resisted. Feeling the pull, Stanford (2015) contends that claims about what exists or about which terms successfully refer are not at issue in debates about realism; instead, antirealist arguments should be construed as targeting only scientific descriptions of the "fundamental constitution and operation of various parts of the natural world." One may naturally wonder here about the relevant sense of "fundamental." Is the intention to target some special part of the spectrum of increasingly refined descriptions of some focus of scientific investigation offered by some realists? This would be puzzling: there is no obvious point at which these descriptions become "fundamental" and, in any case, different species of realism disagree about much description while still belonging to the genus. Perhaps instead, "fundamental" is being used in the way familiar to us from accounts of ontological or explanatory reduction, in which some entities or phenomena are arguably "reducible" to other, more fundamental ones. But this is likewise unpromising, even granting the premise of reductionism, in the absence of some convincing argument to the effect that less fundamental things should not be considered real. If it turns out that superstring theory is true, then it will turn out that particles are modes of vibration of strings, but it is at best unclear how this would make them any less real.

While many realists disagree about the natures of particles and go to great lengths to explain, in conflicting ways, how such talk should be interpreted, they are no less realist *about particles*. This suggests that realism *simpliciter* is something to which one may subscribe along a spectrum of descriptions, from the minimal, as in the case of assertions of reference, to the most refined views of metaphysical natures.[10] Indeed, Stanford (2015, pp. 410–411) acknowledges that in some cases the sheer weight of theoretical and experimental evidence for the existence of something (e.g., atoms) is so great that it is implausible to imagine that future scientists will change their minds. Given that they may change their minds about certain fundamental descriptions, however, and that it is dubious that we are capable of predicting what subset of our current descriptions will be retained in future, realism thus conceived would be "so weak that . . . no historicist opponent will think it worthwhile to contend against it" (p. 416). It is all too easy, though, to place this shoe on the other foot. If the evidence is sufficiently strong as to indicate that we have successfully picked something out in the world, this is music to realist ears—and a justified expectation that this will be preserved across theory change suggests that some significant portion of the theoretical and experimental knowledge justifying this expectation will be preserved as well, furnishing a basis for even more substantial conceptions of realism.

Let us take some final inspiration from those engaged in theorizing and experimenting. Late in his life the great theoretician Werner Heisenberg (1998/1976) betrayed a striking ambivalence between top-down and bottom-up approaches to characterizing the nature of matter: "The question, What is an elementary particle? must find its answer primarily in experiment"; "theory . . . cannot add much to this answer" (p. 211)—but later he could not resist adding, "The particles of modern physics are representations of symmetry groups and to that extent they resemble the symmetrical bodies of Plato's philosophy" (p. 219). To return to earth once more from Plato's heaven, nothing smooths the way better than speaking to an experimentalist. Randal Ruchti is part of the High Energy Physics Group at the University of Notre Dame which participates in experiments at the Large Hadron Collider at CERN, for which they developed a hand-held detector that can be placed in high energy particle beams to yield visual representations of particle interactions. The discourse of

[10] Cf. Magnus 2012, p. 122: "Retail arguments [i.e., arguments stemming from evidence specific to the case at hand] for believing in particular things can give us good reasons to believe that those things exist on the basis of their connections to other things, while leaving questions of things' fundamental nature either unmentioned or unresolved."

experimental particle physics is so rife with collisions, scattering, and detections of "packets" of energy and momentum that realism about particles is a natural default, but Ruchti is quick to add: "don't ask me what they are!"

Acknowledgments

I would like to thank a number of colleagues for thoughts on an earlier version of this paper, including Jonathan Bain, Steven French, Vassilis Livanios, Matteo Morganti, Fred Muller, Kyle Stanford, and Peter Vickers, and audiences at the Universities of Barcelona, Bergen, Indiana-Purdue (Indianapolis), Lund, Miami, Pittsburgh, and Roma Tre, and at the CSHPS annual conference, for stimulating comments on various parts of this project.

References

Bain, J. 2011: "Quantum Field Theory in Classical Spacetimes and Particles," *Studies in History and Philosophy of Modern Physics* 42: 98–106.

Baker, D. J. 2009: "Against Field Interpretations of Quantum Field Theory," *British Journal for the Philosophy of Science* 60: 585–609.

Bauer, W. A. 2011: "An Argument for the Extrinsic Grounding of Mass," *Erkenntnis* 74: 81–99.

Ben-Menahem, Y. 2018: *Causation in Science.* Princeton: Princeton University Press.

Berghofer, P. 2018: "Ontic Structural Realism and Quantum Field Theory: Are There Intrinsic Properties at the Most Fundamental Level of Reality?" *Studies in History and Philosophy of Modern Physics* 62: 176–188.

Bigaj, T. 2018: "Are Field Quanta Real Objects? Some Remarks on the Ontology of Quantum Field Theory," *Studies in History and Philosophy of Modern Physics* 62: 145–157.

Bigelow, J., B. Ellis, & C. Lierse 1992: "The World as One of a Kind: Natural Necessity and Laws of Nature," *British Journal for the Philosophy of Science* 43: 371–388.

Bird, A. 2007: *Nature's Metaphysics: Laws and Properties.* Oxford: Clarendon.

Chakravartty, A. 2007: *A Metaphysics for Scientific Realism: Knowing the Unobservable.* Cambridge: Cambridge University Press.

Chakravartty, A. 2017: *Scientific Ontology: Integrating Naturalized Metaphysics and Voluntarist Epistemology.* New York: Oxford University Press.

Chakravartty, A. 2019: "Physics, Metaphysics, Dispositions, and Symmetries—à la French," *Studies in History and Philosophy of Science* 74: 10–15.

Einstein, A., B. Podolsky, & N. Rosen 1935: "Can Quantum-Mechanical Description of Physical Reality Be Considered Complete?" *Physical Review* 47: 777–780.

Ellis, B. 2005: "Katzav on the Limitations of Essentialism," *Analysis* 65: 90–92.

Esfeld, M. 2009: "The Modal Nature of Structures in Ontic Structural Realism," *International Studies in the Philosophy of Science* 23: 179–194.

Fraser, D. 2008: "The Fate of 'Particles' in Quantum Field Theories with Interactions," *Studies in History and Philosophy of Modern Physics* 39: 841–859.

French, S. 2014: *The Structure of the World: Metaphysics and Representation*. Oxford: Oxford University Press.

French, S., & K. McKenzie 2012: "Thinking Outside the (Tool)Box: Towards a More Productive Engagement Between Metaphysics and Philosophy of Physics," *European Journal of Analytic Philosophy* 8: 43–60.

Galison, P. 1997: *Image and Logic: A Material Culture of Microphysics*. Chicago: University of Chicago Press.

Hanson, N. R. 1965/1958: *Patterns of Discovery: An Inquiry into the Conceptual Foundations of Science*. Cambridge: Cambridge University Press.

Harré, R. 1986: *Varieties of Realism: A Rationale for the Natural Sciences*. Oxford: Blackwell.

Heisenberg, W. 1998/1976: "The Nature of Elementary Particles," *Physics Today* (1976) 29: 32–39. Reprinted in E. Castellani (ed., 1998), *Interpreting Bodies: Classical and Quantum Objects in Modern Physics*, pp. 211–222. Princeton: Princeton University Press.

Jantzen, B. 2011: "An Awkward Symmetry: The Tension between Particle Ontologies and Permutation Invariance," *Philosophy of Science* 78: 39–59.

Jones, R. 1991: "Realism About What?" *Philosophy of Science* 58: 185–202.

Ladyman, J. 1998: "What is Structural Realism?" *Studies in History and Philosophy of Science* 29: 409–424.

Ladyman, J. 2014/ 2007: "Structural Realism," in E. N. Zalta (ed.), *The Stanford Encyclopedia of Philosophy*, http:// plato.stanford.edu/ entries/ structural-realism/ . Stanford: The Metaphysics Research Lab, Center for the Study of Language and Information, Stanford University.

Livanios, V. 2010: "Symmetries, Dispositions and Essences," *Philosophical Studies* 148: 295–305.

Livanios, V. 2012: "Is There a (Compelling) Gauge-Theoretic Argument against the Intrinsicality of Fundamental Properties?" *European Journal of Analytic Philosophy* 8: 30–38.

Livanios, V. 2018: "Hamilton's Principle and Dispositional Essentialism: Friends or Foes?," *Journal for General Philosophy of Science* 49: 59–71.

Lyre, H. 2012: "Structural Invariants, Structural Kinds, Structural Laws," in D. Dieks, W. J. Gonzalez, S. Hartmann, M. Stöltzner, & M. Weber (eds.), *Probabilities, Laws, and Structures*, pp. 169–181. Dordrecht: Springer.

Magnus, P. D. 2012: *Scientific Enquiry and Natural Kinds: From Planets to Mallards*. London: Palgrave Macmillan.

McKenzie, K. 2016: "Looking Forward, Not Back: Supporting Structuralism in the Present," *Studies in History and Philosophy of Science* 59: 87–94.

Morganti, M. 2013: *Combining Science and Metaphysics: Contemporary Physics, Conceptual Revision and Common Sense*. New York: Palgrave Macmillan.

Muller, F. A. 2015: "The Rise of Relationals," *Mind* 124: 201–237.

Saunders, S. 2006: "Are Quantum Particles Objects?" *Analysis* 66: 52–63.

Smart, B. T. H., & K. P. Y. Thébault 2015: "Dispositions and the Principle of Least Action Revisited," *Analysis* 75: 386–395.

Stanford, P. K. 2015: "'Atoms Exist' is Probably True, and Other Facts that Should Not Comfort Scientific Realists," *The Journal of Philosophy* 112: 397–416.

van Fraassen, B. C. 1991: *Quantum Mechanics: An Empiricist View*. Oxford: Clarendon.

Wolff, J. 2012: "Do Objects Depend on Structures?" *British Journal for the Philosophy of Science* 63: 607–625.

Index